Lecture Notes in Electrical Engineering

Volume 674

The book series *Lecture Notes in Electrical Engineering* (LNEE) publishes the latest developments in Electrical Engineering—quickly, informally and in high quality. While original research reported in proceedings and monographs has traditionally formed the core of LNEE, we also encourage authors to submit books devoted to supporting student education and professional training in the various fields and applications areas of electrical engineering. The series cover classical and emerging topics concerning:

- Communication Engineering, Information Theory and Networks
- Electronics Engineering and Microelectronics
- Signal, Image and Speech Processing
- Wireless and Mobile Communication
- Circuits and Systems
- Energy Systems, Power Electronics and Electrical Machines
- Electro-optical Engineering
- Instrumentation Engineering
- Avionics Engineering
- Control Systems
- Internet-of-Things and Cybersecurity
- Biomedical Devices, MEMS and NEMS

For general information about this book series, comments or suggestions, please contact leontina.dicecco@springer.com.

To submit a proposal or request further information, please contact the Publishing Editor in your country:

China

Jasmine Dou, Associate Editor (jasmine.dou@springer.com)

India, Japan, Rest of Asia

Swati Meherishi, Executive Editor (Swati.Meherishi@springer.com)

Southeast Asia, Australia, New Zealand

Ramesh Nath Premnath, Editor (ramesh.premnath@springernature.com)

USA, Canada:

Michael Luby, Senior Editor (michael.luby@springer.com)

All other Countries:

Leontina Di Cecco, Senior Editor (leontina.dicecco@springer.com)

**** Indexing: The books of this series are submitted to ISI Proceedings, EI-Compendex, SCOPUS, MetaPress, Web of Science and Springerlink ****

More information about this series at http://www.springer.com/series/7818

Wesley Doorsamy · Babu Sena Paul ·
Tshilidzi Marwala
Editors

The Disruptive Fourth Industrial Revolution

Technology, Society and Beyond

Springer

Editors
Wesley Doorsamy
Institute for Intelligent Systems
University of Johannesburg
Johannesburg, Gauteng, South Africa

Babu Sena Paul
Institute for Intelligent Systems
University of Johannesburg
Johannesburg, Gauteng, South Africa

Tshilidzi Marwala
University of Johannesburg
Johannesburg, Gauteng, South Africa

ISSN 1876-1100 ISSN 1876-1119 (electronic)
Lecture Notes in Electrical Engineering
ISBN 978-3-030-48229-9 ISBN 978-3-030-48230-5 (eBook)
https://doi.org/10.1007/978-3-030-48230-5

This Springer imprint is published by the registered company Springer Nature Switzerland AG
The registered company address is: Gewerbestrasse 11, 6330 Cham, Switzerland

Foreword

As I first wrote in my 2016 book with the same name, we are living in the era of the Fourth Industrial Revolution (4IR). As previous industrial revolutions, this fourth one is changing all aspects of humankind, from governments and economies, to the way we interact, live and work, and even the way we see ourselves. As before, this revolution also brings with it both challenges and opportunities, which we all need to seek out, and actively contemplate in order to shape our individual and collective futures for the better. But this Fourth Industrial Revolution is on track to be the most impactful one ever, given the scope, scale, and speed of the changes it brings about. All the same, the impact this revolution has is still very much dependent on the choices we make.

The excitement brought about by the potential of exponential advancements in technology to improve the lives of people around the world is however hindered by uncertainties. If we don't actively shape the direction of technology, will it rather help or hurt societies at large? Will the Fourth Industrial Revolution be an opportunity for countries in Africa and elsewhere to "leapfrog", or will it only exacerbate existing development inequalities? And, especially in the current context, will technology be a net positive for a world dealing with triple crises (economic, environmental, and health), or will it fail to live up to its potential?

Addressing the uncertainties and realizing the possibilities, in the Fourth Industrial Revolution, require urgent and in-depth studies of the emerging and future disruptions. They should incorporate both technological and societal analyses. It is in this spirit the book *The Disruptive Fourth Industrial Revolution: Technology, Society and Beyond* has been conceptualized and undertaken. The topics explored in this book are very much representative of some of the pressing issues of 4IR—present and future—and are explored in an insightful and inspiring manner. The book's contextual emphasis of technological disruptions is welcomed, as it not only opens up pathways for future research and innovation, but also prompts dialogues on policy.

In the context of the public health emergency and economic crisis we find ourselves in 2020, I've found particular relevance in reading on topics such as **Metabolomics**, which could help us get to more personalized medicine and faster diagnoses, and **Virtual and Augmented Reality in Surgery**, which could allow us to take the next leap forward in the efficacy and safety of invasive medical treatments. Equally, in the context of much of our life and economic activity now rapidly moving online, it is good do deep dive on emotion recognition in the 4IR, and the importance of cybersecurity. Each of those could and will have a role to play when as we deal with the current crises, and rebuild our systems to be better, be better prepared, and more resilient in the future.

And, knowing that the Fourth Industrial Revolution is another turning point for development, I also enjoyed reading chapters like the one on how to **Prepare Emerging Markets for the 4IR**, highlighting how modern policies and—particularly—higher education play a crucial role in an economy's ability to benefit from this next wave of innovations and growth. In a similar vein, the authors have the right to ask whether the Global South needs to **Decolonize the Fourth Industrial Revolution**. Indeed, countries must be able to shape their own future, while cooperating on their own terms with others that can help along the way.

It was an honor for me to read and write about the *The Disruptive Fourth Industrial Revolution: Technology, Society and Beyond* in this foreword. It is an invaluable contribution to the Shaping the Future of the Fourth Industrial Revolution, which is a crucial challenge for all of us, and a meaningful aide for academic, policymakers, and entrepreneurs who will be contributing to this effort in Africa and the world.

Geneva, Switzerland Klaus Schwab
April 2020

Preface

The Fourth Industrial Revolution (4IR) has already begun. This revolution is not based on a single technology, but rather on a confluence of multiple technologies like Artificial Intelligence, Internet of Things, Machine Learning, Big Data, Additive Printing, Cloud Computing, Virtual and Augmented Reality, etc. 4IR is poised to affect all spheres of society—in the way we interact, conduct business, and govern. As we witness the fast-paced growth of disruptive technology globally, there is a growing sense of urgency to grapple with the far-reaching impacts of 4IR and to prepare ourselves for the future. It is with this sense of urgency that The Disruptive Fourth Industrial Revolution: Technology, Society and Beyond is conceived. The book deals with the latest research in disruptive technology in 4IR and critically analyzes the roles and effects thereof on inter alia economics, society, business, government, labor, and law. A broad perspective of 4IR with a deep-dive technological and societal focus is adopted to offer readers a deeper understanding of recent advancements and future trajectories. Researchers, academics, practitioners, policymakers, and business and industry leaders will form the primary readership of the book.

There are 11 chapters in The Disruptive Fourth Industrial Revolution: Technology, Society and Beyond. The "Robotic Process Automation as a Precursor to e-Government in the Fourth Industrial Revolution" chapter explores the emerging dimensions and models of e-Government in the realm of 4IR. The growing enshrinement of technology in public service delivery is dealt with and a conceptual framework that can be used as a blueprint in the design of automated public services especially considering Robotic Process Automation (RPA) is proposed.

Blockchain and smart contracts have emerged as key disruptive technologies in 4IR. Smart contracts are revolutionary and are dubbed to be the next evolutionary step in contract theory and practice. Chapter "The Disruptive Force of Smart Contracts" considers the disruptive force of smart contracts specifically relating to contract theory and general contract practice. The chapter analyzes historical development of commercial advances and contractual practices, and the development of dispute resolution processes to complete a fully autonomous smart contract ecosystem in commercial and business transactions.

The "Technologizing Infrastructure *for* Peace in the Context of Fourth Industrial Revolution" chapter deals with the growing interest and controversy in the application of various aspects of technology to peacebuilding. More specifically, the chapter focuses on technologizing Infrastructure for Peace (I4P) which refers to the peacebuilding institutional capacity, conflict prevention, and post-war recovery.

One of the foremost concerns of states around the world is the impact of 4IR on cybersecurity. Thus, cybersecurity policy and governance for 4IR are crucial, especially in determining how states under international law should govern cybersecurity globally when faced with the technological disruptions brought about by 4IR. The "State Cybersecurity Governance in the Fourth Industrial Revolution: An International Law Perspective" chapter identifies cyberjustice as the desired foundational normative prescript to manage state cybersecurity governance and policy interventions.

Although the Fourth Industrial Revolution has the potential to perpetuate renewed coloniality at the hands of the more developed parts of the Global South, it has the potential to empower the Global South through decolonisation. Chapter "Does the Global South Need to Decolonise the Fourth Industrial Revolution?" gives a comprehensive history of global colonization and describes the extent to which (economic) coloniality prevails and is being renewed in the Global South. The ways in which the Global South needs to break coloniality in order to reap the benefits of the fourth industrial revolution is also argued.

Chapter "Decentralizing Emerging Markets to Prepare for Industry 4.0: Modernizing Policies and the Role of Higher Education" investigates socioeconomic policies, technological focus, and academic necessities in core competencies and skills development to prepare emerging markets for the technological disruption of the Fourth Industrial Revolution.

Social media has become a leading advertising platform for businesses to leverage market share and yield more significant conversion rates. The "Sell-Bot: An Intelligent Tool for Advertisement Synthesis on Social Media" chapter presents how Artificial Intelligence (AI), Machine learning (ML), Big Data, and Natural Language Processing (NLP) can be exploited for modern advertising.

Chapter "Advancements and Role of Emotion Recognition in the 4th Industrial Revolution" discusses the importance of emotion recognition and how this technology, which uses physical and behavioral expressions—such as facial features, speech features, and their variants—together with Deep Learning (DL), has emerged as one of the prominent pillars of 4IR. A new Speech Emotion Recognition (SER) model is also presented as a case study of ER systems.

One of the key advancements in the era of the Fourth Industrial Revolution is quantum computing, which harnesses quantum mechanical concepts such as entanglement, superposition, and tunneling to perform computation. Although a full-scale quantum computer has not yet been realized, Noisy Intermediate-Scale Quantum (NISQ) computers are already in use. Chapter "Noisy, Intermediate-Scale Quantum Computing and Industrial Revolution 4.0" explores NISQ as a disruptive technology of the Fourth Industrial Revolution and introduces a novel privacy-preserving quantum machine learning scheme.

A new era of systems biology is disruptively emerging, providing holistic descriptions of biochemical phenomena at the cellular and organismal level. Metabolomics is an interface between biology, chemistry, chemometrics, statistics, and computer science that has provided remarkable insights into the mechanisms that underlie various physiological conditions. Chapter "The Disruptive 4IR in the Life Sciences: Metabolomics" discusses positively disruptive force of metabolomics and how its momentum and maturation have revolutionized the life sciences.

Augmented Reality (AR) and Virtual Reality (VR) technologies have received much attention and widespread application. Health care is emerging as one of the leading candidates for application of these disruptive technologies in the Fourth Industrial Revolution. Chapter "Virtual and Augmented Reality in Surgery" deals with the novel applications of AR and VR technologies within the realm of surgery. The chapter clarifies concepts associated with AR and VR in the context of surgery, and addresses technological implementation approaches in surgical preoperative planning and training.

Johannesburg, South Africa

Wesley Doorsamy
Babu Sena Paul
Tshilidzi Marwala

List of Reviewers

All chapters of the book, **The Disruptive Fourth Industrial Revolution: Technology, Society and Beyond**, were peer-reviewed by two (2) or more technical specialists. The review process was conducted anonymously. The editors would like to thank the listed reviewers for their participation in the review process.

Abdul Kader
Amit Mishra
Amlan Chakrabarti
Andréa Campbell
Ghandi Badela
Grace Mwaura
Guy Taylor
Herman Vermaak
Kershree Padayachee
Maria Schuld
Mark Nasila
Michael Zimba
Monica Singer
Olatomiwa Lanre
Pranati Rakshit
Rajarshi Mahapatra
Rajohane Matshedisho
Sudheesh K. Shukla
Sushabhan Choudhury
Tendani Malumedzha
Towela Nyirenda-Jere
Vinod Kumar

Contents

Robotic Process Automation as a Precursor to e-Government in the Fourth Industrial Revolution

Kelvin Joseph Bwalya

Abstract A lot of research has been done focusing on adoption of e-Government by individuals and businesses in the different parts of the world. Despite that being the case, there is a clear gap in understanding the different nuances of ubiquitous e-Government especially in the realm of the Fourth Industrial Revolution (4IR) where automation and intelligence is emphasized. In Africa and other resource-constrained environments, there is a clear knowledge void with regards to unique challenges faced when attempting to design and implement contemporary e-Government. The chapter is anchored on descriptive informetrics as a philosophical orientation. This orientation enables it to employ a detailed bibliographic analysis of work published in scholarly and policy outlets. This enables it to ground the discussion using both formulaic and conceptual definitions of key concepts explored. This chapter is as a result of continuous research in e-Government that has been going on for the past decade. This research intended to explore the emerging dimensions of e-Government in the realm of 4IR. The chapter discusses the emerging model of public service delivery given the growing enshrinement of 4IR in public service delivery platforms. Finally, a conceptual framework that can be used as a blueprint in the design of automated public services especially considering Robotic Process Automation (RPA) is proposed. Fundamental concepts leading to the understanding of 4IR, AI and RPA are adequately explored in the chapter.

Keywords RPA · e-GaaS · Blockchain · Artificial intelligence · Conceptual framework · 4IR · e-Government

K. J. Bwalya (✉)
School of Consumer Intelligence and Information Systems, University of Johannesburg, Melville, Johannesburg 2006, South Africa
e-mail: kbwalya@uj.ac.za

W. Doorsamy et al. (eds.), *The Disruptive Fourth Industrial Revolution*, Lecture Notes in Electrical Engineering 674,
https://doi.org/10.1007/978-3-030-48230-5_1

1 Introduction

Increased adoption of emerging technologies such as Internet of Things (IoT), blockchain, big data and behavioural/predictive analytics, Artificial Intelligence (AI), etc., stands to revolutionalise the way government departments do their businesses. In the context of e-Government, AI emphasises on automatic decision-making, intelligence and connectedness (through mesh networks). In order to achieve automatic decision-making, robots are used in the different governance information systems. Designing stable e-Government systems where all the different aspects of emerging technologies can be appropriately coupled together is a key question in contemporary e-Government systems.

The original motivation of e-Government has been the desire to utilize technology in different government business processes. The introduction of technology was to ensure the achievement of improved efficiencies in the government business processes which will ultimately translate into effective public service delivery as a whole. As anticipated at the onset of e-Government, the level of efficiencies in the public sector delivery platforms will significantly improve in line with future increase in automation. Automating public services will culminate into increased access to information by businesses and individuals thereby encouraging service innovation. The transition towards 4IR-based e-Government will grow the number of start-up businesses coming in with diverse business innovation thereby sustaining the changing technical space of the fourth industrial revolution (Engin and Treleaven 2019). The 4IR will facilitate the creation of possibilities for increased competition among start-ups culminating into heterogeneous technology innovations. For the public sector, the emerging technologies will enable the integration of digital innovative platforms such as bots and robo-advisors to support decision-making at the level of civil servants. For instance, the digital agents will be able to access information from various places in the governance business value chains using distributed ledgers and smart contracts in the blockchains.

Around the world, there have been so many movements towards realizing the massive potential benefits that come with artificial intelligence, process automation and generally the Fourth Industrial Revolution. South Africa has also jumped onto the bandwagon by recently ushering in a special committee to drive 4IR interventions. The 4IR committee aims to find ways of harnessing the different opportunities that come with the 4IR. Being chaired by the President of the Republic of South Africa, there is enough political will and resolve to explore the different 4IR opportunities given the South African context. One of the perceived immediate impact of the work of this committee will be putting on the table the different opportunities presented by 4IR which can be embedded into the general public service business processes. The desired scenario will be for citizens and businesses to communicate with machines and government systems endowed with appreciable levels of automation and intelligence replacing the input of humans. In such a scenario, public services are accessible anywhere and at any time in a truly ubiquitous sense of the word.

In order to design context-aware 4IR-based technology solutions endowed with appreciable degree of automation and intelligence, there is need to clearly understand the underlying information architecture of the system(s). Understanding the information architecture of information systems entails appreciating the workflow models, systems interfaces and integration frameworks. These aspects are going to be cardinal towards achieving desired process automation in the fourth industrial revolution. This chapter aims to explore the different entities that are needed to facilitate automation in e-Government. Principally, the chapter is enshrined upon the concept of Robotic Process Automation. The concepts of automation and intelligence are explored from an information management perspective. The other key aspect explored in the chapter is the information cloud architecture that can be considered to guide the design of e-Government systems capable of facilitating automation in the governance value chains. The proposed architecture is based on e-Government as a Service (eGaaS). It is anticipated the proposed architecture will be used as a blueprint in influencing the conceptualization of context-aware avant-garde technologies that will further open up public service delivery to increased efficiencies.

This chapter is arranged as follows: the next section presents the background which lays the context in which this research has been done. The third section looks at the context of e-Government design in the fourth industrial revolution. In this regard, concepts such as automation and artificial intelligence are introduced. the later parts of the chapter explore design of 4IR-based e-Government systems and the present the concept of e-Government as a Service (e-GaaS). The chapter concludes by articulating the future works and a snapshot of what has been discussed in the chapter.

2 Background

Starting with the decentralisation agenda and the Structural Adjustment Programmes (SAPs) promoted by the Bretton Woods Institutions for implementation in African countries, there have been quite some changes in the public governance models in Africa. With the dawn of rapid advances in Information and Communication Technologies (ICTs) witnessed around the 1990s, many African countries embarked on the digitisation agenda and finally designing the delivery of public services around the concept of e-Government. The motivation to implement e-Government by a majority of African countries was to improve public service delivery, although a handful were motivated just to *keep up with the joneses*. In the 2000s, transformation towards interactive e-Government models surged. As a result, translating into delivery of a majority of public services through ICT platforms.

In the last 5 years, many governments have realised that the anticipated benefits especially with regards to increased efficiencies and reduction in the cost of public service delivery have not necessarily been realised. It was common knowledge that a majority of resource-constrained countries were excluded from implementing meaningful e-Government. As a result, many countries have been motivated to design e-Government solutions based on technology innovations enshrined on the fourth industrial revolution. There is a more realistic chance that automation and artificial intelligence will culminate into reduced cost for delivery of public services and revitalise the public service delivery landscape. The implementation of e-Government in the fourth industrial revolution will revitalise the public governance models used in developing world countries towards increased accountability and transparency (through blockchain in e-Government), reduction in cost of public delivery (use of bots toward increased intelligence and automation), etc.

3 4IR-Based e-Government Design Prospects

E-Government designed upon the principles of the fourth industrial revolution will culminate into a more integrated public information business processes. Process integration is a precursor to the automation and intelligence desired in the public service. The motivation for transcending towards implementation of 4IR-based technology innovations in the public sector given the perceived benefits cannot be overemphasized:

1. Traditional governance systems were endowed with persistent problems with regards to efficiency, effectiveness and appropriateness. Further, tracking of level of service in the governance value chains cannot be monitored due to lack of service integration.
2. Mere use of technology and digitisation of public service business processes in the realm of e-Government has not achieved the anticipated benefits. This has mostly been due to lack of adequate integration of services.
3. Limited empowered public workforce with relatively low technical skills and digital literacy.

Design of contemporary public service delivery platforms take into account the tenets of the fourth industrial revolution. As a result of rapid advances in IoT, the push towards digitising all business processes in the public sector are no longer a mere dream but reality. Several governments around the world have introduced advanced forms of e-Government in their governance value chains to the point of automating some of the mundane processes. A number of countries around the world have jumped onto the bandwagon of automating public services. These include Canada, South Korea, Japan, Australia, etc. In the European Union, Estonia is one of the most cited exemplary reference point for advanced e-Government. Other notable e-Government projects include the Germany's Bundesagentur für Arbeit (https://www.arbeitsagentur.de/), Singapore's SingPass single sign-on system (https://www.singpass.gov.sg/), Korea's KONEPS online e-procurement system (http://www.pps.go.kr/), etc. Given increased interest in designing 4IR-based e-Government systems by countries around the world, there is a clear prospect for innovators and designers to showcase their innovations. Applied computing scientists need to start thinking of how to interrogate the different opportunities brought about by automation in the public sector.

With the onset of Fourth Industrial Revolution, contemporary e-Government demands that there is increased automation in the execution of repetitive or mundane activities in the different public sector business processes. 4IR will blur the boundaries existing between the physical, digital and the biosphere as a result of massive integration of cyber-physical information systems. When the 4IR is eventually achieved in different contextual settings, information will be accessed by Artificial Intelligence (AI) systems integrated into the physical environment. This will enable applications and digital agents to pervasively access information for intelligent decision-making. The 4IR is an advancement of the Third Industrial Revolution which focussed on aggressive use of ICTs to automate production. The 4IR contends that digital agents or robots (*bots*) can now be integrated into heterogeneous business processes and contexts allowing them to make intelligent decisions to the potential of complementing or replacing the input of human beings altogether. This automation needs to be intelligent so that the automatic information management processes are close as possible to mimicking a human being. In order to achieve this, the following is critical:

1. Appropriate machine-to-machine applications where e-Government applications automatically interpret information from different integrated databases to make informed decisions;
2. Integrated open systems with identical or closely-identical data marshalling standards for communication and interpretation of available data (Predictive analytics);
3. Machines and systems optimally trained with local data sets. These machines and systems need to be designed in such a way that they learn by analysing recurring traits in data;
4. Machines and systems able to analyse huge sets of data from multi-dimensional perspectives whether it be structured or non-structured (Big Data analytics).
5. Machines and systems endowed with necessary business rules and public policies in order to make informed intelligent decisions.

The above context depicting the principles that should guide the design of the intelligent automatic business programmes in the context of e-Government is shown in Fig. 1.

Figure 1 shows the modules required to achieve automation in the context of e-Government. Only high-level modules are shown as details can be obtained from

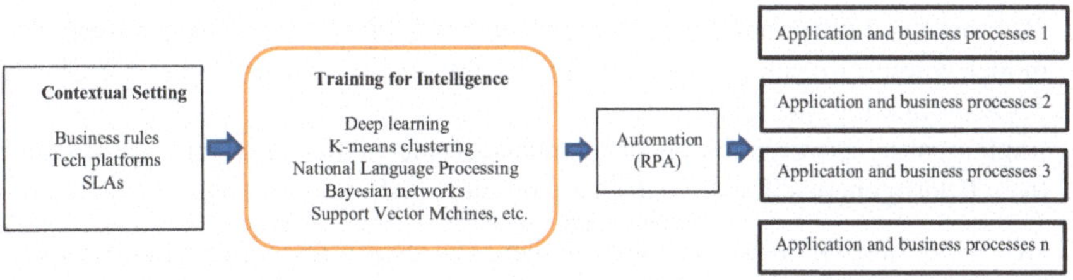

Fig. 1 Contextualising automation in e-Government (*author's own concept*)

reference works. The e-Government platforms are designed with due reference to the context in which they are going to be deployed. Both structured and unstructured data depicting the business levels, service level agreements (SLAs), etc., are utilized in training of the technology innovations. Common machine learning algorithms such as deep learning, Bayesian networks, etc. can be used depending on the context. After training, some level of intelligence is attained al-lowing automation using Robotic Process Automation (RPA, discussed below) can be achieved. The different types of automation are applied in the different business processes.

In order for e-Government to reach its anticipated potential in the era of 4IR, its designs have to be data-centric. In the contemporary world, each individual interacts with a lot of entities out of which data is formed for each of those interactions. Further, the lifetime of human beings culminates into a data profile which is largely unique from one person to another. In order to have automatic and intelligent decision-making in governance value chains, e-Government systems will have to be able to process huge quantities and large dimensions of both structured and unstructured data (Big Data). The phrase "Data is the new oil" puts a lot of emphasis on the importance of data with regards to enriching competitiveness in any given setup. Access to data means that right decisions can be made and a number of opportunities can be unlocked. It is against this background that this research posits that having automation come onto the bandwagon in making data available to humans and applications in the e-Government context promotes universal innovation in the public sector.

The rapid adoption of 4IR in e-Government platforms will culminate into realization of the concept of Open Government which is based on open processes, open applications, open data and open public services. The implementation of 4IR-based e-Government will open ways for realization of Open Government which borders on enhancing transparency, integrity and accountability in the public governance value chains. The level of interaction between human beings and government digital agents will be more interactive as there will be more automation of mundane tasks executed by the agent (Vesnic-Alujevic et al. 2019). Human beings will not have to wait to visit a government office before accessing a service as they can do that ubiquitously at almost the same level of engagement. Open Data and Open Government have been hot topics for over a decade now yet governments have not opened up not even 20% of their data and processes. For corruption-prone environments such as in Africa, Open Data will have positive effect on corruption by bringing about transparency and accountability in the governance processes. Open governance is hinged on two major pillars:

- Opening up of governance processes and encouraging a collaborative governance model where citizens are included in the governance process (policy-making, decision-making, etc.); and
- Implementing governance on open interoperable platforms where information from heterogeneous devices can easily be managed.

Given the fourth industrial revolution, there is a clear orientation towards innovative and Open Government and the concept of user-centricity where citizens' access to public services is a right. With Open Government, the citizen and the public institutions work in a collaborative manner allowing citizens to be more involved in different aspect of governance (Vesnic-Alujevic et al. 2019). Therefore, Open Government advocates for the transformation of the role of the citizens from that of being passive participants or spectators to that of active agents. The expectation is that the active agents will be consulted at all phases of the policy, decision-making processes and the design of the public sector administrative infrastructure. The engagement of citizens as active agents brings them to the center of the innovation value chains. It is worth noting that collaborative innovation is one of the key attributes of the fourth industrial revolution.

Realizing the benefits of e-Government in the 4IR will need careful design and implementation strategies underpinned by systems thinking and stepwise approach. Further, there will be need for careful technology and digital literacy skills education so that individuals and organizations are capacitated with appropriate skills to harness the digital opportunities (Mir Ali 2013). The irony about ignoring this need is that there could be massive digital exclusion of a critical mass of the population thus miss-ing out on innovative and digital opportunities that can be explored. The 4IR implementation in public governance will mean that pervasive access to public services will culminate into mobile workspaces for government workers. This means that public officials will have the opportunity to work at anywhere at any time as long as they have access to the Internet (Adobe 2015).

In the context of this chapter, the Fourth Industrial Revolution points to the integration of automatic business processes in the public sector business processes and intelligence achieved through the adoption and use of various artificial intelligence technologies. Although automation and artificial intelligence have been around for centuries now, the 4IR presents an industrial paradigm where these two are used in large proportions thereby bringing the notion that human beings may eventually be re-placed in different business processes. Therefore, it is clear that the key tenets of the fourth industrial revolution in the realm of e-Government has much to do with automation and artificial intelligence.

4 Automation in the Governance Processes

Understanding the basic principles that underpin automation in governance systems is cardinal to understanding e-Government systems in the fourth industrial revolution (4IR). One of the key requirements in the design of automate governance platforms is the need to understand the information architecture and the workflow processes involved in information exchange between different government units. The anticipated start-point is that the different government information systems need to be integrated in order for them to share information. For this to happen, e-Government designers need to have clear understanding of the different integration and interoperability frameworks that are used to connect information systems.

Many governments around the world have implemented e-Government interoperability frameworks (eGIF). Mir Ali (2013) investigated the effectiveness of office automation system with regards to whether it translated into increased efficiencies, timeliness and accuracy as well as improved decision-making. The results of the study indicated that there is significant increase in levels of efficiency when automation is integrated in the public sector business processes. Although this study indicated substantial increase in efficiency, there is undoubted inventory of cases accentuating the positive role that automation plays in the contemporary information management environments (Mir Ali 2013).

Designing automated business processes starts from understanding the key terms and entities that go into automation. First, automation does not necessarily mean that human beings are kicked out from (Sheridan and Parasuraman 2000). It means that the emerging technology interventions need to be designed in such a way that they complement the efforts of human beings and not replace them completely. Second, automation in environments which followed traditional governance models should be done in a stepwise approach. This means re-engineering of business processes need not be done at once but in stages. Some of the key to be considered included the following:

- Repetitive processes—by training machines and digital agents using correct data, repetitive processes can be fully automated and decisions made with high degree of precision.
- Intelligent automation—entails that digital agents are able to simultaneously analyse huge sets of data, identify patterns and make informed intelligent decisions without the need for input from a human being.
- Business process modelling—dismantling of a business process in order to clearly understand what activities are embedded therein in order to make decisions whether they can be automated or not.

In order to achieve automation certain requirements have to be met. Automation requires requisite planning in terms of re-engineering of government processes. The level of automation in an information-intensive system such as an e-Government system is dependent upon the ability of the system to manage information through a given information lifecycle. The information lifecycle details the capability of a system to collect (capture), do information representation (data visualisation), analyse, interpret, act or share information.

The Sheridan Framework articulates the different nuances of automation degrees in different contexts. The framework indirectly posits that automation needs to be approached with a stepwise mind framework where not all business processes in the public sector need to be automated at once. During the maturity life-cycle of automation, there needs to be incremental levels of control between computers/digital agents or robots and human beings. Thus, 4IR-based e-Government systems need to be designed in such a way that they are complementary with traditional e-Government systems where human beings and machines are used complementary (Smith et al. 2010). In order to achieve automation, different Automated Decision

Making Systems (ADMSs) are used. ADMSs are grounded upon the following type of technologies:

- **Workflow applications** (software programmes for information-intensive interconnected processes);
- **Rule engines** (technologies that execute conditional statements in a bid to observe the business rules in an environment);
- **Intelligent systems** (systems that search and perceive conditions, learn and act intelligently in decision-making);
- **Enterprise systems** (systems for connecting and managing information flows or transaction processes within the organisations (Sheshasaayee and Bhargavi 2017).

These are advanced intranet systems with a wider functionality for connecting the different units of the organisation into one information space to guide decision-making); industry-specific packages (technology packages designed for a specific industry for guiding the automation of the different business processes); statistical or numerical algorithms (focussing on management of quantitative data within the automated information spaces); etc.

With increased need for automation in the public services, Business Process as a Service (BPaaS) will become an unavoidable requirement for innovative e-Government systems. BPaaS is an architectural arrangement where the data for a given business process is stored in the cloud so that it can be accessed pervasively by the different agents of the business. BPaaS allows most of the data and information needed in the business processes to be stored in the cloud so it can fuel automation and intelligence in the public sector. BPaaS further allow the organisation to concentrate on their core business whilst utilising third-experts in supportive aspects of the business such as market analysis, supply chain, etc. The BPaaS allows an opportunity for rapid introduction of technology in the business processes replacing the legacy information systems towards an absolute agile environment where technological innovation can easily be introduced in the business.

One of the key issues that need to be addressed towards automation of public services is the integration of systems in the governance business process. Integration entails that the different systems in the public service value chains are connected and are interoperable thereby able to share information, applications and able to simultaneously access distributed databases.

It is anticipated that AI and automation coupled with the envisaged 5G technology will have significant effect on how information is managed in the public sector. Therefore, it is anticipated that e-Government will slowly transcend into automation of mundane tasks. It is worth noting that with the increasing level of automation, the role of a human being (operator) will slowly diminish and the computer system/robotic agents become to occupy a more pronounced role in decision-making and takes all the necessary discretionary actions (Smith et al. 2010).

Although a lot of ground has been covered with regards to the development of capabilities for digital agents in the automation lifecycle, it is important automation in the e-Government environments be guided by a complemented set of capabilities of bots and human beings. In order to design a strategy for automation, one needs to understand the levels of automation anticipated. The scale of degrees of automation can be seen in Sheridan (1992; 358). Automating service delivery platforms in the public sector have been informed by the Sheridan Automation Framework (Parasuraman et al. 2000; Smith et al. 2010). PwC estimates that automation will not be achieved in one step but will follow a stepwise process passing through three different waves:

- **First wave**: will be about algorithmic transformation where automation will only target simple computational tasks and analysis of structured data;
- **Second wave**: will be an augmentation one ushering in a paradigm for dynamic interaction with technology to improve efficiency in clerical duties; and
- **Third wave**: is a pure autonomous one culminating into automation of physical labor and problem-solving in real-world environments.

In the context of this chapter, automation may involve the use of various control systems for operating machines or the use of digital agents so as to minimise input of human beings in a business process. In the contemporary edge, at the pinnacle of business process automation is the concept of Robotic Process Automation (RPA). The following section discusses RPA in depth, as applied in e-Government systems.

5 Robotic Process Automation

Robotic process automation (RPA) is a new concept that is being aggressively utilised in the fourth industrial revolution. RPA makes use of software robots (bots) that use rule-based reasoning (RBR) to automate transactional processes thereby mimicking typical reactions by human beings. RPA entails employing software agents to do work that was previously done by people. Digital agents are trained using advanced machine learning algorithms, pattern recognition, natural language processing, neural networks (deep learning), etc., are able to mimic the likely behaviour of a human being in a given situation (Adobe 2015). RPA will enable workflow automation in the business processes of the public sector business processes. Workflow automation for repetitive tasks prone to human error such as collecting basic patients' information (such as age, patient name, weight, etc.) or medical history can be done by digital agents thereby increasing efficiency in managing patient information. It is important to note that RPA allows the design of systems and applications that are able to meet enterprise IT agility needs such as auditability, scalability, security, change management, etc.

In order to implement RPA, it is important to clearly understand the type of activities that can be executed using RPA. Such activities are repetitive in nature, involve data manipulations and are generally data intensive, electronically executed, have high error rate, involve reconciliations, etc. It is worth noting that the cost for the utilisation of RPAs in public service business processes is relatively lower than using permanent employees or outsources human resources. Further, with advances in technology, it is becoming increasing clear that robots at the centre of automation are easy to train and give relatively higher degrees of automation. Therefore, bots make fewer errors, fair judgement and do not involve emotions as they make decisions. Ultimately, the cost of public services is significantly reduced (Adobe 2015).

Envisaged automation in e-Government in the realm of RPA can be achieved owing to unique features of RPA. Application based on RPA are easy to design as RPA uses drag-and-drop designs and therefore does not require extensive programming skills as required in many business re-engineering processes; is non-invasive i.e. can be designed to sit on top of already-existing systems (the invasiveness allows new application designs to access business logic and data layers in the different layers of the existing IT architecture stack (Lacity and Willcocks 2016). Another potential espoused by automation is the use of RPA in healthcare diagnosis and disease management. The digital agents are able to search through cloud-based platforms to obtain historical medical information that can be used to inform the current or future medicines to be given to the patient.

Automation will be achieved in different domains of the public sector with careful implementation of RPA-based technology innovations. The prospects for automation in the health care will culminate into improved efficiencies and increased capabilities in the treatment of ailments that humans find too complicated. This is because robots are able to reach complicated and intricate contours of the human beings thereby performing complicated medical procedures. The public sector can take advantage of the opportunity to utilise RPA in delivering a myriad of public service scenarios. Several public services will potentially be automated in the near future. Typical public services include transactions for applying for services offered by government agencies such as passports, Identity Documents (IDs), social benefits (walefare), drivers' licences, motor vehicle registration, medi-care cards, voters' cards, customs security checks, payment for services, etc. Such type of services forms a mundane set of repetitive activities which can easily be automated in the realm of e-Government (Adobe 2015). Other potential areas where automation can be implemented include: Anti-fraud checks, credit checks, permit applications, crime reporting, claims processing, case management, medical diagnostics, student finance management, tax calculations, license applications processing, etc.

6 Automation Potential in the South African Public Sector

There are some steps towards automation in the public sector of South Africa. For example, the department of e-Government in Gauteng has been mandated to coordinate the integration of back-end processes. The integration of back-end processes is a precursor to automating the public sector. It cannot be overemphasised that automation promises cost savings, a general improvement in the efficiency and effectiveness of public services, and improvement in revenue collection. These opportunities brought about by automating public services can be missed if not designed carefully.

The Province of Gauteng is leading the country in as far as advancing the public service delivery. E-Government and eventual automation of the services have been identified as key drivers for bringing intelligence and efficiency in the public service delivery domains. One of the first key strides towards achieving this goal was the automation of self-service for suppliers where they could do self-registration in the database of suppliers. This self-registration is based on the automation of the Source-to-Pay (S2P) processes. S2P processes are conceptualised and based on the Oracle Procurement Cloud. The S2P integrates all the players and events involved in the procurement process into one information space. The different events in the process are executed automatically. Another key intervention in Gauteng has been the erection of the multi-billion rand Gauteng Broadband Network (GBN) which is poised to transform Gauteng government into a smart establishment. One of the key focuses of the GBN is to achieve connectedness of the government information systems which will enable different government platforms to share information in real-time. Having the GBN fully implemented will culminate into increased innovations in the governance value chains.

A good example of automation in the South African context is the implementation of e-filing system by the South African Revenue Service (SARS)—the system is working relatively well and individuals are now able to file tax returns online. It is expected that, in the long run, this system will culminate into massive savings on the part of the government (increased efficiencies, reduced workforce for processing returns, reduced cost, etc.) and individuals (elimination of notorious long queues at the revenue offices, time savings, convenience, etc.). Gathering aggregate information through the e-Filing system makes it possible for deeper economic trend analysis to be done on a tax payers and therefore give an indication of the performance of the economy. The only obvious downside to the implementation of the e-Filing system is that there is a possibility that the huge information gathered through the system may be susceptible to privacy and data protection issues.

At various border controls in South Africa, biometric systems are being implemented using a phased-approach. These systems enable the information of individuals crossing the border controls to be checked against various databases to enable on-spot identification. In government information systems, there are a number of potential database interface connections that could be created: the Interpol database, home affairs database (police, crime prevention, etc.), bank database systems, etc. Checking individual information obtained against these databases would enable border officials to immediately identify individuals who would pose potential threats to national security. It is anticipated that the integration of different disparate database systems, for instance on a cloud infrastructure, to form one information space will culminate into the formation of a highly capable integrated government information system which be able to execute a great deal of automated tasks.

7 Regulatory Environment and Digital Leadership in e-Government

The adoption of robots or digital agents in the public sector business processes is currently being monitored the world over. The monitoring is important because it gives an opportunity to understand the level of digitisation of public services throughout the world. Intelligent agents and robotics (bots) can be adopted into the different public business processes if there is a clear understand of what potential can be unlocked once implemented carefully. This means that careful understanding of fit-for-purpose need to be understood first before robots or intelligent agents are adopted. South Korea has arguably the best *robot density ratio* in its industry including the government South Korea is the global leader in innovation as it leads on the global Automation Readiness Index (http://www.automationreadiness.eiu.com/static/download/PDF.pdf). South Korea has invested a lot in terms of developing capacity in technology innovation, implementation and leadership over the years given the dedicated by the government and leading tech companies such as LG and Samsung Electronics.

Countries intending to jump onto the bandwagon with regards to accessing the different opportunities that come with 4IR will have to ensure that they have requisite policy encouraging the enshrining of 5G into the economy. Desired automation and intelligence cannot be achieved and sustained if there is no adequate legal, institutional, regulatory and policy framework. As the information landscape changes, it is important that the different frameworks guiding its management change in tandem with the changing technology environment. This means that in order for requisite innovation to be achieved in any public sector arrangement, it is important that there is agile policy resume which can take a dynamic nature throughout its implementation cycle. In the fourth industrial revolution, an entrepreneurial mind is needed to drive innovation and service/product innovation. There is need for policies and leadership that is going to promote the entrepreneurship at all levels of the socio-economic establishment. In a highly-connected and integrated world, more people are able to easily share ideas for consolidation into intelligent offerings. This means that innovative ideas will have a shorter lifespan as new ideas keep popping up every time (WEF 2019).

Requisite leadership is one of the key requirements for the revitalisation of public governance in any given context. In the contemporary e-Government environment, adaptive leadership is a must. Adaptive leadership enables individuals who are high innovative, agile and open to new ideas with a flexible mind allowing them to aggressively explore emerging themes in e-Government. Adaptive leaders will employ a people-centred approach in the design of intervention and responses to the fourth industrial revolution. Innovative leaders in this space will aim for creating data economies that will open the vast amounts of opportunities brought about by implementing the key principles of the fourth industrial revolution. With a clear political will and leadership capacity, South Africa stands on a fertile ground to launch aggressive programmes for stepwise exploration of 4IR implementation in the public sector.

The existence of a requisite institutional, legal and regulatory frameworks coupled with a willing government to introduce innovation in the public sector governance systems presents a good opportunity for the implementation of automation in the South Africa's public sector.

8 Integration of e-Government Systems

Contemporary e-Government cannot be considered as a single entity but as a multi-faceted phenomenon that demands amalgamation of different concepts in order to achieve its purpose in its entirety. For example, need for intelligent transport systems in the realm of smart cities, integrated medical and health management systems (provision historical data for management of diseases, etc.). Integration is important for making a broader governance digital eco-system where all the heterogeneous government departments and systems are integrated. With a key focus on connectedness demanded by the fourth industrial revolution, the need contextual integration paradigms cannot be overemphasised. Connectivity will drive innovations as more information will be easily shared in information networks. In the fourth industrial revolution, innovation will be encouraged by a variety of enabling technology platforms such as the following:

- The emergence of enabling technologies like the Internet of Things (IoT) will enable drag-and-drop design consoles to be integrated into the core of design bureaus thereby replacing the traditional objected-oriented programming modules. The need for programming skills in such an environment will be wiped away culminating into increased participation into design attempts.
- Within environments where network infrastructures are enshrined upon the IoT and IPv6, the concept of Self-Sovereign Identity (SVI) enables any entity in the environment will be given an IP address. This will enable all the entities in the environment to be given an identity and therefore connected. When entities are connected objects, sharing information among them can easily be achieved.
- Application Programming Interfaces (APIs) and suitable integration platforms will make government organisations share data and information easily and effectively.

The fourth industrial revolution entails that e-Government will be designed in such a way that it will have the following attributes:

(a) Flexible information architecture; and
(b) Designed upon the cloud and fog computing models. E-Government as a Service (e-GaaS) espoused upon the cloud computing model enabling ubiquitous access to public services. The e-GaaS is designed in such a way that the major e-Government infrastructure is espoused the cloud computing model.

Important issues to be considered in order to achieve complete pervasive computing will be digital profiling, were citizens will have digital identity cards, passports, etc. Digital profiling will be achieved by ensuring that every citizen has a digital identity (made possible by the emergence of IoTs). These emerging technologies will be embedded with chips which collect information and updates information stored in the government's public or private clouds. The future is already being set by Russia leading the pack towards introducing digital passports by 2022.

Deployment of any technology innovation in the public sector begins from understanding the informational architecture of the business. The architecture is correctly understood by first understanding the synthesis of the underlying business process. For example, Building Information Modeling (BIM) has been used to create physical models of information systems (Conradie 2009).

9 Design of 4IR-Based e-Government Systems

The fourth industrial revolution brings about plethora of opportunities. This is because the 4IR enables digital, physical and biological connectedness (WEF 2017). Given that a lot of citizen innovation happens on mobile platforms, there is need to investigate potential inculcation of semantic Web (Web 3.0) onto e-Government design (Mir Ali (2013). In the fourth industrial revolution, desired e-Government applications should allow innovation at the demand side of public services. Consumers of public services need to be given an opportunity to inculcate their contextual characteristics into the e-Government modules (Singh 7 Chandel 2014). With the aggregation of heterogeneous technologies such as grid distributed computing, service-oriented computation, and parallel computing, cloud computing is a potential technological orientation that can go a long way in the design of contemporary e-Government applications (Patil et al. 2017).

Cloud computing is implemented to achieve on-demand self-service, broadband network access, rapid scaling and elasticity, resource pooling, and a dedicated service. These characteristics, for example resource pooling, allows e-Government to use the cloud computing model to have public services be accessed by heterogeneous devices and applications simultaneously (Singh 7 Chandel 2014). The use of cloud computing is the root for realising e-Government as a Service (e-GaaS). The use of e-GaaS reduces the cost of data storage in the e-Government environment. A number of countries are already taking advantage of cloud computing to come up with different contextual versions of e-GaaP. The following are some of the examples:

- Singapore has implemented a private government cloud (G-Cloud). This cloud platform provides the storage console for all e-Government data and enforces the different business rules with regards to public service delivery and data access. The G-Cloud is also embedded with the Basic Assurance Zone which is shared with public cloud users (Patil et al. 2017).
- Japan has the Kasumigaseki Cloud which provide a platform for data storage, information sharing, systems integration, resource sharing (including IT programmes and infrastructure) and therefore consolidate the government

services. the Kasumigaseki Cloud is a bid distributed computing infrastructure focussing on both information and application integration. The idea behind putting in place such an infrastructure is that information systems and applications will be integrated so that they can share data and realise the goal of automation of public services (Patil et al. 2017)., etc.

Successful implementation of cloud computing in e-Government environments will enable an improved access to information by citizens and business for decision-making. This chapter proposes an information architecture that can be used in conceptualising physical e-Government designs based on the e-GaaS model. Figure 2 shows the proposed conceptual model for e-GaaS which can be implemented to take advantage of the opportunities brought about by the fourth industrial revolution.

In Fig. 2, amalgamation of different technology modules culminated into an integrated e-Government information system based on the cloud (e-GaaS). The first module represents the macros which basically carry out the scripting functionality in the business processes. Macros are important in the recording of transactions in the e-Government environment—for purposes of audit trail. The macros record the process state at any given time in the business process. The second module is the Software Development Kit (SDK) which are a set of support kit that offers IT development community the capability to build RPAs according to the context in which they operate. The SDK allows experienced programmers to build single robots based on their knowledge with pre-requirements, best practice, methodology, etc. The third module is the cloud computing space which store information or application states at any time in the business process. The cloud configuration allows other functionalities of the e-Government system to access information ubiquitously. The fourth module

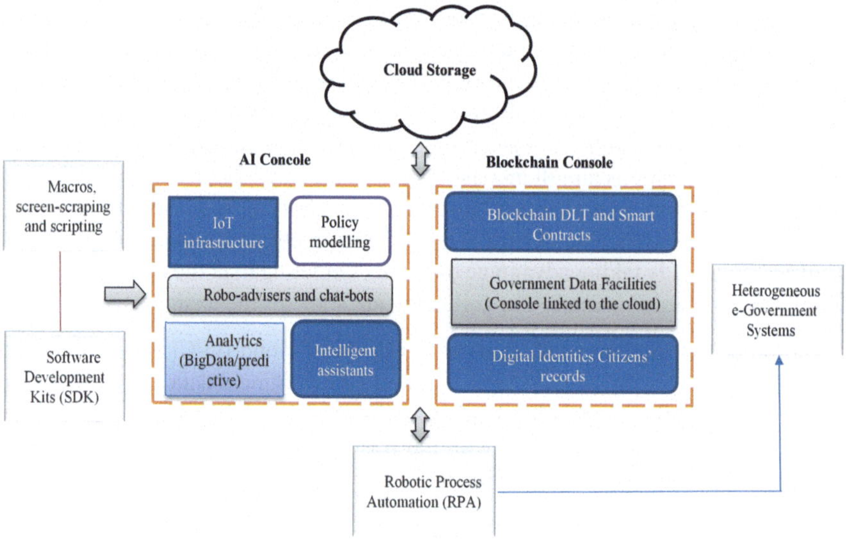

Fig. 2 e-GaaS integrated modules (*author's own concept*)

is the RPA which is connected to the SDK, the macros, the artificial intelligence console as well as the blockchain console. RPA allows e-Government systems to achieve automation and a level of intelligence thereby complementing or replacing the human being in repetitive mundane tasks.

The fifth and sixth consoles are the AI and blockchain consoles, respectively. The artificial intelligence console has different component in order to coordinate actions and achieve an acceptable level of intelligence. The intelligent agents are able to simultaneously analyse multi-dimensional structured or unstructured data (bid data) and utilising predictive analytics to advise digital agents of future scenarios. The robo-advisers render advice to civil servants interacting with e-Government platforms or autonomously execute business processes without the intervention of a human being. In order to achieve intelligence in e-Government environments, it is important that appropriate policies are in place supporting digital innovations. There should also be requisite IoT infrastructure to support e-GaaS. The blockchain console uses distributed ledger technology and smart contracts to ensure that information is adequately shared among the stakeholders in the e-GaaS environment to promote efficiency, accountability and transparency. Government departments are now transcending towards Social, Mobile, Analytics and Cloud (SMAC) integration which brings about all the entities in one information space. SMAC is used to make available to users simplified services and user experiences.

10 Future Prospects

There are a lot of opportunities that can be unlocked by designing e-Government applications with reference to the key tenets of the fourth industrial revolution. In order to amass these opportunities, it is important that public organisations be ready to re-engineer their business processes towards having in place more agile processes. Putting in more scalable business processes which can be redesigned anytime given the dynamic changes in business process reconfiguration given short lifecycle of technological innovations.

Future models of automation and intelligence in the ambit of e-Government will include e-Voting systems (vote casting, automatic counting, and declaration of results of the poll), permit allocation and administration at the Home Affairs—individuals upload their photos and other information in online systems and digital agents check against requirements and quickly make decisions on allocation of permits (residence permits, passports, identity documents, etc.), etc.

11 Conclusions

There are many opportunities that public organisations can take advantage of in the fourth industrial revolution. Because progressive 4IR-based information systems are being developed with reference to robotic process automation, many technology vendors have come up with diverse RPA software solutions that public organisations can take advantage of and customise according to their unique contexts. Agility is a key requirement in 4IR-based information systems. 4IR-based e-Government systems enable automation and intelligence to be achieved in the public governance value chains better than what a human being can do. The ability of digital agents or bots to analyse huge sets of information in a relatively short period of time is a huge advantage towards automation and intelligence.

References

Adobe (2015) Digital government transformation. https://offers.adobe.com/content/dam/offer-manager/en/na/government/Digital%20Transformation%20in%20the%20Public%20Sector%20eBook.pdf. Accessed 17 July 2019

Conradie D (2009) Building information modelling (BIM). https://researchspace.csir.co.za/dspace/bitstream/handle/10204/3364/Conradie_2009.pdf?sequence=1&isAllowed=y. Accessed 16 July 2019

Engin Z, Treleaven P (2019) Algorithmic government: automating public services and supporting civil servants in using data science technologies. Comput J 62(3):448–460. https://doi.org/10.1093/comjnl/bxy082. Accessed 20 July 2019

Lacity M, Willcocks L (2016) Robotic process automation: the next transformation lever for shared services the outsourcing unit, working research paper series paper 16/01, pp 1–35. http://www.umsl.edu/~lacitym/OUWP1601.pdf (Accessed 21 July 2019)

Mir Ali MR (2013) Review office automation system effect to improve the staff managers decision making process of Sabzevar' municipal. Global J Manage Bus Res Adm Manage 13(10-A):1–5. https://journalofbusiness.org/index.php/GJMBR/article/view/1064. Accessed 18 July 2019

Parasuraman R, Sheridan TB, Wickens CD (2000) A model for types and levels of human interaction with automation. IEEE Trans Syst, Man Cybern Part A: Syst Hum 30(3):286–297

Patil P, Patil S, Chaudhari N (2017) Use of cloud computing & automation for e-Government in India: challenges and benefits. Int J Eng Manage Res 7(1):316–331

Sheridan TB (1992) Telerobotics, automation, and human supervisory control. MIT press

Sheridan TB, Parasuraman R (2000) Human vs. automation in responding to failures: an expected-value analysis. Hum Factors 42:403–407

Sheshasaayee A, Bhargavi K (2017) A study of automated decision making systems research inventy. Int J Eng Sci 7(1):28–31

Singh VJ, Chandel A (2014) Evolving e-Governance through cloud computing based environment. Int J Adv Res Comput Commun Eng 3(4):57–74. https://doi.org/10.5539/ijbm.v9n9p57

Smith ML, Noorman ME, Martin AK (2010) Automating the public sector and organizing accountabilities. Commun Assoc Inf Syst 26:1–16. https://doi.org/10.17705/1cais.02601

Vesnic-Alujevic L, Stoermer E, Rudkin J, Scapolo F, Kimbell L (2019) The future of government 2030+: a citizen-centric perspective on new government models. EUR 29664 EN. Publications Office of the European Union, Luxembourg. ISBN 978-92-76-00165-2. https://doi.org/10.2760/145751, JRC 115008

WEF World Economic Forum (2017) Harnessing the 4th industrial revolution for sustainable emerging cities, November 2017 fourth industrial revolution for the earth. http://www3.weforum.org/docs/WEF_Harnessing_the_4IR_for_Sustainable_Emerging_Cities.pdf. Accessed 16 July 2019

WEF World Economic Forum (2019) Industrial revolution, putting people at the centre in collaboration with accenture leading through the fourth. http://www3.weforum.org/docs/WEF_Leading_through_the_Fourth_Industrial_Revolution.pdf. Accessed 16 July 2019

The Disruptive Force of Smart Contracts

Michele M. van Eck

Abstract The fourth industrial revolution hints at technological advances that could disrupt entire industries. Already glimpses of the disruptive force can be seen with the advent of technologies such as cryptocurrencies, blockchain, adaptive and predictive algorithms and the use of big data. The smart contract has emerged from these technological advances as a sophisticated self-executing computer program with the capability of automatically adapting itself throughout the contract life cycle to achieve the contractual objectives of the parties. Smart contracts are fully autonomous, self-evolving and dynamic processes in the formation, interpretation and management of contracts. Smart contracts could also remove the need for trusted third party intermediaries in commercial transactions, and thereby fundamentally alter the traditional approach to contract theory, practices and dispute resolution processes. Smart contracts will likely be incorporated into commercial transactions in some way, shape or form but the extent of such integration remains uncertain. This new paradigm of contractual theory and practice certainty will bring about different challenges and opportunities as the legal and commercial world grapples with the ethical and philosophical issues of controlling a contracting system that, at its core, is intended to liberate transactions from central control.

Keywords Fourth industrial revolution · Disruptive force · Smart contracts · Online dispute resolution processes · Technology

1 Introduction

The legal system, for the most part, has been described as a slow and somewhat tedious process in the administration of justice in the fields of business and commerce. Even in the literary classic of Henry VI, Shakespeare uses satire and rhetoric by

M. M. van Eck (✉)
Department of Private Law, Faculty of Law, University of Johannesburg, Aukland Park Kingsway Campus, Johannesburg, South Africa
e-mail: mmvaneck@uj.ac.za

W. Doorsamy et al. (eds.), *The Disruptive Fourth Industrial Revolution*, Lecture Notes in Electrical Engineering 674,
https://doi.org/10.1007/978-3-030-48230-5_2

stating that "[t]he first thing we do, [is] kill all the lawyers" (Henry VI, Part 2, Act IV, Scene 2). This does not point to literal murder but rather is a plea for a different approach to the law and more efficient systems, processes and ways of doing things.

The law is heavily steeped in tradition and is almost proud of its heritage and dependence on precedence. This exact process has provided the necessary stability to foster growth, development and eventual maturity in commercial and contractual engagements from the height of legal thought in the *Pax Romana*, through the Age of Reason and, finally, to the present day. However, a few occasions may be pinpointed in history that have firmly taken hold of the rudder of progress and steered the legal system, thoughts and approaches into a new course of development. Examples are to be found in the codification of the Roman law in the Twelve Tables in 450 BC (Goodman 1995: 131; Thomas et al. 2000: 26) as well as the modern codification of the civilian legal systems in Europe during the nineteenth and twentieth centuries, which charge was led by the French *Code Civil* of 1804 and the German *Bürgerliches Gesetzbuch* of 1900.

At the dawn of digitalization and the third industrial revolution the plea for more efficient and accessible legal systems, processes and ways of doing things became more prominent through advanced global communication capabilities that called for a breakaway from the old traditional manners of communication (which still is popularized by the modern Plain Language movement) and international commercial engagements. The disruptive force of further change has been felt in new business models such as Airbnb, Uber and the like that have taken hold of the commercial world with radically-transformed businesses and commercial structures, and have left the established giants of industry scrambling for financial survival and relevance in the new digitized world of commerce.

Society finds itself again approaching the precipice of change as the developments of human interaction with technological advances threaten to disrupt the very convention upon which the old and established traditions are founded, and now demand new models and approaches altogether. Existing structures continue to reinvent themselves as a new wake of technological developments, such as blockchain, cryptocurrencies, artificial intelligence, adaptive and predictive algorithms and big data, challenge the traditional conventions of business and commercial transactions. Flowing from these advances, the advent of what has been called the "smart contract" has brought about the possibility of contracting in a way never before imagined. Contracts traditionally have been concluded either tacitly, verbally or in writing (and only recently digitally), and in their very nature have been reactional to the needs and requirements of the contracting parties. Smart contracts have the potential of being proactive in nature with the capability of monitoring compliance of obligations and adapting to the needs of the contracting parties in real time.

The potential of these technological advances is but a glimpse of the imminent fourth industrial revolution in which some predict an extinction-level event for the classical model of contract theory and practice. The disruptive force on contracts could be revolutionary and possibly could be the next evolutionary step in contract theory and practice. This chapter considers the disruptive force of smart contracts specifically relating to the South African contract theory and general contract practice by considering (i) the close link between the historical development of commercial advances and contractual practices; (ii) the functioning of the self-driving smart contract; (iii) the constraints, developments and opportunities of smart contracts; (iv) new skills and knowledge that would be acquired; and (v) the development of dispute resolution processes to complete a fully autonomous smart contract ecosystem in commercial and business transactions.

2 Commercial Advances and Contractual Practices

"The law of contract is of fundamental importance in the modern world, because it is woven into and is inseparable from every form of economic activity" (Christie et al. 2011: 1). The complexity and development of commercial structures and contract law have been, and continue to be, interlinked, as contract theory and practice develop in-step with that of commercial transactional requirements (Savelyev 2016: 7).

Commercial transactions of the most primitive of societal and commercial structures were concomitant to honor-based contracting, religious oaths and the use of the bartering system. Initially, little differentiated the law and religion (Thomas et al. 2000: 26), and society's belief system had a significant impact on the way in which parties commercially interacted with one another. There also was a heavy reliance on the bartering system for commercial transactions, which required little trust between the contracting parties as it allowed for the immediate exchange of something of value for another thing of similar value. The complexity of contract law at the time was in-step with such bartering commercial transactions which required the immediate exchange of value.

The first real sophistication in commercial and economic activity was seen with the advent of what one may call future promises. In this structure goods and services were provided on the strength of a promise of future payment, and vice versa. As there was no immediate exchange of value, transactions required contracting parties to trust that the agreement would be performed as promised. Contract theory again developed to cater for the enforcement of such future promises. However, agreements generally existed in the form of pacts or *pacta* in which parties had to trust that the agreement would be performed (Buckler 1985: 5). The Romans later recognized specific categories of contracts, being real contracts (*contractus re*); verbal contracts (*contractus verbis*); written contracts (*contractus litteris*); and consensual contracts (*contractus ex consensu*) (Thomas et al. 2000: 217; Justinian's Digest 1985: 44.7.1.1; Justinian's Institutes 18893.8.2; Gaius 1904: 3.89). The verbal contract was the most popular (likely due to the illiteracy in society; see Maine 1906: 302) and generally was based on forms of verbal exchanges clothed in ceremony and formalities that often included the use of specific predetermined words that formed a valid contract. The exchange was based on trust that the contracting parties were honor-bound to fulfil their promises. A typical example was that of the *stipulatio* in Roman law. Although the popularity of verbal contracts prevailed during this time, there was a gradual increase in written documents in order to record the agreements and create some level of legal certainty regarding the terms of the commercial transaction. A typical example of a written contract was found in a ledger system in which the creditor would record the financial transactions and the ledger would constitute proof of the debts of a debtor (Thomas et al. 2000: 262; Gaius 1904: 3.129–131).

The late twentieth century flowing into the twenty-first century signaled the start of the digital age which brought different, faster and more efficient ways of transacting. Notwithstanding the different manners of executing transactions, commercial activity remained based on the fulfilment of future promises. This period of history brought about the Internet, digital commerce and, in particular, e-mail exchanges, which birthed new branches of law, such as telecommunications law, cyber law and data protection law. These additions to the legal framework also recognized digital forms of contracts, and legal rules were developed to regulate the global digitalized world of commerce.

However, it was only in 2002 that South Africa officially recognized data messages as a form of contractual engagement through legislative intervention in the Electronic Communications and Transactions Act (ECTA). The purpose of the ECTA, among other things, was to "provide for the facilitation and regulation of electronic communications and transactions; to provide for the development of a national e-strategy for the Republic; to promote universal access to electronic communications and transactions and the use of electronic transaction[s]". The ECTA steered South African law into recognizing not only tacit, verbal and written contracts but also the validity of digital contracts (Sections 11 and 22 of the ECTA), as well as electronic signatures (Sections 12 and 13 of the ECTA and the Supreme Court of Appeal case of *Spring Forest Trading v Wilberry*). From digital contracts stemmed further development in specific forms of contractual rules in respect of online trading and subcategories, such as clickwrap contracts (terms being accepted by clicking on an acceptance button) and browsewrap contracts (terms being binding by merely having such terms displayed on a website) (Koornhof 2012: 44–46). Despite the digital nature of such contracts, they often were not fully digitized and many contracts required ancillary paper work to execute, perform and monitor the performance of contractual obligations (Savelyev 2016: 12).

Contracts have developed throughout history and continue to be reactional in nature, being focused on ensuring that the innocent party has the necessary remedial action to enforce contractual promises (Caruso 2018: 1795). Traditional contract law primarily is focused on ensuring that there are remedial actions available to the contracting parties, and is aimed at resolving disputes in respect of *ex post* performance and not that of *ex ante* performance (Werbach et al. 2017: 318). Further technological advances in blockchain and consequentially smart contracts focus on the *ex ante* performance and have the potential of ensuring that the performance of contractual promises essentially is guaranteed. Thereby, smart contracts would have the ability of removing the trust requirement between the contractual parties for future promises (Caruso 2018: 1795).

This technology has the potential of disrupting the very nature and characteristics upon which commercial and economic structures rest. Due to the close link between commercial activity and contracts, the smart contract technology would likely disrupt the legal rules governing such commercial activity. The next phase of commercial activity appears to be moving away from future promises to that of self-executing and self-sustainable contractual ecosystems. The paradigm for smart contracts has the potential of transcending the traditional contractual enforcement mechanisms and that of inter-contractual party trust necessary for the performance of contractual promises. These future disruptive changes, however, do not imply that the developments of the past should be disregarded, but rather that the advances in new technology should be used to develop faster, better and more efficient manners of contracting (Szabo 1997), thereby ensuring that contract theory and practice remain in-step with current commercial and business activities steered by technological advances.

3 Self-driving Smart Contracts

A smart contract may be described as a self-automated and self-executing computer program, which operates in a closed and secure system and ensures contractual performance by means of blockchain technology. In 1994 Nick Szabo first alluded to the concept of smart contracts in the context of future technological developments. At the time the concept of a smart contract was a practical impossibility due to the immaturity of technological capabilities. It was only in 2014 in the Ethereum White Paper of Vitalik Buterin and the recent advances made in blockchain technology that the smart contract was resuscitated into a viable commercial transacting methodology.

The functionality of a smart contract has often been described as the modern version of a vending machine (Szabo 1997). The vending machine, as in the case of the smart contract, is a self-contained and automated method of contracting between parties. Anyone can participate in transacting with a vending machine provided that they have the appropriate currency. Once the currency (in this case coins) is placed in the vending machine, the vending machine automatically dispenses the goods (Szabo 1997). The smart contract operates similarly through the use of blockchain technology, in which a series of pre-programmed protocols automatically execute certain actions (being the contractual performance) when pre-determined events occur. The smart contract, therefore, removes the need for trusted intermediaries through the use of blockchain technology. As in the vending machine example, this is a rather straightforward transaction in which a thing of value is exchanged for another thing of value. This would be the most basic of functionalities of a smart contract and could be described as a shallow-smart contract (Kraus et al. 2019: 109). However, it would be possible to have more sophisticated smart contracts that are based on "related, sequential and conditional" performances of the contractual promises (Bacon et al. 2018: 45), which could be described as a deep-smart contract (Kraus et al. 2019: 109). The practical applications of smart contracts could theoretically be used in any type of commercial transaction, whether simple or complex in nature, such as leasing a car, renting a property or even in financial instruments to facilitate the transfer of property, goods or services from one person to another.

3.1 Blockchain Technology

Blockchain technology was initiated in 2009 and essentially is a sophisticated ledger system that originally was developed for cryptocurrency transactions that do not require traditionally-trusted intermediaries (Kolber 2018: 216). The technology achieves so-called decentralized consensus through various nodes in the blockchain that verify the accuracy of the data in a blockchain. Blockchain's key features firstly is data security and is achieved through its hash functions that create a system that avoids tampering with its records (Bacon et al. 2018: 6). The second key feature of the technology is that of identity authentication which is achieved through the public key infrastructure that is authenticated by the various nodes in the blockchain (Bacon et al. 2018: 6). These features of the blockchain technology render the system immutable and secure.

The disruptive force of blockchain technology has rendered many of the traditionally-trusted intermediaries in transactions virtually obsolete (Kolber 2018: 216), and continues to threaten the operation of established industries such as the banking and insurance sectors. It has become apparent that the blockchain's distributed ledger technology is not limited to cryptocurrency transactions but, with emerging technologies, could infiltrate many other sectors of society. Blockchain technology, for example, could be used for crowdfunding, voting systems, cloud file storage, ownership verification and documentation control (Werbach et al. 2017: 326), and the potential now exists to use the technology in smart contracting ecosystems.

Blockchain can take one of three forms (Mckinney et al. 2018: 320–321) being (i) public blockchains that permit anyone to read and send transactions and are fully decentralized (a typical example of a public blockchain is Bitcoin and Ethereum); (ii) consortium blockchains that consist of a number of preselected nodes that would verify the blocks in the chain, and are partially decentralized in nature (a typical example of consortium blockchains is found in financial institutions that use decentralized blockchain systems for secure financial transactions; this type of blockchain would also be the likely candidate for smart contract arrangements); and (iii) private blockchains in which transactions are controlled by a single party, are completely centralized and are often used for private record keeping purposes.

Blockchain technology is attractive for commercial transactions as it has the ability to accurately keep records, is self-validating through "decentralized consensus" processes and renders such transactions "tamper-proof" (Cong et al. 2018: 8). Smart contracts, being self-regulating and self-automated transactional ecosystems, have now been considered a potentially viable contractual mythodology.

3.2 *Self-driving Smart Contracts*

Once it has been formed the smart contract requires no further human intervention (Kolber 2018: 208). The contracting parties merely need to agree on the macro-objectives of the contract, being the outcome that they wish to achieve by entering into the transaction. The smart contract would automatically execute various micro-objectives that include methods, requirements, processes and steps to achieve the macro-objectives of the contract. The smart contract thereby monitors and automatically adapts the *ex ante* performance of the contract.

The micro-objectives in traditional contractual arrangements normally would be agreed up-front by the contracting parties and the contracting parties thereby would ensure certainty in the execution of contractual performances. However, this has proven to be practically difficult to achieve as the contract would have to cater for an infinite number of possibilities and eventualities that could impact the achievement of the macro-objectives of a transaction. The traditional contractual arrangement is inflexible and requires the consent of the contracting parties for any amendments to the contract. This likely is due to the traditional contract's main purpose being the regulation and execution of *ex post* performances. The traditional contract therefore may be described as reactional in nature.

The smart contract, on the other hand, has the capabilities of automatically adapting its micro-objectives in order to achieve the macro-objectives of the contracting parties. The smart contract would automatically monitor and adjust the micro-objectives by means of data integration from a number of sources to achieve the macro-objectives. Therefore, the smart contract may be described as an intuitive, proactive approach to the monitoring and execution of contractual performances.

Smart contracts only require the contracting parties to agree to the ultimate objective of the contract and, by adapting the micro-objectives, could write its own terms by "filling in the gaps". This could be achieved (Casey et al. 2017: 2–7) by (i) the contracting parties agreeing to the macro-objectives and leaving the micro-objectives to the smart contract to work-out; (ii) using machine-driven analytics and artificial intelligence to achieve the macro-objectives through adaptive and predictive algorithms; or (iii) using external collected data to monitor the performance and execution of contractual obligations of the contract (adapted from Casey et al. 2017: 2). Thereby, smart contracts would not require the contracting parties to agree to all terms but only to the transaction's ultimate objectives.

3.3 *Future of Smart Contracts*

The future development of smart contracts stands at a proverbial fork in the road. On the one hand, the smart contract could be reduced to a mere tool to execute the obligations of the traditional contract. This is a popular view as resistance persists in accepting the possibility of a fully-integrated legally-recognized smart contract (see for example Kraus et al. 2019: 108; O'Shields 2017: 185). The smart contract would then be viewed as a function to fulfil the legal contract between the parties, and not the legal contract itself (Bacon et al. 2018: 46). If technology cannot adapt to the challenges that smart contracts face, it would be impossible for the smart contract to evolve further, and thereby could be reduced to merely an effective tool for the fulfilment of the performance of a traditional contract.

On the other hand, the possibility exists that technology could develop to enable a fully autonomous, self-fulfilling, self-regulating and legally-recognized form of a smart contract which is more than a tool for the execution of contractual performances. If this is the direction of the smart contract's development, then the technology would have the potential of augmenting both the contract practices themselves as well as the enforceability of contractual promises (Cong et al. 2018: 8). To this end, decentralized autonomous organizations (DAO), being self-sustainable and self-regulating bodies or entities, would establish such rules and codes to regulate self-autonomous transactions that could be used in the execution and governance of smart contracts.

Currently the development of smart contracts are hampered by the immature technological capabilities in collecting accurate data for the monitoring of contractual performance. Smart contracts are premised on the collection and use of accurate data, which could be achieved through so-called oracles. Oracles essentially collect data and information for a specified period of time which is then fed into the smart contract (Lauslahti et al. 2017: 17). Oracles, therefore, may be described as the access point between smart contracts and information available in the rest of the world. If the accuracy of the information collected by oracles can be verified and authenticated (Mik 2017: 23), and should oracles be successfully integrated to most areas of society, then such data could be used to determine whether performance or conditions were fulfilled and whether the smart contract would be able to accurately determine the execution of its pre-determined protocols (Lauslahti et al. 2017: 17).

The smart contract is characterized by its ability to adapt and, therefore, this technology need not be limited to the mere collection of data. The smart contract could also interpret the data of possible future outcomes through the use of adaptive and predictive algorithms, which would predict possible outcomes and adapt the micro-objectives to ensure that the macro-objectives are fully realized and achieved.

The practical possibilities of smart contracts are endless and have the potential of disrupting the very basis on which modern society contracts. We have already seen the first generation of self-driving smart contracts in the form of "self-pricing contracts" in the auto-insurance industry (Casey et al. 2017: 3). In these contracts the contract price automatically adjusts based on computer analytics (Casey et al. 2017: 3), which often is linked to how well or badly a person drives. This information would be provided to the smart contract and would then directly influence and adjust the price of auto-insurance. Further advances in predictive technology would create more sophisticated and sustainable models of self-driving smart contracts (Casey et al. 2017: 3), both in the insurance industry and also in other sectors. Another example of the practical use of smart contracts is found in 2016, where a smart contract was created to provide insurance for passengers affected by delays on

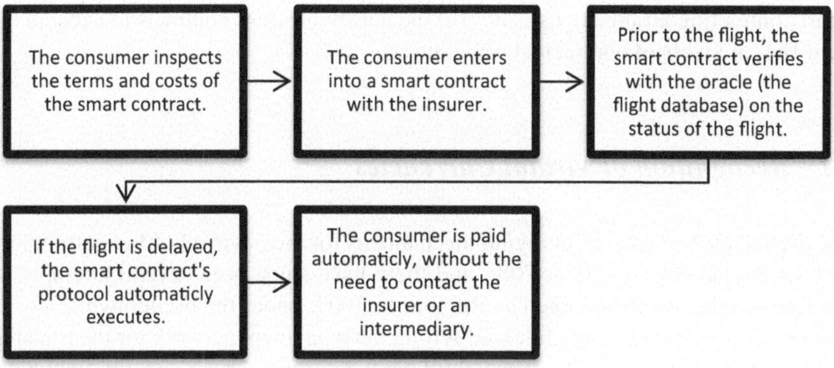

Fig. 1 The workflow of smart contracts used in delayed flight insurance

certain flights (Kolber 2018: 209). This may be illustrated in the following workflow (Fig. 1) (adapted from Kolber 2018: 209).

A person could inspect the terms and cost of the insurance and enter into the contract if they were comfortable with the terms. Prior to the flight, the smart contract would verify delays in the flight and if there were delays the smart contract would execute the protocol and automatically settle the claim by paying the insured (Kolbe 2018: 209). The database of the flights would function as the oracle, in other words, the database would be the connection point between the raw data and the smart contract. Furthermore, all payments in this scenario would be executed in real time without the need to contact an intermediary at all.

4 Constraints, Limitations and Opportunities

Change is the product of progress, and progress inevitably stems from technological advances. Change brings about questions and inescapable uncertainties, but also is the breeding ground of opportunities that previously were unavailable. This is evidenced in the wake of the first three industrial revolutions that brought radical economic and societal changes and a number of opportunities for progress. If past experience is the indicator of the future, then it would not be unreasonable or unlikely to anticipate similar industry-wide changes and opportunities in the advent of the fourth industrial revolution.

The current technological capabilities of smart contracts and blockchain technology have not fully matured and still have some constraints and limitations that would hamper the full integration of the technology into modern society and commercial transactions. The constraints, limitations and possible opportunities are explored in three general aspects that have a direct bearing on smart contracts, being (i) the issue of recognition of virtual currencies; (ii) the need for regulatory oversight in smart contracting arrangements; and (iii) the ability of smart contracts to execute on the different layers of contractual obligations.

4.1 Recognition of Virtual Currencies

The digital age brought about digital mechanisms for the payment of fiat currencies, such as PayPal (Spruyt 2018: 708), and there have since been several attempts at creating virtual currencies, such as Bitcoin. In 2009, under the pseudonym Satoshi Nakamoto, a publication was released setting out a payment network for the transfer of virtual currency without trusted third party intermediaries, and this payment network became known as blockchain technology (Spruyt 2018: 709). These virtual currencies, such as Bitcoin, are mechanisms of payment that are underpinned by the functionality of blockchain technology, and consequently that of smart contracts.

In 2017 the South African Reserve Bank advised that the development of a national cryptocurrency would be too risky for South Africa (Haig 2017), and it continues to view virtual currencies and cryptocurrencies not as a valid currency in South Africa (South African Reserve Bank Position Paper 02/2014).

The apparent hesitancy in recognizing virtual currencies and cryptocurrencies largely is due to the potential disruptive force on the banking sector and concerns relating to consumer protection, such as (South African Reserve Bank Position Paper 02/2014) that (i) the unregulated nature of virtual currency and cryptocurrency contains an inherent risk of financial losses to the end user; (ii) security breaches, theft and user error could result in loss and financial exposure to end users; (iii) risks associated with fraud and fraudulent activities are of particular concern in light of the irreversibility of transactions regulated by blockchain technology; and (iv) no insurance cover exists for lost wallets or services becoming unavailable.

The recognition of cryptocurrencies is an important, but not necessarily fatal, part to the successful application of smart contracts. Cryptocurrencies, on the one hand, may be seen as forming the basis of payment for automated electronic commercial transactions based on blockchain technology. On the other hand, cryptocurrencies in smart contracts certainly would speed up the process of transacting, however, smart contracts can function without the use of cryptocurrencies. As the South African Reserve Bank currently does not recognize virtual currencies or cryptocurrencies as valid currency in South Africa, the use of such virtual currencies are at the consumer's risk (South African Reserve Bank Position Paper 02/2014).

Notwithstanding this, the South African Reserve Bank embarked on Project Khokha in which a proof-of-concept was successfully trialed in distributed ledger technology to manage wholesale payment settlements between banks (South African Reserve Bank Press Statement 2018). Project Kokha is not, strictly speaking, a cryptocurrency, but is a real time interbank settlement system which was successfully implemented and which illustrates the powerful and diverse application of technology. The popularity of virtual currencies continues to increase and eventually some form of regulation would be introduced. Until then, those who use virtual currencies in smart contracts may very well be left without contractual legal recourse should there be an error, incorrect goods or services provided, or the like. Fortunately and depending on the circumstances, there may be recourses available under South African law, for example, in the field of undue enrichment. However, the enforcement of such action in an international sphere may render such recourses ineffective or impractical.

Some countries, such as China and Thailand, have banned the trading of virtual currencies outright (Spruyt 2018: 719). South Africa, on the other hand, has been slow to set out regulations for virtual currencies (Spruyt 2018: 719). In order for smart contracting to become an effective option for commercial transactions, virtual currencies would have to be recognized within the legislative and regulatory framework of South Africa. Until then, a significant limitation remains in place for the full development and integration of smart contracts in commercial transactions in South Africa.

4.2 Regulatory Oversight

An attractive feature of smart contracts is the potential of removing trusted third party intermediaries and having agreed regulations applicable only between the contracting parties. This technology has been described as "[t]rustless public ledgers, maintained by no one but available to everyone" (Fairfield 2014: 36). It seems that contracts of the future would be regulated by what one may call the "consensus protocol" of the blockchain (Fairfield 2014: 36). In such instances the code has been described as the law between the contracting parties. This, perhaps, is not too far removed from the traditional approach as contracts, after all, have been described as ad hoc legislation *inter partes*, which effectively means that the contractual terms agreed by the parties are as good as legislative requirements (Van Eck 2015: 23). The extent to which contracting parties could contract without trusted third party intermediaries or governmental oversight remains uncertain. Traditionally, the ad hoc rules between contracting parties have always been limited by and subjected to the legislative and regulatory framework, the values of society as a whole and the rule of law. Smart contracts appear to attempt the removal of such oversight and control in individual contractual engagements.

Contract theory has recognized that not all contractual engagements can be considered valid or enforceable. Examples include a person entering into a contract that lacks the necessary contractual capacity, such as those suffering from mental illness (Hutchison 2017: 154–155). Further, the law has established rules to protect the vulnerable and disenfranchised of society by allowing them to escape contracts that were entered into due to a material mistake (*error*) (Hutchison 2017: 84); duress (*metus*), which includes threats and intimidation; as well as fraud (*dolus*) (Hutchison 2017: 141, 145, 147). Contract theory has also established rules to ensure that contracts that are illegal or against public policy are not enforced (Hutchison 2017: 182). The most prominent of these examples would be contracts for murder or slavery that would be considered illegal and consequently would not be enforced in law. However, there does not appear to be similar regulatory intervention contemplated in the functioning of smart contracts, as the function of the smart contract is regulated solely by the intention and needs of the contracting parties.

Smart contracts consider only the expression of what the parties intended, denoting that the smart contract is a manifestation of the consensus reached by the parties (Savelyev 2016: 19). Further, due to its nature smart contracts do not take into account consequences of contracts that are based on duress, undue influence or a misrepresentation by either party. In such instances the smart contract would still execute its protocols regardless of the circumstances, and the contracting parties would be unable to terminate or reverse the transaction without the consent of both parties (which could be achieved by entering into a new smart contract in order to correct the initial transaction). It is for these reasons (and because smart contracts should conform to the traditional requirements of a valid contract) that critics have alluded to the fact that smart contracts are not in themselves complete and legal contracts (see for example Kraus et al. 2019: 108; O'Shields 2017: 185).

The current state of smart contracts appears to be similar to that of the declaration theory in traditional contractual engagements to establish contractual liability (Christie et al. 2011: 25). In terms of the declaration theory (which is not accepted as the sole basis for liability in South Africa; see Christie et al. 2011: 1) the physical manifestation or expression of the parties' actions, conduct or words indicates whether a contract exists or not. Unless qualified in some form, this approach has the potential of leading to unjust and sometimes absurd results (Hutchison 2017: 16). The declaration theory and that of smart contracts would function similarly. Take, for example, a contract entered into by the parties for the purchase of goods. The parties in this example intend to buy and sell the goods for a certain amount of money but the contract incorrectly records the purchase price. In terms of the declaration theory, and the current functionality of the smart contract, the parties would be bound to the incorrectly-recorded purchase price solely because the act of signing (or execution) of the contract indicates that the parties have agreed to the content of the agreement (irrespective of their true intent). As a result of its irreversible nature, if the smart contract remains unqualified it could lead to unjust results. This is particularly relevant in instances where a smart contract is entered into as a result of some dubious conduct (such as duress, undue influence or even a misrepresentation by the supplier) or there is a flaw in the programming of the code which results in the incorrect execution of the smart contract. The smart contract would execute irrespective of the reason for entering into the contract and would not take into account any programming error. The smart contract itself, therefore, could be open to vulnerability should the code be incorrectly programmed. For example, in 2017 it was reported that an end user had accidently exposed a flaw in a smart contract called *Parity* (Bacon et al. 2018: 55) in which a substantial sum of value had been frozen in a wallet, thereby becoming inaccessible to the end user (Bacon et al. 2018: 56).

It goes without saying that notwithstanding the allure of having a system of contracting without third party intermediaries or regulatory influence, there is a real possibility of unfair and unjust results in a smart contracting system without qualification. Whatever form smart contracts take in the future, it would require some form of regulatory oversight and that traditional contractual theory, to some extent, would remain intact in instances where the computer code diverges from the contracting parties' intent (Werbach et al. 2017: 322). A further consideration is that South African contract law does not allow parties to enter into an agreement that attempts to "deprive the courts of their normal jurisdiction" (Christie et al. 2011: 365; see also *Schierhout v Minister of Justice*: 424), which essentially is one of the core characteristics of smart contracts.

4.3 Execution of Contractual Layers

Being a computer program, a smart contract is based on code that uses Boolean logic. The code is not open to interpretation, as is the case with traditional contracts, and contains a form of mathematics that is based on true or false values that fit within the binary numbering system of computer science (Savelyev 2016: 12). The smart contract therefore cannot adapt to change and must execute as its program demands.

At this stage this Boolean logic limits the smart contract's ability to fully integrate all the types of obligations that exist in traditional contracts. Contractual engagements have a number of variables and may be divided into three general layers of performance, namely, (i) boilerplates, which generally house the administration of the contract, such as the applicable law and jurisdiction; (ii) the non-commercial layer, which includes obligations that are of a non-commercial nature, such as restraint of trade, confidentiality, intellectual property protection and *force majeure* provisions; and (iii) the commercial layer, which includes aspects relating to the performance of commercial activities such as services, the delivery of goods and that of payment, penalties and interest.

Current technological capabilities enable a smart contract to generally execute the commercial layer of a contract, such as payment and a delivery release, based on the available data at the time. What remains questionable is the smart contract's ability to successfully execute the non-commercial and boilerplate layer of a contract, as well as the smart contract's ability to effectively incorporate natural language into its code (Mik 2017: 15). The technology has not yet developed to fully monitor, integrate and execute the boilerplate and non-commercial layers of a contract. This area of the technology must still reach a level of maturity to ensure a fully-integrated smart contract in commercial transactions. Should technology develop further to accommodate the boilerplate and non-commercial layers of a contract, then smart contracts would be a serious contender for the recognition of a new contracting form and methodology. Until such time the smart contract would remain an effective tool for the performance of a traditional contract.

5 New Skills and Knowledge

The traditional approach to contractual engagements is for lawyers to draft and prepare contractual paperwork by expressing ideas, concepts and the inner workings of the minds of the contracting parties in natural language (mostly in English), which often is open to ambiguity and interpretational issues. Ambiguous and vague provisions can result in lengthy and costly litigation processes in which an adjudicator would attempt to establish what the contracting parties' intentions were at the time of the conclusion of the contract.

These traditional approaches are misaligned to the heartbeat of international commercial transactions powered by new and developing technologies such as blockchain and, consequently, smart contracts. Innovation, speed and agility are characteristics of these technological advances. However, the legal system and "[t]raditionally educated lawyers are not usually known for key characteristics like agility and the capacity to innovate" (Fenwick 2017: 353). The legal fraternity is considered to be a trusted intermediary for contractual engagements but "the counseling, deal-making, matchmaking, gatekeeping, and enforcing roles are increasingly performed by technology" (Fenwick 2017: 362). The drive to introduce smart contracts as a viable alternative to traditional contractual practices is likely to lead to the need for different legal skillsets.

The extent to which a lawyer would be required to adapt and learn new skillsets would depend largely on the direction in which smart contracts develop and are incorporated into commercial transactions. If, for example, smart contracts become a tool for the enforcement of traditional contracts, a lawyer may not need to learn any new skills as the programming of a smart contract could simply be outsourced to an individual who is trained in coding. In instances where smart contracts become the predominant manner of contracting, there is a real possibility that lawyers may need to acquire skills to program the appropriate code for the smart contract. Notwithstanding this, there are still some that argue that a written form of a smart contract would be required for the contracting parties to appreciate and agree to the terms of the smart contract (Mik 2017: 16). The third possibility is that the traditional contract drafter could become obsolete, which may occur in instances where blockchain suppliers include the functionality and flexibility for individuals to simply choose the attributes and qualities from a variety of standard provisions for their smart contract (Mik 2017: 17). In doing so, there would be no need for the contract drafter to act as an intermediary and would thereby be removed from the process altogether. This is visible on the *Ethereum* platform which allows individuals to "draft their own contracts" (Surujnath 2017: 273) and in *Open Law*. Naturally, such options may limit an individual's ability to create bespoke transactions, however, nothing prevents contracting parties from engaging with third party coders for the creation of bespoke smart contracts.

The drive to reduce the time for legal processes has been referred to as the "law lag" (Reyes 2017: 435). The "law lag" may be described as the typical time taken to assess the law, the consideration of a person's legal position, the preparation of the necessary documentation for signature and any enforcement processes for commercial transactions (see Reyes 2017: 435). Advances in technology would either remove or drastically reduce the "law lag" in transactions by enabling smart contracts to be immediately updated and to have real time feedback through distribution ledger technology (Reyes 2017: 435). The code can be changed more quickly than natural language and commercial transactions could then be in-step with the fast-paced world of technology (Reyes 2017: 436).

The crypto-legal structure has been described as "the law in action" (Reyes 2017: 435). Already new branches of the law have started to develop such as *lex cryptographa* or crypto law that would ultimately regulate the creation, enforcement and adjudication of smart contracts (Reyes 2017: 401). Instead of viewing the self-regulation of smart contracts outside the constructs of the law, it is rather becoming an emerging field of the law, which would ultimately require the international adoption of a single set of rules for any effective application. Once these rules have been fully developed, lawyers would be required to apply such laws to commercial transactions.

As technology develops, so the law must develop to ensure that commercial activity is regulated and that there is a consistent application of rules governing commercial transactions. Not only would there be a requirement (as argued above) for some legal or regulatory oversight in smart contracts, but there would also be a need for the rules and laws to be established for the operation of smart contracts. Due to the specialized nature of crypto law, it is to be expected that specialized trained lawyers would be required not only to understand the technology but also to apply the legislative and regulatory framework that governs smart contracts (Evans 2018: 286).

6 Dispute Resolution Processes

"Contracts are like hearts. They are made to be broken", but smart contracts do not break that easily (quote by Ray Kroc, played by Michael Keaton, in the 2017 Filmnation Entertainment production of *The Founder*). Throughout history, the law has attempted to regulate the enforcement of contractual promises and has developed a complex set of rules to ensure that persons keep their promises. The law has regulated and continues to effectively regulate trust and honesty between contracting parties for future promises and the transfer of value through various dispute resolution and enforcement mechanisms. These mechanisms provide the necessary assurances that contractual promises are kept, thereby achieving certainty in commercial transactions. However, technological advances in blockchain technology and, consequently, smart contracts now boast that "trusting people is easy again" (see the website of Confideal), which hints at the core issue of contract theory and that of contract enforcement mechanisms.

Smart contracts disrupt the conventional operation of contracts and ensure the execution of performance, thereby allowing transactions without trust or even knowledge of the identity of the other contracting party. This autonomous and self-executing computer program seemingly has the power of preventing a party from breaking (or breaching) a contract. As smart contracts ensure performance and possess unique security features and self-executing attributes, it is understandable that one could conclude that the need for the traditional dispute resolution process or other forms of enforcement mechanisms would be removed altogether from blockchain transactions. Unfortunately this is not the case. No matter how perfect the system, where there are humans there is bound to be some form of dispute and, consequently, a need for appropriate mechanisms to resolve such disputes. Also, the smart contract would be incomplete if it did not cater for instances where, for example, the services, products or goods are defective or incorrectly delivered. Therefore, a major element for a completely autonomous system of smart contracts are effective dispute resolution mechanisms and processes that are in-step with current and future technological advances.

6.1 Traditional Dispute Resolution Processes

The traditional and natural approach to resolve disputes is to refer the matter to court. Access to court is considered the default dispute resolution mechanism, and court access is available to all persons in terms of Section 34 of the South African Constitution. Litigation has historically proven to be inherently lengthy and costly to the average person, and over time various alternative mechanisms have developed to resolve disputes between parties. The purpose of these alternative dispute resolution mechanisms is to allow for quicker, more efficient and affordable ways of resolving contractual disputes between parties.

The first type of dispute resolution may be classified as an informal mechanism (Van Eck 2015: 268). These informal mechanisms often exist in the contractual provisions to ensure that the parties attempt to resolve the dispute through an informal manner before taking the dispute to court. This includes, for example, a contractual provision that requires good faith negotiations between the parties in an attempt to resolve the dispute (Van Eck 2015:268-269).

The natural progression and evolution from such informal dispute resolution mechanisms are more formalized mechanisms such as mediation and conciliation, in which case a third party would facilitate the discussions between the parties to assist in reaching a resolution to the dispute. If these mechanisms fail, the parties could still refer the matter to court.

On the other end of the spectrum are formal mechanisms such as arbitration. Arbitration is regulated in South Africa by the Arbitration Act and is recognized as an alternative to court processes. There have been much debate around the advantages and disadvantages in resolving disputes by means of arbitration. However, it is worth noting that arbitration remains reliant on the courts to enforce any form of arbitration award (see Section 31(1) of the Arbitration Act, which allows an arbitration award to be made an order of court).

These dispute resolution mechanisms have ultimately proven ineffective to the average consumer who often can afford neither the cost of going to court nor the costs of such formal alternative dispute resolution mechanisms. Due to the vulnerability of the consumer in the market the legislature intervened by introducing not only specific legislative protections for the average consumer (such as the National Credit Act and the Consumer Protection Act), but also various types of mechanisms to resolve disputes. These include (i) ombudsman referrals for specific consumer sectors (for example the Motor Industry Ombudsman of South Africa, the Consumer Goods and Services Ombudsman, the Ombudsman for Banking Services, the Credit Ombudsman, the Ombudsman for Financial Service Providers, the Ombudsman for Long-Term Insurance, the Ombudsman for Short-Term Insurance, the Tax Ombudsman and the Health Ombudsman); (ii) special courts for specific and specialized areas of the law (for example the circuit courts, the Special Income Tax Courts, the Equality Courts, the Small Claims Court, the Child Justice Courts, the Maintenance Courts, the Labour Court, the Competition Commission, the Electoral Court and the Land Claims Court); (iii) tribunals and commissions established for specific consumer matters (for example the National Consumer Commission, the National Consumer Tribunal and the National Credit Regulator); and (iv) alternative dispute processes specifically catering for domain name registration disputes in terms of Section 69 of the ECTA and its regulations governing alternative dispute resolutions. These traditional dispute resolution processes are summarized in Fig. 2 (adapted from Van Eck 2015: 267-278).

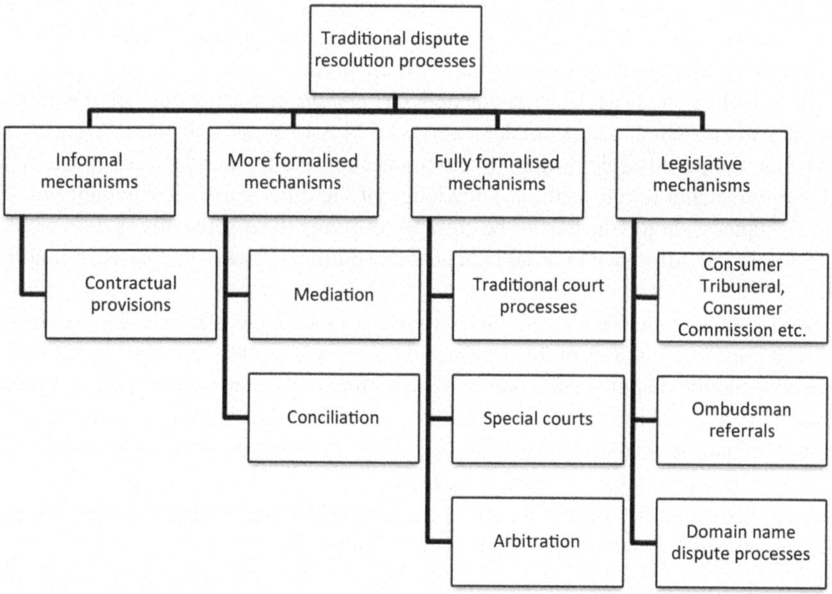

Fig. 2 Traditional dispute resolution processes

It is questionable whether these dispute mechanisms truly are effective in resolving disputes relating to international and online trade (Van Arsdale 2015: 110). Considering the international nature of smart contracts, it seems almost deficient to rely on the traditionally-established manners of resolving disputes. The drive to find mechanisms to address disputes between suppliers and consumers for international trade already is underway. Dispute resolution mechanisms premised on current technological capabilities are known as online dispute resolution processes (ODRP) and flow directly from their traditional dispute resolution counterparts (Albornoz et al. 2012: 43). Currently, there appear to be several available options for more effective dispute resolution processes between contracting parties as society moves into the fourth industrial revolution.

6.2 Technological Developments in Informal Resolution Processes

The developments in informal dispute resolution software programs have already been undertaken in an attempt to automate and assist in resolving routine disputes. Condlin (2017: 725) describes these types of disputes as "simple, small-stakes, high-volume, standardized, commercial transactions over the Internet". The developments in these dispute resolution mechanisms closely resemble the traditional informal dispute resolution mechanisms in which parties attempt to negotiate and find a resolution between themselves without engaging a third party at all. The modern version, however, intervenes via technology and uses a system or program to facilitate the discussions and ultimate resolution between the parties.

Blindbidding such as *Smartsettle* and *Cybersettle* are examples of resolution mechanisms in which parties make a series of bids to settle a dispute. Algorithms determine the permissible range for settlement bids and the parties must make their bids within such a range, without knowledge of the other party's settlement bid. In some instances, should the parties be close to resolving the dispute, the system would simply split the difference in order to achieve resolution between the parties (Condlin 2017: 725).

Square Trade is another example of a system that provides settlement proposals between parties. However, in this instance the system would function as the intermediary to settle the dispute. This, however, is neither a fully automated nor a complete system for dispute resolution as a failure to resolve the dispute would result in the system automatically referring the dispute to an online mediator (Condlin 2017: 725). It is anticipated that in future these technological intermediaries would become more sophisticated by means of robots, avatars or even holograms taking the reins of resolving such disputes between parties (Condlin 2017: 725).

6.3 Technological Developments in Formal Resolution Processes

This type of dispute resolution mechanism closely resembles the traditional alternative dispute resolution mechanism of arbitration, but functions in a more automated manner. There are two major areas of development in this type of dispute resolution mechanism. The first area of development allows contracting parties to execute a smart arbitration contract based on blockchain technology. This effectively is an electronic version of an arbitration agreement and would still be subject to some form of arbitration rules and process. The arbitration process often is regulated and subject to applicable legislation, as is the case in South Africa's Arbitration Act and, therefore, smart arbitration contracts would be required to conform to national and international legislative requirements.

Smart arbitration contracts have already been used in the system *Confideal* which utilizes *Ethereum* blockchain technology for their smart contracts and includes an optional feature for the appointment of third party arbitrators in instances of disputes between the contracting parties (Francis 2017). These third party arbitrators are qualified professionals in the field and have been vetted by having submitted all the necessary paperwork to the system (Francis 2017). The arbitration process on *Confideal* conforms to the UNCITRAL international arbitration regulations, which gives the necessary structure to the dispute resolution process (Francis 2017). This process, however, is not fully integrated or automated, as once the arbitration process had been completed the matter would move to the normal court process for the enforcement of the arbitration award (Francis 2017). Therefore, the alternative dispute resolution system still requires the traditional court process and structures for the enforcement of any arbitration awards.

The second area of development is that of an online alternative dispute resolution platform. This seemingly functions as a substitute for parties being physically present in one room and allows parties to participate and contribute to the arbitration process regardless of the distance or their location. Such a system already is in existence on the European Commission's online dispute resolution platform, designed to resolve disputes regarding online trade within the European Union, Norway, Iceland and Liechtenstein for consumer complaints, and in instances where the supplier wishes to make a complaint, it would apply where the consumer is based in Belgium, Germany, Luxemburg or Poland. This system, however, is not automated and both the consumer and the supplier must agree to the online dispute resolution process and the dispute resolution body that would be appointed to resolve the dispute.

Both these areas of development are similar to the traditional arbitration processes and the parties must still identify the necessary authority and processes to adjudicate the dispute. The enforcement of the outcome of such an adjudication process ultimately rests with the traditional court system and, as is the case with contemporary arbitration awards, these types of online dispute resolution awards would only be executed in the court that has jurisdiction over the matter. For example, South

Africa signed the Convention on the Recognition and Enforcement of Foreign Arbitral Awards of 1958, more commonly known as the New York Convention, which forms the basis for the recognition and execution of some foreign arbitration awards in South Africa.

While these developments are progressive, they remain costly and time-consuming for the average consumer, but may still be a valuable option for disputes between established business entities.

6.4 *Disruptive Force and Innovative Dispute Resolution Mechanisms*

The creation of a different dispute resolution mechanism exclusively through the use of blockchain technology already is underway. This type of dispute resolution mechanism has been called crowdsourcing or an outsourced dispute resolution mechanism (Van Den Herik et al. 2012: 99). Although not strictly recognized as a dispute resolution process, such a crowd-sourced online dispute resolution process (CODRP) could be described as a *sui generis* form of alternative dispute resolution process (Van Den Herik et al. 2012: 100).

CODRP would be achieved by adding additional protocols and contingencies to the smart contract itself, making it a complete autonomous mechanism not only for the performance of obligations but also for the resolution of any disputes that flow from contractual performance. There already are developments in this area in the mechanisms offered by *eJury* and *VirtualJury*, which serve as more of an opinion poll for lawyers (Van Den Herik et al. 2012: 100), while systems such as *Jury.Online* and *Kleros* are more progressive towards autonomous dispute resolution processes. In these instances a dispute would be referred to the online dispute resolution process and a pool of third party jurors would be appointed. A general appointment would allow any person to become a juror, as seen in systems such as *SideTaker* and *iCourthouse*. Alternatively, a closed online jury system could be used in which jurors are appointed based on pre-set criteria that would often require certain knowledge, expertise or experience to qualify as a juror (Van Den Herik et al. 2012: 105). These jurors would then decide on the outcome of the dispute after considering the evidence. The majority decision of the jurors would indicate who won the matter.

The process of resolving a dispute without consulting a lawyer is appealing. Such an online dispute resolution process would remove the need to refer matters to third parties (whether they be lawyers or adjudicators, which can be costly and time-consuming). It certainly has the potential of being a viable alternative to effectively resolving disputes for both online and offline disputes (Albornoz et al. 2012: 45–46), thereby side-stepping and disrupting the traditional legal processes and rendering them obsolete. This type of dispute resolution process could also disrupt the need for lawyers and institutional adjudicators as such positions would become redundant under a new dispensation of dispute resolution processes.

However, there are serious risks to the rule of law and justice in using online jury systems in their current form. The manner in which these systems are built is on the premise that the participating jurors would obtain some sort of monetary benefit for a decision that conforms to the majority vote. Theoretically, due to the use of a secure system facilitated by blockchain technology, there should be no risk of bias, corruption or prejudice in reaching a decision. However, blockchain technology is not yet fool-proof and has been vulnerable to attacks and breaches in recent years (O'Neal 2018; Orcutt 2019; Petersson 2018; Risberg 2018), thereby raising the concern as to whether the method of reaching a decision in an online jury systems would in all cases be truly unanimous, unbiased and without third party influence.

Even if the blockchain technology becomes completely secure without the possibility of breach, vulnerabilities or third party influence, the question remains as to whether a majority decision would be based on law, a certain set of predetermined rules, or whether it would be determined by an individual juror's personal sense of fairness, or even their opinion alone. If a decision is based on a person's sense of fairness or solely on their opinion, these online jury systems would simply be a court of public opinion, rather than one of justice and law. As the CODRPs are effectively described as opinion polls (Van Den Herik et al. 2012: 105), the widespread use of such systems could lead society down a road of so-called mob justice, littered with inconsistencies and uncertainties resulting from decisions based on the volatility of human emotions rather than achieving justice through established rules and values. Van Leeuwen (1820: 1.1.3) mentions that a person's right (and consequently the enforcement thereof) is closely linked to justice, which effectively is giving every person what is due to them in accordance with what is right and lawful. The establishment of what is right and lawful is not determined by the individual's sense of right or wrong but rather by established rules that are reflective of the community as a whole.

Notwithstanding this, the CODRP could have a positive impact on resolving disputes between individuals provided that it is entrenched within requisite regulations and structures. Although many aspects remain that must be resolved and questions to be answered in the process of CODRP, it certainly is the required next step in ensuring that smart contracts operate as self-sustainable ecosystems with the ability to effectively resolve disputes. There already are a variety of CODRPs in existence, pointing towards a possible future state of online dispute resolution processes. If the right balance can be achieved in the resolution of disputes and the maintenance of law and order, these developments could be a valuable tool for completing the function and role of smart contracts in society. Such technological advances would ultimately disrupt the general localized dispute resolution processes by replacing them with quicker and more efficient manners of solving disputes in the global commercial sphere.

7 Conclusion

Smart contracts have the potential not only of radically changing conventional thinking and approaches of commercial transactions but also the legal framework of such transactions. The disruptive force of technological advances, such as blockchain, cryptocurrencies, artificial intelligence, adaptive and predictive algorithms and big data, could reinvent contract theory and practice. However, it is the current immaturity of these technologies and the resistance to change that hamper the full integration of smart contracts into commercial transactions. Many questions remain unanswered regarding the practical and legal implications of smart contracts, and whether the removal of trusted third party intermediaries would require a form of regulatory oversight or compliance requirements.

Change is the product of progress and often is uncomfortable as it requires one to walk away from the conformity of the familiar. However, it is clear that smart contracts will in some way, shape or form be incorporated into future commercial transactions, although the extent of such integration remains uncertain. Smart contracts certainly will require a change in mindset, approach and ways of doing things. As with any change, the impeding integration of smart contracts into commercial transactions can be either ignored, resisted or embraced. Either way, the new paradigm of contractual theory and practice is certain to bring about different challenges and opportunities as the legal and commercial world continues to grapple with the ethical and philosophical issues of controlling a contracting system that, at its core, is intended to liberate transactions from central control. Society can be either reactive by living in the wake of this progress, or proactive by riding the wave of change and embracing the potential disruptive force of the technological advances of smart contracts.

References

Albornoz MM, Martín NG (2012) Feasibility analysis of online dispute resolution in developing countries. Univ Miami Inter-Am Law Rev 44(1) (Fall):39–61. Last accessed on 28 May 2019 from https://www.jstor.org/stable/23645500

Arbitration Act No. 42 of 1965

Bacon J, Michels JD, Millard C, Singh J (2018) Blockchain demystified: a technical and legal introduction to distributed and centralised ledgers. Richmond J Law Technol 25(1):1–106

Buckler WH (1985) The origins and history of contract in Roman law down to the end of the Republican period. CJ Clay & Sons, Cambridge University Press, Warehouse, London

Caruso D (2018) Then and now: Mark Pettit's modern unilateral contracts in the 1980s and in the age of blockchains. Boston Univ Law Rev 98:1789–1796

Casey AJ, Niblett A (2017) Self-driving contracts. J Corpor Law 43(1):1–33

Christie RH, Bradfield GB (2011) The law of contract in South Africa, 6th edn. LexisNexis, South Africa

Consumer Protection Act No. 68 of 2008
Constitution of the Republic of South Africa Act No. 108 of 1996
Condlin RJ (2017) Online dispute resolution: stinky, repugnant, or drab. Cardozo J Confl Resolut 18(3):717–758
Cong LW, He Z (2018) Blockchain disruption and smart contracts. National Bureau of Economic Research Working Paper Series, No. 24399, pp 1–41. Last accessed on 28 May 2019 from http://www.nber.org/papers/w24399
Confideal. Last accessed on 28 May 2019 from https://confideal.io/
Cybersettle. Last accessed on 28 May 2019 from http://www.cybersettle.com/
eJury. Last accessed on 28 May 2019 from https://www.ejury.com/
Electronic Communications and Transactions Act No. 25 of 2002
European Commission's online dispute resolution platform. Last accessed on 28 May 2019 from https://ec.europa.eu/consumers/odr/main/index.cfm?event=main.home2.show&lng=EN
Evans J (2018) Curb your Enthusiasm: the real implications of blockchain in the legal industry. J Bus Entrepr Law 11(2):273–300
Fairfield JAT (2014) Smart contracts, bitcoin bots and consumer protection. Wash Lee Law Rev Online 71(2):35–50. Last accessed on 28 May 2019 from https://scholarlycommons.law.wlu.edu/wlulr-online/vol71/iss2/3/
Fenwick M, Kaal WA, Vermeulen EPM (2017) Legal education in the blockchain revolution. Vanderbilt J Entertain Technol Law 20(2) (Winter):351–384
Filmnation Entertainment (2017) Production of the founder
Francis J (2017) Built-in arbitration system is the hallmark of confideal. Last accessed on 28 May 2019 from https://bitcoinist.com/built-arbitration-system-hallmark-confideal
French Code Civil of 1804
Gaius (1904) (Trans by Poste E). Gai Institutiones or Institutes of Roman Law by Gaius. Clarendon Press, Oxford
German Bürgerliches Gesetzbuch of 1900
GNR.1166 of 22 November 2006: alternative dispute resolution regulations (being the regulations to the Electronic Communications and Transactions Act No. 25 of 2002)
Goodman E (1995) The origins of Western legal tradition from Thales to the Tudors. Federation Press, Sydney
Haig S (2017) South African officials consider National cryptocurrencies "Too Risky". Last accessed on 5 May 2020 from https://news.bitcoin.com/south-african-officials-consider-national-cryptocurrencies-too-risky
Hutchison D, Pretorius C (eds) (2017) The law of contract in South Africa, 3rd edn. Oxford University Press, Cape Town
iCourthouse. Last accessed on 28 May 2019 from http://www.icourthouse.com/main.taf
Jury.Online. Last accessed on 28 May 2019 from https://jury.online/
Justinian (Digest) (1889) (Trans by Watson A). The digest of Justinian, Vol IV. University of Pennsylvania Press, United States of America
Justinian (Institutes) (1889) (Trans by Moyle DCL). The Institutes of Justinian, 2nd edn. Clarendon Press, Oxford
Kolber AJ (2018) Not-so-smart blockchain contracts and artificial responsibility. Stanf Technol Law Rev 21(2):198–234
Kleros. Last accessed on 28 May 2019 from https://kleros.io/
Koornhof PGJ (2012) The enforceability of incorporated terms in electronic agreements. Speculum Juris 2:41–65
Kraus D, Obrist T, Hari O (2019) Blockchains, smart contracts, decentralised autonomous organisations and the law. Edward Elgar Publishing, Cheltenham UK and Massachusetts USA
Lauslahti K, Mattila J, Seppälä T (2017) Smart contracts—how will blockchain technology affect contractual practices? ETLA Reports No. 68, pp 1–27. Last accessed on 28 May 2019 from https://pub.etla.fi/ETLA-Raportit-Reports-68.pdf
Maine HS (1906) Ancient law. Hendry Holt and Company, New York

Mckinney SA, Landy R, Wilka R (2018) Smart contracts, blockchain, and the next frontier of transactional law. Wash J Law Technol Arts 13(3):313–347
Mik E (2017) Smart contracts: terminology, technical limitations and real world complexity. Law Innov Technol 9(2):269–300 (last accessed on 28 May 2019 from https://ink.library.smu.edu.sg/sol_research/2341)
National Credit Act No. 34 of 2005
O'Neal S (2018) SIM swapping: how hackers stole millions worth of crypto via victim's telecoms operator. Last accessed on 28 May 2019 from https://cointelegraph.com/news/sim-swapping-how-hackers-stole-millions-worth-of-crypto-via-victims-telecoms-operator
Open Law. Last accessed on 27 November 2019 from https://www.openlaw.io/
O'Shields R (2017) Smart contracts: legal agreements for the blockchain. N Carol Bank Inst 21:177–196. Last accessed on 28 May 2019 from https://scholarship.law.unc.edu/cgi/viewcontent.cgi?article=1435&context=ncbi
Orcutt M (2019) Once hailed as unhackable, blockchains are now getting hacked. Last accessed on 28 May 2019 from https://www.technologyreview.com/s/612974/once-hailed-as-unhackable-blockchains-are-now-getting-hacked
Petersson D (2018) Thefts, hacks and surveillance: whose side is blockchain on? Last accessed on 28 May 2019 from https://www.forbes.com/sites/davidpetersson/2018/11/26/thefts-hacks-and-surveillance-whose-side-is-blockchain-on/#175bf40c33be
Reyes CL (2017) Conceptualizing cryptolaw. Nebr Law Rev 96(2):384–445
Risberg J (2018) Yes, the blockchain can be hacked. Last accessed on 28 May 2019 from https://coincentral.com/blockchain-hacks
Savelyev A (2016) Contract Law 2.0: "Smart" contracts as the beginning of the end of classic contract law. In: Working papers, series: law, WP BRP 71/LAW/2016, National Research University Higher School of Economics, Basic Research Program, pp 1–24. Last accessed on 28 May 2019 from https://wp.hse.ru/data/2016/12/14/1111743800/71LAW2016.pdf
Schierhout v Minister of Justice 1925 AD 417
SideTaker. Last accessed on 28 May 2019 from https://www.sidetaker.com/
Smartsettle. Last accessed on 28 May 2019 from https://smartsettle.com/
South African Reserve Bank (2014). Position Paper Number 02/2014. Last accessed on 28 May 2019 from https://www.golegal.co.za/wp-content/uploads/2018/01/Virtual-Currencies-Position-Paper-Final_02of2014.pdf
South African Reserve Bank (2018) Media relations—press statement. Last accessed on 28 May 2019 from https://www.resbank.co.za/Lists/News%20and%20Publications/Attachments/8753/Project%20Khokha%20press%20statement%2006%20September%202018.pdf
Spring Forest Trading 599 CC v Wilberry (Pty) Ltd t/a Ecowash & Another 2015 (2) SA 118 (SCA)
Spruyt W (2018) An assessment of the emergent functions of virtual currencies. Tydskrif vir die Suid-Afrikaanse Reg 707–725
Square Trade. Last accessed on 28 May 2019 from https://www.squaretrade.com/
Surujnath R (2017) Off the Chain! a guide to blockchain derivatives markets and the implication on systemic risk. Fordham J Corp Financ Law 22(2):257–304. Last accessed on 28 May 2019 from https://ir.lawnet.fordham.edu/cgi/viewcontent.cgi?article=1440&context=jcfl
Szabo N (1997) Smart contracts: formalizing and securing relationships on public networks. First Monday 2(9). Last accessed on 28 May 2019 from http://firstmonday.org/article/view/548/469
Thomas PHJ, Van der Merwe CG, Stoop BC (2000) Historical foundations of South African private law, 2nd edn. Butterworths, Durban
Van Arsdale S (2015) User protection in online dispute resolution. Harv Negot Law Rev 21(1) (Fall):107–142
Van den Herik J, Dimov D (2012) Towards crowdsourced online dispute resolution. J Int Commer Law Technol 7(2):99–111
Van Eck MM (2015) Drafting of contracts in South Africa. University of Pretoria. Unpublished Doctoral Thesis. Last accessed on 28 May 2019 from https://repository.up.ac.za/handle/2263/52340

Van Leeuwen S (1820) Commentaries on the Roman Dutch Law (translation from Dutch). Joseph Butterworth & Son, Dublin
VirtualJury. Last accessed on 28 May 2019 from http://www.virtualjury.com/
Werbach K, Cornell N (2017) Contracts ex machina. Duke Law J 67(2):313–382

Technologizing Infrastructure *for* Peace in the Context of Fourth Industrial Revolution

Francis Onditi and Robert Gateru

Abstract This chapter builds on Ayse Ceyhan's '*technologization of security*' thesis to suggest alternative ways of harnessing technology for peacebuilding beyond physical security. We argue that in as much as technology is key to the various dimensions of security, its current application remains overly a 'hard security' issue (surveillance, biometrics and military intelligence). Yet, the emerging peace and security architecture is deeply rooted in human security facets that requires multidimensional approach. Application of technology to peace and security has also been found to generate new social challenges. For example, in politically charged ethnic societies, researchers have established a significant correlation between cell phone coverage and occurrence of violence. The concern of this chapter is therefore to explore ways of harnessing the 'constructive' attributes of technology for peace, at the same time define the 'negative externalities' resulting from 'fusion of technology' to the infrastructure of peace (I4P) within the fourth industrial revolution (4IR) environment.

Keywords I4P · 4IR · Technologization · Disruption · Constructive agents · Security

1 Introduction

The growing interest and controversy in the application of various aspects of technology to peace-building is a reflection of the changing nature of conflict and war in the new world order. Consequently, this shift in conflict and war landscape demands

---

F. Onditi (✉)
School of International Relations and Diplomacy, Riara University, P.O.BOX 49940-00100, Nairobi, Kenya
e-mail: fonditi@riarauniversity.ac.ke

R. Gateru
Vice Chancellor, Riara University, P.O.BOX 49940-00100, Nairobi, Kenya
e-mail: rgateru@riarauniversity.ac.ke

W. Doorsamy et al. (eds.), *The Disruptive Fourth Industrial Revolution*, Lecture Notes in Electrical Engineering 674,
https://doi.org/10.1007/978-3-030-48230-5_3

a reflection on conflict prevention and management systems and infrastructure for peace (I4P), in terms of mechanisms, approaches and innovative responses. Infrastructure for Peace refers to the building of institutional capacities for peace-building, prevention of violent conflict and recovery from post-war violence (Lederach 2013). Although I4P has proved to be an important strategy for tackling conflicts (Van Tongeren 2011), ambiguity and lack of appropriateness of I4P structures in different contexts have swamped stakeholders into unnecessary strife and rivalry. As a matter of illustration, this chapter evaluates two dialectical attributes of technologizing I4P: (1) technology as a 'constructive' agent; and (2) technology as a 'disruptive' agent. This relationship highlights the pivotal role played by technology in the ever-changing peace and security architecture.

In view of this dynamic peace and security architecture, we argue that these changes in conflict environment, methods of warfare, adaptation of human beings to the use of sophisticated technology and the camouflaging character of actors call for changes of tact in responding to their impact on human society. Owing to the technological nature of contemporary (in)security and conflict threats, the 4th industrial revolution (4IR) presents an opportunity to address some of the world's sophisticated conflicts. Indeed, there exists significant potential for using 4IR to enhance the effectiveness of interventions across conflict spectrum-prevention, management, mitigation and resolution. A typical example is the growing culture of multilateralism and multistakeholderism in internet governance (Weinstein 2013), which can be utilized to resolve the differences that often occur in a multiple-actor ecosystem. But, as peacebuilding processes continue to evolve within the broader framework of 4IR, it is imperative that stakeholders exploit the different modalities that interface technology, including cooperation and norm development as well as the management of the negative externalities of the technology of things. On this account, we have coined the notion of *concentric interface model* (CIM) to represent a conceptual representation of the dialectical interface between technology, security and I4P.

Before we delve deeper into the intersection between technology and peacebuilding, it is important to understand the I4P concept. In sub-Saharan Africa, I4P application is more evident mainly in South Africa, Kenya and Ghana. The Great Lakes Region (DRC and Burundi) and Zimbabwe have also developed I4P, albeit at a rudimental level. The epitome of I4P was demonstrated through the successful diffusion of the otherwise destructive violence in South Africa that preceded the historical democratic elections in 1994. Other examples, include, the role played by the Wajir Peace and Development Committees in the troubled northern Kenya in 1993 and the formation of the National Peace Council (NPC) in Ghana in 2005. The NPC in Ghana was particularly instrumental in ensuring a peaceful election in 2008 (Ojielo 2010). Typologies of I4P vary according to context; in the Kenyan society, Peace Committees and Community Diaries would constitute an I4P. The model has since evolved and spread across the country with mixed outcomes and iterations.

How should the technologized I4P look like? The point of departure for this chapter is to explore strategies for *technologizing infrastructure for peace* (T-I4P) as a rejoinder to Ayse Ceyhan's *technologization of security* (2005, 2006a, b, 2008) thesis. This chapter approaches this issue in two dimensions: First, T-I4P as a possible 'disruptor' of peace—for example, the failure of technology to deliver peace messages timely or the deployment of technology to spread 'hate' speeches and other derogatory messages that threaten peace. There are also instances where technology has been used to deconstruct people's identity, degrade confidence and marginalize communities leading to 'e-resistance' movements. Second and finally, T-I4Ps as a peace 're/constructor' i.e. the ability of technology to create and sustain peace, enhances a well-functioning government, enables free flow of information and fosters good state-societal relations (Global Peace Index 2015). The notion of technologization of security as applied by Ceyhan (2005, 2006a, b, 2008), attests to the fact that technology remains a security phenomenon. This study, however, is not designed to interrogate the 'hard' state-centric security dynamics. Rather, it presents an 'exploratory study' of how technology could enhance the effectiveness of the I4P within the broader framework of 4IR. This approach to studying significance of technology in day to day life could help deal with societal challenges, which in essence forms the research question for this chapter; can 4IR technologies enhance infrastructural capacity for peace and foster sustainable peace in the African society?

In order to address the above question, the development of this chapter utilized a range of techniques including the documentary research method, field observations and interviews. The documentary research method provides useful information that is linked to the theoretical framework (peace indicators) (Atkinson and Coffey 1997). This design enabled the researchers to deconstruct the typologies for I4P, leading to the coining of the notion of "technologization of infrastructure for peace (T-I4P) ecosystem". In order to establish the *authenticity* and *truthfulness* of data, we complimented the documentary review approach with in-depth field interviews from communities and experts involved in conflict monitoring and peace-building processes from selected conflict 'hot spot', in the Kenyan towns of Naivasha, Wajir, Isiolo, Eldoret and Kisumu.

Section 2 of this chapter proffers a response to Ayse Ceyhan's thesis of 'technologization of security,' by interrogating the two dimensions of technologization of security-constructive and disruptive. Section 3 goes beyond Ceyhan's thesis by introducing the notion of peacebuilding and its linkages to the various facets of technology. Section 4 identifies structural and systematic challenges of technology when applied to peacebuilding processes. Section 5 develops a new model of 'thinking' the nexus between technology-security-peacebuilding. Finally, the conclusion emphasises the need to put precautionary measures in place to mitigate any negative externalities emerging from technologization of I4P.

2 Responding to Ayse Ceyhan's Thesis

Ayse Ceyhan is one of the leading scholars on the intersection between technology, modernity and security (Ceyhan 2004, 2005, 2006a, b, 2008). She has explored several strands of technology and security, including the technology–security nexus and context dynamics, drivers of uncertainties, governance and security. Within this framework, technology is believed to provide considerable contribution towards early warning systems. Ceyhan has also cautioned that application of technology to different aspects of the society raises fundamental questions; ethical, legal, philosophical, sociological and political problems that require research attention in light of human relations induced by technology (Ceyhan 2006a, b). Ceyhan's analysis leaves no doubt about the double-edged role of technology in society. Double-edged, because it presents both 'disruptive' as well as 'constructive' impacts to human relations. The two conceptions of how technology is revolutionizing peace and security in this chapter, fits well into Ceyhan's (2008) 'technologization of security' thesis. She constructs 'security' first as an 'enabler' of war against terror and secondly as a 'facilitator' of insecurity/terrorism (Ceyhan 2008; Lyon 2003).

As aforementioned, the Global Peace Index reiterates that building peace in the current digital society requires access to technological infrastructure. However, in the global south this is hampered by poor access and lack of deliberate efforts to create what we call a 'peace ecosystem' that is responsive to the rapid technological advancement. The contribution of this chapter is therefore timely, because the world over is experiencing technological revolution; the first industrial revolution characterized with mechanized production, the second is characterized with mass production, the third with automation (Sachs 2015). The 4IR is driven by variety of devices, including the internet, artificial intelligence (AI), drones, and 'big data'. The most fascinating aspects of the 4IR is its capability to create a seamless intersection between the physical, digital and biological spheres. It is important to note that technologization of peace and security is not a new phenomenon. It began with the transition of the International Telecommunication Union (ITU) into one of the United Nation's (UN) specialized agencies in 1947 (International Telecommunication Union 2012). As a result, the significance of technology on international peace and security matters came into focus. In the advent of 4IR, the field of international relations, especially the human relations such as peace and security can no longer operate without technology.

The adoption of open data, open standards, and open source and innovation could improve the effectiveness of peace intervention. For instance, the potential of 'big data' remains untapped—this could allow decision makers to track peace processes and improve social protection (Letouze 2012). The fact remains, however, that in Africa as everywhere in developing nations, technology is deployed mainly in 'security'-surveillance, military operation and government espionage. Therefore, the next part of this chapter would be dedicated to analysing the Ceyhan's typology of technology-security-modernity nexus from an enabler perspective. It is this analysis which would later on be utilised to construct the relationship between technology and I4P, either, as 'constructive' or 'disruptive' agents.

2.1 *Technology as a 'Constructive' Agent*

Ceyhan (2008) traces his argument from the global terror acts such as the 9/11 attack on New York and Washington D.C. This terror attack was executed by the Al Qaeda terror group. There is no doubt that the terrorists operated under a meticulous intelligence and counter-intelligence support mechanism aided by a highly sophisticated technology. Since this incidence in the US, hard security-focused technologies, such as identification, surveillance and risk assessment have become the centrepiece of strategic security. Ceyhan's conception agrees with Lyon's (2003) understanding that technology in security has been constructed in two perspectives; enabler of war against terror and potential facilitator of terrorism (Lyon 2003). This binary role of technology can be explained in three logics: (1) logic of intercepting the so-called risky people or objects; (2) logic of managing flow of goods and people; and (3) ambient intelligence - integration of microprocessors in the daily life of individuals to make them more comfortable. This classification of roles plays a facilitative task in the management and control of state security and can actually defend the state from internal unrest as well as external aggression.

Ceyhan believes the state can attain "certainty" by enforcing technology-enabled security systems. The constructive attribute of security is therefore a precursor to effective surveillance, management, and monitoring movements. Haddad and Binder (2019) acknowledge that technology can be developed beyond 'hard security' to improve governance, serve civilian needs-including promotion of social co-existence and crowdsourcing information through social media. Paradoxically, technologization of security produces 'dialectic control', which essentially advances new criminological concerns, particularly in conflict zones, in which technology of security is knightly tied to politics, law, social relations and functions of the state (Shalhoub-Kevorkian 2011). Nevertheless, the right to security has been recognized by several countries leading authorities to deliberately mainstream ICT into security systems as a means to managing risks and other external threats.

On the same issue of security and risks, Ceyhan (2008) agrees with Collier and Lakoff (2014) that appropriate application of technology can be a means of achieving security by linking up infrastructure with thoughts. In other words, technology is a security "enabler". On a similar note, Jasonoff and Kim (2013) introduce T-S (sociotechnical) imaginaries. In the world of intelligence, Kim's T-S notion can explain how the imaginaries of terror attacks influences technology-based counter-insurgency strategies. T-S is imperative in understanding the nexus between technology, modernity and day to day life. In this regard, technology can be a factor for "constructive" social change, including changes in the international system. A study by the International Peace Institute (2016) noted that, just as technology has been applied in sustainable development, the UN should lead the process of applying the same to other areas of society, such as peace and security. In Uganda, for example, the women-led 'situation room' facility utilized the power of internet to prevent violence against women and children during the Presidential elections in 2016. But the application of technology to human activities does not always construct the society positively. The controversies surrounding invention and use of devises such as drones raises moral questions: How *disruptive* is technology *disruptive to the society*? How does the disruption affect human interaction, dignity, privacy and well-being?

2.2 Technology as a 'Disruptive' Agent

Ceyhan (2008) coins the notion of "uncertainty" as relates to the state, institutions and political regimes. Ceyhan expands the 'uncertainty' into a more comprehensive framework of human security, which encompasses the political, the environmental, and the society. In this framework, the 'uncertainty' situation and '*nemesis*' is no longer limited to the communist ideology as was the case during the Cold War era. In this realm of human security, security threats can be constructed from various human–environmental interactions and other human-human interaction activities such as migration, urban movement, transportation and trade. Little wonder, therefore, in his own words, Lipschutz (2000: 1) describes the contemporary security environment as "security demands certainty". In other words, social insecurity, risks and reflexivity is a function globalization, which eventually reproduces risky societies at various levels—local, national and global. Also, geopolitics play a significant role in generating an increase in perceived uncertainties and insecurities as states interact to fulfil their national obligation. Ceyhan endorses this juxtaposition, albeit without qualification, by observing that technologies that support biometrics, video cameras, smartphones, scanners, databases, are capable of mitigating personal, state and human security risks.

However, there are ethical and philosophical questions: Should technology replace human operators? To what extent does technology contribute to formation of negative political identities and perception? Does technology replace the human cognitive capabilities? Despite these concerns, processes of technologizing security using biometrics and cameras continues to inform management of refugees and control of immigrants. This has contributed to the new identities of immigrants, leading to some sort of "securitized immigrants". The 'push factor' side of the technologization of security has produced social norms; immigration, borders formation, identity, welfare, crime, and terrorism (Buzan 1991). Taner and Meyer's (2015) ethnographical study has revealed the 'disruptive' side of mobile phones and other portable devices. These devises have tilted the working environment. As people try to adapt to the living conditions, risk images are created and recreated. A deeper challenge, however, remains that of evaluating the efficiency and beneficialization of technology to individuals, state and society.

Ceyhan interprets the intersection between technology, security and society the same way as Weberian in terms of globalized and transnationalized processes. The survival and maintenance of state authority is dependent on how well it can control its citizens. However, the state needs to be reconceptualized in a post-Weberian framework for which the globalization of biometrics contributes to the deterritorialization of control. In other cases, the role of technology in the securitization of human interactions such as migration is overbearing. Barry (2013) theorizes securitization in the context of Actor-network Theory (ANT) and considers ANT as a transition zone in international relations. It, therefore, implies that technology is the greatest variable in state security, power and sovereignty. On a similar note, Binder (2016) observes that although dual-technologies are useful to civilians, the challenge is the varied interpretation (boundary objects) of these technologies. This subjectivity of emerging technologies poses new ethical questions: How should authorities regulating the use of technologies ensure that they are only used for human development? Who determines the type of technology to be deployed in what context?

Another 'disruptive' agent is the technological marginalization and exclusion. Ceyhan's impassioned denunciation of the role of technology in border controls cautions that subjecting human beings to technological identification and surveillance is an act of marginalization. Marginalization occurs through social infrastructures, but also physical structures. Gokalp's (1992) notion of Large-Technological Systems (LTS) links technology-based infrastructures to human activities and society at large. LTS has evolved to become an international instrument of control; however, this trans-boundary evolution of the technology–security interface raises fundamental governance questions.

Shalhoub-Kevorkian (2012) takes the debate (technology, security and society) beyond Ceyhan's technologization of security thesis. Shalhoub-Kevorkian presents a case of how technology is a peace 'disruptor' in the Israel–Palestine conflict system. He demonstrates how the use of technology to control civilian's movement can degrade human dignity; leading to restricted movement, limited access to basic services and contestation of space. The friction between the authorities' resolve to use security technologies that are overbearing and the resistance from the population leads to the formation of new identities and social norms: discrimination and exploitation; marginalization; construction of unwanted individuals, illegalities and insecurity.

On the issue of spatial restriction, Shalhoub-Kevorkian (2012) writes that the 'victims' of technologization of security are likely to develop new peaceful resistance acts. Similarly, communities that have been exposed to violation of their rights through technologization of security are capable of developing resistance to the "disruptive" attributes of technology. In this chapter, this phenomenon is coined as '*immune to technology-based securitization.*' The resisting/immune population might adopt the various options available; for example, mobilising supporters through social media to stage civil unrest. This results in 'technologies of resistance'. In this chapter, our conception of the "*future of technology and society*", the "technology of resistance" is an expression of an evolving digital society, expressing itself in different ways such as frustrations through cyberspace. This is probably the most effective way the modern society expresses its 'soft' power within the broader framework of 4IR. The removal of the long-serving President of Sudan, Omar Hassan al-Bashir from power in 2019 was, to a large extent driven by social media civil unrest.

From the analysis it is clear that the various forms of technology present both 'disruptive' and 'constructive' attributes, and that the two variables reveal the relationship between technology and security as argued by Ceyhan. However, in order to understand the foreseeable impact of technology on I4P, and hence, T-I4P, the following section evaluates the application of technology to peace-building processes in various contexts.

3 Scaling Ceyhan's Argument Beyond Security

As seen from the previous sections, various technologies have been used to mobilize citizens and institutions to promote security systems. However, in Ceyhan's thesis, it is apparent that human population have borne the brunt of negative securitization at the behest of technology. Bringing together classical use of technology in security sector and peacebuilding platforms, and reflecting on the threats and opportunities of 4IR technologies, this section presents the possible applications of 4IR technologies in peace efforts. It also reveals that the opportunities and threats are 'two sides of the same coin.' On one hand, technology can create opportunities for enhancing peace, while on the other hand, technology can inhibit peace efforts by among others, increasing the feelings of vulnerability among individuals and communities.

Notwithstanding the challenges posed by technology, human relations have benefited from technological advancement. Technology has, for example, empowered people to participate in localized conflict management and peace-building, (Kahl and Larrauri 2013). Technologies have also been known to make peace-building efforts more effective, for example, by bridging the gap between warning and response. Kahl and Larrauri (2013) correctly points out that digital *engagement tools* strengthen social cohesion and effective governance. Innovative technologies when integrated into existing civil society initiatives can enhance horizontal flow of information (Mancini and O'Reilly 2013), hence, enabling stakeholders and individuals to report and react more rapidly for timely intervention. We give a few examples shortly.

For example, *gaming technology*; an emerging behavioural change field, can enhance elements of peace to the citizenry, as a subtle alternative to other actions. With the deep interests among the youth to gaming technologies, peace actors can work with gaming technology experts to develop games that inculcate positive habits of the mind among the youth with a view to breaking the chain of negative contributors to conflict. The development process for such games must proactively incorporate all aspects of behaviour, including experts in psychology, child behaviour, early childhood education and learning patterns among others. A focus on games to structure the players' behaviour portends powerful tools for positive behavioural change (Kahl and Larrauri 2013). Building on successes of previous gaming initiatives such as *The Hillary Project* in the United States (Bardall 2013), diverse stakeholders can develop targeted web-based video games that enhance positive behaviour change/development by for example awarding points to a player, each time they demonstrate a desired change of behavior or attitude towards "an opponent(s)", and by working collectively to achieve collective good. Attributes of opponents have to be specifically selected by behavioural psychologists to avoid entrenching the attributes the gaming technology seeks to deter.

In looking at the functional characteristics of technology as one of the contributing factors to development of I4P, it is useful to illustrate examples which include the various dimensions of technology that brings into focus the constructive and disruptive nature of technology in four countries-Brazil, Colombia, Mexico and Kenya. These countries are obviously shaped by different contextual landscapes, however, they all have experienced internal conflicts emanating from, ethnicity, drug-related organized crime and political quandary.

In Brazil, the *Infocrim* system, *narcoblogs* and *narcotweets* are utilized by the country's civil society groups to prevent violence of different scales and magnitude (Mancini and O'Reilly 2013). In both Colombia and Mexico, drug trafficking remains the main driver of violence. However, through the power of digital media, information is regularly published on crime related events. At the same time, advocates of peace sustain interactive peace networks through appropriate technology and social media platforms. In Mexico, this is often at considerable risk and personal cost as criminal groups (including drug cartels) often infiltrate online reporting networks to identify and harm activists. Citizens have also been known to falsely report information and generate panic or lead civilians into meting vigilante justice against alleged criminals (Muggah and Diniz 2013). Despite the threats from drug dealers and perpetrators of violence, peace advocates have utilized the power of technology to diffuse those threats. Twitter, for instance, has been used in Brazil and Mexico to promote anti-drink-driving campaigns.

In such cases, 4IR technologies such as the use of Internet Bots present immense opportunities for deeper engagements with communities. Internet bots (Bots) are miniature computer programs that are "trained" to behave like humans but not directly linked to any human being. In areas of conflict, where fear of reprisal is very real, Bots would enable citizens to for examples post tweets anonymously, devoid of the fear of becoming targets for enemies of peace. Kenya's use of information technology in managing conflict and violence continues to rise. In part due to the country's high mobile and internet penetration rates. Since the outbreak of the electoral-related violence in 2007/08, the use of social media to diffuse tensions has become a norm. However, despite the many applications of innovative technologies, the gap between the early warning indicators and response remains a challenge. This disconnect becomes obvious when non-state actors at the local level are often not able to access information submitted to various situation rooms located at the ministerial level.

Technology-enabled social media platforms have been used widely to facilitate information sharing: (1) the *Ushahidi* platform is part of the crowdsourcing mapping initiatives in the country. The key feature that allowed the platform to map electoral violence in Kenya was the integration of mobile technology. The original system gathers data from crowds in a progressive manner: the crowd sent a text message which is then received by an administrator who can approve or not approve the message; an approved message is in turn loaded on the digital map for public view and information. The strength of digital-based initiatives such as *Ushahidi* and *Infocrim* lies in their ability to connect with the local population (Mancini and O'Reilly 2013: 2). *Uwiano@108*-this is a peacebuilding platform for peace established by the National Steering Committee (NSC) to engage the public through various media platforms at the national level. The objective of Uwiano@108 is to maximize the inclusivity of citizens in peace-building and conflict mitigation activities in the country.[1]

Depending on the nature of information received from such technologies, the relevant state agencies would be better equipped to take charge of reported cases. The platform's related social media pages are used to spread peace messages to social media consumers; (3) The National Cohesion and Integration Commission (NCIC). Through this arrangement, weekly reports are submitted from all the counties. Through the system, Cohesion monitors working for the NCIC are deployed on the ground across all counties with recording devices (video cameras and voice recorders) to monitor hate speech. Police officers have also recently been issued with recorders and video cameras to better monitor and record incidences of hate speech and others that may trigger conflict. The reports are submitted through various media channels such as WhatsApp groups. WhatsApp groups are used to gather information on incidences.

There are internal groups used specifically by NCIC agents and staff. When Cohesion monitors gather intel, they submit to the internal groups for action by concerned departments. The Integrated Public Referral Mechanism (IPRM) is an online system established by the NCIC for reporting complaints regarding the Commission, and incidences in regions where the Commission lacks physical presence. It is a multi-agency system consisting of the NCIC, the Kenya Human Rights Commission, the Ethics and Anticorruption Commission, the Office of the Ombudsman, Transparency International and the Kenya Justice Network among others. Incidences can be reported to any of these agencies who will in turn report them to the Commission through the IPRM.

In such cases, the authors propose the use of Big Data technologies to amalgamate data emanating from the diverse "unstructured sources in the filed" such as social media platforms, short text messages, among others, and using artificial intelligence to corroborate this data with data from trusted sources such as Chiefs and Community Officers of the State, in order to derive patterns and map possible conflicts. This would not only enable early detection of possible conflicts, but also inform early responses. With the use of Geographical Information Systems, Satellite Mapping, as well as other technologies, state actors in collaboration with non-state actors are able to better map potential and actual hotspots and rapidly deploy appropriate responses.

The use of Drones would enable the military and police personnel to provide better surveillance on conflict hotspots in real time. Drones would then relay real-time data to response centres enabling more targeted interventions. In cases where conflicts have alienated specific communities or groups of people, drones would be instrumental in delivering crucial humanitarian assistance, such as food items, medical aid, clothing, surveillance of their conditions, among others. In extreme cases, specialised Robots and remote sensors (data gathering devices) can be deployed as advance agents of the State to respond in areas where sending human agents would be considered too risky to attempt. Robots have been used in times of war and it is expected that they will be used more widely in rescue operations and other peacekeeping activities.

4 Limitations and Lessons for the Future

Despite their success, application of technology to peace-building can pose both access and usability challenges (Martin-Shields 2013). Indeed, the same technology can be used by governments as tools of hegemony and oppression to control people, enhance surveillance or promote and reinforce existing power asymmetries. In summer 2011 when South Sudan gained independence and conflicts erupted, the Government of South Sudan identified IDP (internally displaced person) camps and burned images in near real-time and documented evidence for war crimes and crimes against humanity (Wang et al. 2013). It is unfortunate that effective application of satellite imagery as a tool to support international justice mechanisms is hindered by the absence of a procedural and methodological framework to standardize and scale up its application by non-governmental actors (Wang et al. 2013). In the same trail of ideas, there are three criteria for determining whether and how satellite imagery could be deployed to document alleged war crimes against humanity: feasibility, data reliability and legal admissibility. It is however outside the purview of this chapter to discuss in details these legal methodologies.

With inadequate technological facilities, the failure to provide actionable intelligence to the relevant authorities can hamper efforts in preventing further escalation of violence (Wang et al. 2013). This information gap, which often leads to the failure of authorities to act, can be corrected through the application of big data analytics. Emmanuel (2012) defines big data as any capability that turns imperfect, complex, often unstructured data into actionable information. With big data, communication tools have allowed stakeholders (especially citizens), regardless of their geographical location, to engage in peace-building efforts by strongly registering their voice on issues. The application of ICT to peace-building efforts draws on big and small data mining techniques, including from official and unstructured social media data. Unstructured data associated with big data can identify patterns associated with conflict and patterns of violence. Law enforcement agencies can identify patterns in data from emergency incidence and make decision for further action. On the other hand, academics and civil society actors can apply big data to predict social unrest and riots before it happens by tracking food prices or how people are feeling about a certain topic and correlating patterns with previous events. However, to effectively utilize big data, significant system-wide hurdles must be overcome in order to reliably inform conflict prevention and management.

Technology has also been used to perpetuate violence. In Baringo, Kenya, for example, whereas social media is used to report incidents, for example sighted attackers, it is also used for tribal connotations and incitements. Residents here have WhatsApp Groups and Facebook pages for each tribe that are used for incitements.

Like Wajir, most of the people readily believe much of what is on social media. It is common for people to keep asking "what's on Facebook?" as a way of keeping up with news or trends—including tribal issues about attacks or incitements against other tribes. In recent times, this has caused trouble in Baringo, Pokot, and Marakwet. However, unlike Wajir which has coordinated WhatsApp groups used by security agencies and peace agents, there is no coordination among agencies in Baringo. This often leads to escalation of attacks, even those which have been forecasted through various citizens' social media channels.

Technological infrastructures include online forums, such as the *Wajir WhatsApp Group* whose membership comprise of the County Secretary, and all peace and regional security officials. An additional Wajir Countering Violent Extremism (CVE) Forum (WhatsApp group) also exists with membership including all CVE actors in the county—both governmental and non-governmental. Membership in the group is restricted due to the fragility of CVE and includes only trusted key actors; it is a very active group. The groups are mainly used to relay information, and it is the duty of actors to respond or not. There are often follow up forums to check on the kind of action agencies took on the information relayed. The main challenges are: unregulated social media which is used to fuel conflict: the challenge of bloggers; government inaction, and slow response despite early warnings; collapse or inaction (incident reports are read but not acted on); and lack of confidentiality, there are often sympathizers in the group who may end up reporting prescribed action or reporting on members for victimization. In addition to the lack of resources, ad hoc interventions (and their lack of continuity) by both government and donors present sustainability challenges as interventions often end immediately the problem appears to be solved.[2]

There are also lessons learnt in integrating technology in both security and peace-building that are useful for the society. From the analysis, it is clear that the application of 4IR technologies to peace-building, and especially I4P, plays not just an infrastructural role but also as an enabler for industrial revolution and system changes. It is also true that the potential of technology in peace-building in the global south remains marginalized and under-explored, sometimes due to a lack of knowledge on how technologies could be applied to peace-building (Tellidis and Kappler 2016). Reviews of various studies present some lessons, but also good practice on how to leverage technology for averting violence and building peace. On this note, Mancini and O'Reilly (2013) suggest that when seeking to leverage on new technologies authorities and civil groups should consider conducting assessment to ascertain conditions for the Do-No-Harm.

The Wajir Counter-violence Forum (the WhatsApp group for CVE) was instrumental in thwarting a potential attack from Al-Shabaab. Actors in Mandera sighted Al-Shabaab militants, their numbers and even the area they were targeting and shared the information through the group. Necessary response action was taken. Messages in the forum are shared and visible to all agencies in the area. Even though one security agent died while the militants were being pushed back into Somalia, mass casualties were avoided. Technology can also be used to instigate violence, mostly during the electioneering period when some residents post information that may trigger conflict through social media. Challenges with deploying technology include the inability to authenticate the information—often is challenging to confirm the information posted or justify the source; and confidentiality of the information. Organizational policy may require that some information be kept confidential (including that which may help avert a potential conflict). But if information is not shared due to institutional confidentiality policies may cause occurrence of or worsen a conflict which may have been averted had the information been shared.[3]

Integrating technology in I4P has the potential to contribute to the broader agenda of the 4IR by fostering intersections of various aspects of human society, including migration and identity, climate and environmental determinism, urbanism, as well as creation of wealth through digital applications. All these represent the characterization of the new peace and security architecture that requires a broader resolution approach. The importance of local content and perception in technologizing of I4P is crucial. On this modality, Banks (2013) argues that innovation should leverage tools already being used by the community in context and focus on user behaviour, keep technologies 'real' since sometimes a user-centric technology is more effective than hyped new technologies, and mainstream the technology with varying sectors of development (Banks 2013). Transforming the "*disruptiveness*" of technology to "*constructiveness*" largely depends on the local input, but also, the international partners for standardization and increasing access of internet, this is particularly important for communities engaged in utilizing I4P for sustainable peace.

Technology has been used as early warning reports. Wajir is one of the counties in marginal areas of Kenya with limited access to formal education, hence it utilizes traditional forms of communication, mainly SMS or telephone conversations. SMS and phone calls were recently used to prevent a potential clan conflict. In the region of Ademasajida, clans were quarrelling over water points where, due to drought, the clans had built small individual reservoirs. But conflict arose over one clan or clans building these reservoirs at the expense of others. Peace actors were notified of the emerging conflict, and a response team was sent, and they successfully de-escalated the conflict. In addition to illiteracy, another challenge facing the deployment of technology is the lack of communication infrastructure. Network connectivity remains a challenge, as most clans live in rural and disadvantaged areas with limited networks. There is also the challenge of external actors who have become significant others. These are often sympathizers of conflicting groups. For example, they use SMS or mobile calls to warn warring clans that security agencies or conflict mitigators are coming—sharing the target locations, the people sent and strategies. In the end, they sabotage the peace-building processes.[4]

Conflicts in Wajir often result from disagreements over natural resources. Several organizations have been established because of WPDA (Wajir Peace and Development Agency), including (i) Al Fatar Council of Elders—which was formed after the signing of Al Fatar Accord in 1995. The Council remains operational today. However, Al Fatar was not enough as it was mainly town-based and lacked touch with grassroot communities. Peace Committee was then formed, and later spread to other communities and counties. Other similar platforms include (ii) Interfaith, which brings together the clergy and Imam to champion peace; and (iii) Peace Diary, which monitors activities on the ground, and records and reports incidences (this is however no longer functional due to lack of funding to motivate volunteers). These are some of the main infrastructures utilized at the community level in Wajir. These community-based initiatives link to the regional and national systems in various ways, including the CEWARN (Conflict Early Warning and Response and Mechanism).[5] CEWARN is an IGAD (Intergovernmental Authority on Development) initiative established to coordinate collection of information and analysis on conflict and development events in support of evidence-based intervention.

The conceptual and policy limitations and lessons of the existing framework for explaining technologization of security and I4P, calls for a new thinking for understanding the triad interaction between technology, security and I4P.

In a nutshell, the challenges of technologizing peacebuilding activities lie in the following broad categories:

Capacity: The peace sector has largely been dependent on direct human interventions with limited investments in technological interventions. In these cases, capacities of the various actors to shift to technology-assisted environments will continue to be inhibited at individual, organizational, national, regional, and indeed global levels, unless deliberate efforts are made to continually but rapidly build this capacity.

Access: With peace disruptions normally affecting large sections of the population, the more the general populace has access to smart technologies, the more effective will be the technological interventions that rely on big data. This challenge is also directly related to the cost of such technologies. There is need therefore to make appropriate technologies easily accessible to the populace.

Ethics and Trust: In the era of the internet and the advent of the "fake news" phenomenon, trust will become a major issue for the global community, more so in peace efforts. Fake news has the potential to cause interventions that would cause deep conflicts, even where none existed before the fake news. There is need therefore for continued investment in building ethics and trust within any and all envisioned technological interventions.

Security: Related to trust and ethics, security of technological I4P will continue to be of paramount importance to all actors. The I4P is premised on among others, the fact that information is the new 'currency.' In this regard, the security of such information and systems must be proactively safeguarded by all.

Control versus liberalism: Whereas some countries continue to adopt liberalism in relation to the internet and access to smart technologies, others are stuck in market control. It is expected that even countries that adopt liberalism will continue to invest in surveillance technologies, the quest for "ultimate control" will continue to be a major issue of concern for all actors globally. It is still too early to predict not just who will ultimately win this contest, but if winning is at all possible or even feasible. The next section explores possibilities of a new framework for rethinking a digital society.

5 Exploring a New Concentric Interface Modelling for a Digital Society

In this chapter, we have argued that 4IR technologies portend immense opportunities for Peace Actors to enhance their activities and interventions. Technology also produces both disruptive and constructive effects on human relations when applied in two different environments—security and I4P. The new configuration brings out the concentric relationship in a style we have referred to as '*Concentric Interface Modelling.*' The two platforms are *contrasted* in order to allow informed decisions on which aspects of human relations (security vs I4P) should the emphasized within the broader framework of the 4IR in Africa and beyond (see Fig. 1).

Fig. 1 A Concentric Interface Model (CIM) of the dialectical representation of technologization of security and I4P. *Source* Author's theorization of the dialectical interface (disruptive and constructive agents) between technology, security and I4P in an African peace and security environment

From Fig. 1, it is clear that while the notion of *technologization of security* is key in understanding threats and risks to peace, the focus is overly focused on criminology, surveillance, cyberspace and spatial control of security phenomenon (Shalhoub-Kerorkian 2011). Even in situations where the GPI indicators are key to the peace process (Omanga 2015), it narrowly focuses on the role of social media[6] technology such as 'Twitter' on chieftaincy, with limited possibilities of demystifying the potential of integrating technology in peace processes. The body of knowledge on the linkages between technology and security has made the former the centrepiece of security systems, leading to a blatant securitization of human interactions including migration and registrations of persons (Ceyhan and Tsoukala 2002). However, securitization of human interactions using technology does not necessarily imply peace and stability to individuals, communities or states. In fact, in circumstances where technology serves the 'disruptive' role (see Fig. 1), technologization of security has been found to increase vulnerability, victimization and militarization of communities (Shalhoub-Kevorkian 2012). As such, even though peace committees are core to the success of I4P, they are particularly vulnerable to political turbulence (Lederach 1995), as well as technological legitimacy. Besides the politics of things (PoT), the intersection between technology and peace processes (as illustrated in Fig. 1), is influenced by several systemic and structural factors: capacity, conceptual ambiguity and plurality of I4P, inclusivity and scalability.

The first issue is *capacity*. The effects of the lack of I4P in a given community may depend on whether the model is functionally active or lacks the pre-requisite capacity and institutional arrangement to inculcate proactive, participatory, inclusive and transformative approaches to conflict transformation (Lederach 2003; Lemay-Hebert 2009). In deeply divided societies, the main objective of I4P committees may be to create trust and reconciliation between communities to prevent violence and resolve disputes. However, in countries with fairly established peace-building architecture, the challenge for peacebuilders is to ensure harmonization and coordination of various I4P structures, institutions and capacities. Moreover, while 'diverse' stakeholders can be a 'dividend' for the success of I4P, if not well harnessed they could create tensions on who plays what role with whom? The significance of I4P, moreover, extends far beyond *harmonization* and *coordination* of multiple actors and factors. The economic factors and governance of the internet, for example, should be organized to ensure equity, and facilitate equal access for all (World Summit on the Information Society 2003).

The second issue is the *conceptual/typological* orientation of I4P. The typologies of I4P have been examined in scholarly literature (Garnett 2016). Some have questioned the role of I4Ps in the formation of peace committees, as to whether it upholds inclusivity (Kovacs and Tobias 2016). There are three typologies of I4P: *state-controlled peace structures, independent peace structures and hybrid structures*. Although the state-centric approach guarantees resources for the implementation of I4P, the approach may be too prescriptive; in many circumstances technologization of security by the state may lead to victimization and marginalization of communities. The hybrid conception of I4P could flexibly allow integration of constructive technologies. Scholars agree that the plurality of I4P environment is key in the evolution of a comprehensive I4P. Chalmers (2007) agrees that the involvement of civil society in I4P ensures capacity, institutional arrangement, coherence, coordination and inter-dependence of its actors.

The third issue is *inclusivity* and *dialogue*. Although Ryan (2012) examines institutionalization and systematic capacity building, the question of *building synergy* across various administrative levels in an inclusive manner remains unresolved. Inclusivity and dialogue are the basic foundations of I4P. But this can only thrive in an environment endowed with not only human and financial resources, but also digital capabilities. In this regard, the constructive attributes of technology can serve as a *multiplier*, to improve the geographical spread, hence increasing access.

Finally, *scalability*. Although empirical evidence from countries such as Afghanistan, Colombia, DRC, Kenya, South Sudan and Uganda have applauded the usefulness of I4P, the 'how' of *scaling* successful utilization of this approach to other areas remains problematic; on this note, there exists no systematic study to demonstrate what works (lessons), where (context) and more importantly, for this chapter, the how (technology) of making the I4P efficient and accessible.

Moreover, studies have shown that most I4P structures involve both men and women, but women predominate in the structures that also integrate livelihood activities such as the Village Savings Loan Association (VSLA) schemes (Bouman 1995; Vanmeenen 2010). However, more recent studies show that women predominate informal schemes of I4P (Chivasa 2015). Although the previous findings on gender dimensions shed some light on patriarchal moves in peace and security architectures in Africa, future research should focus on the implications of women-led I4Ps on sustainability of conflict transformation. Access to appropriate technology remains a problem in the global south, especially in marginalized communities.[7] Thus, exploratory research will help in revealing the informal structures that are responsible for increasing or decreasing community's access to appropriate technologies in peacebuilding activities.

6 Conclusion

This chapter sought to demonstrate the potential use of some of the 4IR technologies to enrich infrastructures for peace (I4P). We have also explored some of the 'disruptive' attributes and challenges of these technologies and proposed several mitigating factors. Whereas most of these technologies are at their infancy stages, especially in relation to peace-building, it is indeed time that the Peace Sector makes a deliberate but calculated shift towards tapping the potential for technologies such as Big Data, Robotics, GIS/Satellites, Internet Bots, among others. In stressing the significance of the constructive attributes of technology, the chapter confirms just how much application of relevant technologies can mitigate peace and security threats, including those emanating from insurgency situations. But also, authorities and scholars studying *technology and society* should be cautious that if sophisticated technologies land in the hands of extremists and other radicalised groups, the damage to the very value of the society will be incomprehensible. This brings to the fore the importance of the governance, ethics as well as accountability of the use and disuse of technology. In view of this technology–security-society triad, systematic integration of technology in the I4P ecosystem will require both in-depth theorization (hence the proposed new thinking logic dubbed CIM in Fig. 1), as well as evidence-based decisions as to what type of technology is best suited for promoting peace, social relations and development. Technological interventions in I4P will therefore require continued dialogue, cooperation, and support of various stakeholders, from peace and security experts to technology experts, as well as other experts in psychology, diplomats, human behaviour, monitoring and evaluation, migration, human rights, community development, social innovation, among others. This notwithstanding, our analysis reveal that the positive attributes of 4IR technologies far outweigh the negative externalities, and that it is in the best interest of humanity that all proponents of global peace work together to build the requisite infrastructure, human and social capacity to remain a head of the threats to peace using 4IR. The converse, where the "enemies of peace" run a head in capacity for 4IR, is incomprehensible.

We would like to caution, however, the CIM model we propose in this chapter is not an absolute cure for the peace and conflict challenges facing the globe, but it is an alternative thinking on how stakeholders can enhance the effectiveness of structures, capacity, institutions and mechanisms for the I4P in the contemporary digital society.

Notes

1. The UWIANO platform has a live map system (Reports Map (https://nscpeace.go.ke/108/)) where the public can see reported (verified) incidences, geographical location, number of incidences, and the nature of the incidences from all UWIANO platforms. UWIANO was responsible for pushing out peace messages by leveraging on voter education. Social media platforms like Facebook, Instagram and Twitter alongside television and radio were used to send out aggressive informercials that contained peace messages (UWIANO Long Term Election Observation report 2017).
2. Interview with Farah, Principal Peace Officer, Office of the Governor, Wajir County. Peace falls under the responsibility of the national government, and most agencies only supplement efforts of the national government. The mandates of the peace committees are to support peacebuilding efforts. Most conflicts in Wajir are clan-based and result from sharing natural resources e.g. water. WhatsApp groups are used for incidence reporting's, and as avenues for peace actions. There used to be a fee but it is no longer there. Obstacles hindering adoption of technology include lack of funding; lack of skilled manpower; lack of authenticity of information which is often unverified; and lack of donor coordination on interventions—interventions often overlap, there is no single coordinated platform beyond the WhatsApp groups as every donor seems to bring there on platform.
3. Interview with Dekow Hassan, Team Leader: Danish Demining Group (DDG). Peace forums act as avenues for sharing information. They help in coordination of information by acting as early warning systems on CVE, clan conflicts, and

resource-based conflicts that afflict the area. Online forums include the Wajir Counterviolence Group (WhatsApp) for CVE, telephone calls, emails and the INSO system which gathers information on security and sends updates on incidences within the region or county for the concerned actors to respond. The International NGO Safety Organization (INSO), an international charity that supports the safety of aid workers in high risk contexts, has established an interactive online map for relaying information of conflict incidences to humanitarian agencies. The INSO is used by humanitarian organizations to alert their staff and communicate to the people in the identified area.

4. Interview with Wajir County Government Office. Hussein Adan Muhamud, County Director of Peace, Cohesion and Integration (former CEO of Wajir Peace Development Agency (WPDA), and current Patron. Wajir, 10th May 2019. Grassroots community Peace Committees are constituted by the National Government. The county office mainly complements the work of the national government on alternative dispute resolution (ADR) work through county structures. In addition, the County government facilitates training by availing forums which result in social contract or agreements.
5. Interview with the representative of the Wajir Peace and Development Agency, Abdinasir Saman, Chief executive officer. The CEWERU is an online platform for submitting reports to National Steering Committee (NSC) and IGAD mostly on border issues—governance, livelihood, environments and peace—for the North Eastern region.
6. National Cohesion and in Integration Commission (NCIC). Interview with Salim Omar Salim Omar Project Officer, NCIC. The National cohesion and integration commission (NCIC) established a multi-agency working group to deal with hate speech in public spaces including social media, ahead of the 2017 general elections with an aim of minimizing online hate speech. A key partnership was with The Kenya Information Communication Association Network (KICTANeT), a major social media player and multi-stakeholder platform involved in ICT policy regulation. This partnership resulted in the creation of positive conversations in social media spaces.
7. Wajir Women for Peace. Interviewed Ahmed Idle, M&E officer—0728930279, Wajir. Women are the pioneer peacemakers and pillars of peace in Wajir. The organization was created by women from various clans and is one of the peace clusters in the county. Women are always at the forefront of peace in Wajir. When conflicts arise, they visit the affected areas using a women caravan. Wajir Women for Peace sometimes uses social media, mostly Facebook, to send peace messages.

Acknowledgements This paper was written as part of the first author's fellowship with the African Institute of South Africa (AISA) of the Human Sciences Research Council (HSRC), Pretoria, South Africa in 2019. We would like to sincerely thank the HSRC and AISA for granting the first author the fellowship, which facilitated field work that informed the paper. Finally, we would like to thank Moses Ogutu, a Mandela Rhodes Scholar for his invaluable research assistantship during the field work in Kenya.

References

Atkinson P, Coffey A (1997) Analysing documentary realities. In: Silverman D (ed) Qualitative research: theory, method and practice. Sage, London, pp 45–62

Banks K (2013) Reflections on a decade of mobiles in development. Stab Int J Secur Dev 2:1–5. http://dx.doi.org/10.5334/sta.cm

Bardall G (2013) Gender-specific election violence: the role of information and communication technologies. Stab Int J Secur Dev 2(3):60, 1–11. http://dx.doi.org/10.5334/sta.cs

Barry A (2013) The translation zone: between actor-network theory and international relations. Millenn J Int Stud 41:413–429

Binder C (2016) Science, technology and security: discovering intersections between STS and security studies. EASST Rev 35. Available at https://easst.net/article/science-technology-and-security-discovering-intersections-between-sts-and-security-studies/. Accessed 18 Feb 2020

Bouman FJA (1995) Rotating and accumulating savings and credit associations: a development perspective. World Dev 23(3):371–384

Buzan B (1991) People, state, and fear: an agenda for international security studies in the post-cold war era. Lynne Rienner, Boulder

Ceyhan A (2004) Sécurité, frontières et surveillance aux Etats-Unis après le 11 Septembre. Cultures Conflits 56:113–145. Available at https://journals.openedition.org/conflits/1001?lang=en. Accessed 18 Feb 2020

Ceyhan A (2005) Policing by dossier: identification and surveillance in an era of uncertainty and fear. In: Bigo D, Guild E (eds) Controlling frontiers: free movement into and within Europe. Ashgate, London, pp 209–232

Ceyhan A (2006a) Enjeux d'identification et de surveillance à l'heure de la biométrie. Cultures Conflits 64:13–47. Available at https://journals.openedition.org/conflits/2176. Accessed 18 Feb 2020

Ceyhan A (2006b) Technologie et sécurité: une gouvernance libérale dans un contexte d'incertitudes. Cultures Conflits 64:11–32

Ceyhan A (2008) Technologization of security; management of uncertainty and risk in the age of biometrics. Surveill Soc 5:102–123

Ceyhan A, Tsoukala A (2002) The securitization of migration in Western societies; ambivalent discourses and policies. Alternatives 27(Special Issue):21–39

Chalmers M (2007) Spending to save? the cost-effectiveness of conflict prevention. Defence Peace Econ 18:1–23

Chivasa N (2015) Peacebuilding among Shona communities in transition: a participatory action research. Unpublished PhD thesis, University of KwaZulu-Natal, Durban

Collier JS, Lakoff A (2014) Vital systems security: reflexive biopolitics and the government of emergency. Theory Culture Soc 0(0):1–33

Emmanuel L (2012) Big data for development: challenges and opportunities. UN Global Pulse, May 2012. Available at www.unglobalpulse.org/sites/default/files/BigDataforDevelopment-UNGlobalPulseJune2012

Garnett J (2016) Grassroots environmental education: developing environmental peace infrastructure in the nascent democracy in Myammar. Peace Conflict Rev 9:80–91

Global Peace Index (2015) Measuring peace, its causes and economic value. Institute for Economics and Peace. Available at http://economicsandpeace.org/wp-content/uploads/2015/06/Global-Peace-Index-Report-2015_0.pdf. Accessed 18 Feb 2020

Gokalp I (1992) On the analysis of large technical systems. Sci Technol Human Values 17:57–78

Haddad C, Binder C (2019) Governing through cybersecurity: national policy strategies, globalized (in) security and sociotechnical visions of the digital society. Osterreichische Zeitschrift fur Soziologie 44:115–134

International Peace Institute (2016) The impact of new technologies on peace, security and development. Discussion paper. Available at https://www.ipinst.org/2016/05/new-technologies-peace-development. Accessed 18 Feb 2020

International Telecommunication Union (2012) Impact of broadband on the economy: the impact of new technologies on peace, security and development. In: Discussion paper. Independent commission on Multilateralism, p 4. Available at https://www.itu.int/ITU-D/treg/broadband/ITU-BB-Reports_Impact-of-Broadband-on-the-Economy.pdf. Accessed 18 Feb 2020
Jasanoff S, Kim S-H (2013) Sociotechnical imaginaries and national energy policies. Sci Culture 22:189–196
Kahl A, Larrauri H (2013) Technology for peacebuilding. Stab Int J Secur Dev 2:1–15. http://dx.doi.org/10.5334/sta.cv
Kovacs AB, Tobias P (2016) Questioning peace infrastructure and peace formation. Peace Conflict Rev 9(1). Available at https://www.academia.edu/30877018/Questioning_Peace_Infrastructure_and_Peace_Formation. Accessed 18 Feb 2020
Lederach JP (2013) The origins and evolution of infrastructures for peace: a personal reflection. J Peacebuild Dev 7:8–13
Lederach JP (2003) The little book of conflict transformation, intercourse. Good books, PA
Lederach JP (1995) Preparing for peace: conflict transformation across cultures. Syracuse University press, Syracuse, NY
Lemay-Hebert N (2009) Statebuilding without nation-building? legitimacy, state failure and the limits of the institutionalist approach. J Interv Statebuild 3:21–45
Letouzé E (2012) Big data for development: challenges and opportunities. UN Global Pulse, May 2012. www.unglobalpulse.org/sites/default/files/BigDataforDevelopment-UNGlobalPulseJune2012.pdf, p 6
Lipschutz R (2000) After authority: war, peace and global politics in the 21st century. State University of New York Press, Albany
Lyon D (2003) Technology vs terrorism: circuits of city surveillance since September. Int J Urban Reg Res 23:666–678
Mancini F, O'Reilly M (2013) New technology and the prevention of violence and conflict. Stab Int J Secur Dev 2:1–9. http://dx.doi.org/10.5334/sta.cp
Maple C (2017) Security and privacy in the internet of things. J Cyber Policy 2:155–184
Martin-Shields C (2013) Inter-ethnic cooperation revisited: why mobile phones can help prevent discrete events of violence, using the Kenyan case study. Stab Int J Secur Dev 2:1–13. http://dx.doi.org/10.5334/sta.cu
Muggah R, Diniz G (2013) Digitally enhanced violence prevention in the Americas. Stab Int J Secur Dev 2:1–23. http://dx.doi.org/10.5334/sta.cq
Ojielo O (2010) Designing an architecture for peace: a framework of conflict transformation in Ghana. In: Paper presented at the first biennial conference and general assembly of the society for peace studies and practice in Abuja, 22 Jan 2007. Legon Centre for International Affairs, University of Ghana, Lejia
Omanga DM (2015) Chieftaincy in the social media space: community policing in a twitter convened Baraza. Stab Int J Secur Dev 4:1–16
Ryan J (2012) Infrastructure for peace as a path to resilient societies: an institutional perspective. J Peacebuild Dev 7:14–24
Sarkar S (2015) Use of technology in human trafficking networks and sexual exploitation: a cross-sectional multi-country study. Trans Social Rev Soc Work J 5:55–68
Shalhoub-Kerorkian N (2011) E-resistance and technological in/security in everyday life: the Palestinian case. Br J Criminol 52:55–72
Shalhoub-Kevorkian N (2012) E-Resistance and technological in security in everyday life. Br J Criminol 52:55–72
Tanner S, Meyer M (2015) Police work and new security devices: a tale from the beat. Secur Dialogue 46:384–400
Tellidis I, Kappler S (2016) Information and communication technologies in peacebuilding: implications, opportunities and challenges. Cooperation Conflict 51:75–93
Van Tongeren P (2011) Increasing interest in infrastructures for peace. J Conflictol 2:16–26

Vanmeenen G (2010) Savings and Internal Lending Communities (SILC): voices from Africa. Catholic Relief Services, East Africa Regional Office

Wang B, Raymond N, Gould G, Baker I (2013) Problems from hell, solution in the heavens? Identifying obstacles and opportunities for employing geospatial technologies to document and mitigate mass atrocities. Stab Int J Secur Dev 2:1–18. http://dx.doi.org/10.5334/sta.cn

Weinstein JM (2013) Transforming multilateralism: innovation on a global stage. Stanf Soc Innov Rev (spring). Available at https://ssir.org/articles/entry/transforming_multilateralism_innovation_on_a_global_stage. Accessed 18 Feb 2020

World Summit on the Information Society (2003) Declaration of principles, UN Doc. WSIS-03/GENEVA/DOC/4-E, 12 Dec 2003, pp 48–49

State Cybersecurity Governance in the Fourth Industrial Revolution: An International Law Perspective

Jentley Lenong

Abstract The state remains the primary role player, which will determine cybersecurity policy and governance for the 4IR. The purpose of this chapter is to determine how states under international law should govern cybersecurity globally when faced with the disruptions that the 4IR will bring. The chapter uses the perceived future 4IR disruptions, prominent international law policy documents and the diversity of state practice to discern the prevailing normative order of state cybersecurity governance. The chapter identifies cyber justice as the desired foundational normative prescript to manage state cybersecurity governance and policy interventions. It further identifies two critical disruptions for state cybersecurity governance under a 4IR paradigm. These are the redundancy of customary legislative and regulatory intervention to legal and policy challenges and the threat to the notion of the state and state sovereignty through an evolved interpretation of cyber sovereignty as uncoupled from state territorial integrity. The research in the chapter is prescriptive. It provides a novel contribution for normative modelling of state cybersecurity governance under international law.

Keywords International law · Fourth industrial revolution (4IR) · State cybersecurity governance · Cyber justice · Cyber sovereignty · Normative governance

1 Introduction

At around 12:30 p.m. on 12 May 2017, computers of about ten hospitals across the United Kingdom were suddenly frozen and local files remotely encrypted by

J. Lenong (✉)
Department of Mercantile Law, University of Johannesburg, Corner University and Kingsway Road, Aucklandpark, Johannesburg, South Africa
e-mail: jentleyl@uj.ac.za

W. Doorsamy et al. (eds.), *The Disruptive Fourth Industrial Revolution*, Lecture Notes in Electrical Engineering 674,
https://doi.org/10.1007/978-3-030-48230-5_4

malware. Bewildered hospital staff were confronted with the malignant demand—pay $300 in Bitcoin in order to undo the cyber-attack.[1] The malware responsible for this cyber-attack was named "WannaCrypt/Wannacry" (Steyn 2017). WannaCry has been without a doubt the most widespread and devastating cyber-attack the world has up to date collectively experienced. It is estimated that between 400,000 and 1 million computers were affected by the malware (Venktess 2017). A global conference of skills pulled together and eventually, a patch for its vulnerability was released by Microsoft on the 14th of March 2017 (Pascariu et al. 2017), but not after the malware spread havoc across more than 150 countries. None of the more than 150 states affected by Wannacry had a rapid, decisive or concerted response to this cyber-attack.

This begs the question, who is burdened with the responsibility when it relates to transboundary emanating cyber-attacks? Under international law, the state remains primarily responsible for its citizens when faced with transboundary threats (Iyi 2016). Consequently, state cybersecurity governance[2] needs to be a central mechanism through which national cybersecurity policies and laws become operational. However, internationally state cybersecurity governance was particularly exposed by the Wannacry cyber-attack, with justice for the victims of these cyber-attacks - some of whom potentially could have lost their lives, still not having been achieved.

Under this reality, will the Fourth Industrial Revolution (4IR) compound the challenges of state cybersecurity governance? With anticipated significant disruptions at all spheres of society, the 4IR demands an interrogation of the traditional international law approach to cyber challenges. The horizontal integration of technologies under a 4IR paradigm would potentially carry particularly challenges for policy and governance of state cybersecurity. Policy decisions of states will ultimately determine the social and political disposition of peoples of the world in relation to the development that the 4IR will enable. The philosopher Dooyeweerd aptly remarked that: "[p]erhaps there is no other organized human community whose character has given rise to such a chaotic diversity of opinions in modern social philosophy and social science as the State." (Freeman and de Jongste 1984). Consequently, states need a communal response to the novel capabilities that 4IR technologies will bring. State governance of these technologies, their risks and benefits need to occur in a manner that will protect its populace and preserve the integrity of the state as well as its cyberinfrastructure. The focus of this research is consequently to answer the question, what governance approach should states undertake under international law to produce state cybersecurity governance that is just in the 4IR.

---

[1] Bitcoin is one of the most famous digital currencies used online. It is also known as cryptocurrencies or e-currencies. It employs a decentralised banking method to record and maintain transactions. The solving of mathematical problems through computational solutions is how new units of the currency is produced.

[2] The employment of the term state cybersecurity governance is deliberate to distinguish it from instances of cyber security governance that involves natural persons. See (European Union 2016). The focus of this chapter is how the state governs its own security in anticipation of the disruptive technologies of the fourth industrial revolution.

Contemporary international law does not yet have a codified response to cybersecurity in general or state cybersecurity in particular. This chapter will proceed by presenting the contemporary international law, state cybersecurity governance approach. The chapter then interrogates the two overarching disruptions that the 4IR will bring to state cybersecurity governance. The methodology that is followed uses the perceived future 4IR disruptions, prominent international law policy documents and the diversity of state practice to discern the prevailing normative order of state cybersecurity governance. The possibility of forms of normative modelling is then postured in order to respond to the primary research question. The chapter concludes by postulating a theory of cyber justice as critical to a 4IR future for states.

2 The Fourth Industrial Revolution and the State

Unlike the previous three industrial revolutions, the 4IR cuts across the globe and holds the potential for equitable development and technological growth. The process advances of the three industrial revolutions differ significantly, not only in the specific technologies convoluting but also in the resulting forms of enterprise and in the nature and role of the actors involved (Colli and Nicoletta 2013). Consequently, predicting the winners and losers is not predicated on the state as a player or even current levels of development of a particular nation. The implications of private individuals affecting the global balance of power will be far-reaching for human development. This withstanding, "[t]he state, however, remains the principal actor in the international arena, and the *raison d'être of the international legal system.*" (Dugard et al. 2011).

The question could be asked, why overarching state intervention is even necessary in a 4IR world of cyberspace. In response, lessons have been learned from the Third Industrial Revolution. Even though states did not play a critical role in the technological expansion of the Internet, but rather private corporate entities—they were critical to the efficiency and dynamical proliferation of this technology. The success of the United States' internet equipment producers is still very dominant globally precisely because they were the early movers in these new high-tech fields (Colli and Nicoletta 2013). Consequently, policy intervention by states in the very early stages of the Third Industrial Revolution meant the difference between success and failure (Colli and Nicoletta 2013).

The advances in technology brought about by the 4IR are focused around three clustered megatrends that exploit the proliferation of digitalization and information processing power. They are primarily physical,[3] digital[4] and biological[5] (Schwab 2016). Collectively, these technological advances will challenge cybersecurity via their pervasive characteristics. The fundamental characteristics of 4IR technologies that will disrupt the traditional regime of state cybersecurity are four-fold. These characteristics are 1. Interconnectivity; 2. The potential of hyper-communication; 3. uber-intelligence; 4. independent self-learning (Schwab 2016; Groscurth 2018; World Economic Forum 2019). This is not a closed list (*numerus clausus*) of characteristic but is representative of the core challenges that cybersecurity faces. The consequences of this transformed creature on state cybersecurity governance are fundamental disruptions to the traditional functioning of states collective policy response. The pivotal disruptions should be observed on a national level and an international level. Nationally, cybersecurity governance will be disrupted in the manner that it is brought about – via legislative and regulatory processes. Internationally the disruption will challenge the legitimacy of the enforcing state itself, through the notion of cyber sovereignty and the accompanying challenges of jurisdiction. These disruptions are critically interrogated below. The interconnectivity that for example the 'internet of things' technology will produce will fundamentally disrupt state control over cybersecurity protocols. Further, three main factors have been observed in recent years has served as catalysts for the proliferation of 4IR technologies, they are:

1. "Digital components such as sensors, actuators, cameras and microphones nowadays are so small and can be produced so inexpensively that we can use them to teach things to see, hear and feel. By the way, several German producers are leaders in the world market for such products.
2. Since the 2010s, an internationally applicable protocol, IPv6, has existed which enables almost everything to be supplied with its own internet address. This enables a device to establish contact to other devices and people as well as send and receive data.
3. Finally, information science as an engineering discipline has matured and is on the way to become the most important discipline of all. It is used to help networked, sensitive things to act in a sensible and increasingly autonomous manner." (Sendler 2018)

These technological advances have removed much of the control over processes and stakeholders that the state traditionally enjoyed. This produces novel cybersecurity threats for both the state and its citizenry. International law, in turn, has not

[3]The main known examples of these tangible technologies are self-driving or autonomous vehicles, advanced robotics and 3D printing.

[4]The digital revolution will be driven by the interconnectivity of things via what has been described as the 'internet of things'. The internet of things technology will connect the digital worlds with the physical realm.

[5]The advances in computing processing power have opened a door to the biological building blocks of humanity. Consequently, the 4IR allows for a world where human genes can be sequenced, activated and edited.

been able to respond to this new dawn. Consequently, cybersecurity law and governance has not evolved or developed to contend with the challenges of the 4IR. In holding with traditional international law, the objective of state cybersecurity law and governance remains the maintenance of international peace and security. However, the 4IR brings with it a paradigms shift and the developments in human rights law needs to filter through cybersecurity law and governance's notion of justice. In order to secure this objective, state cybersecurity law and governance needs to evolve in order to produce justice under a 4IR paradigm. It is not an understatement that the 4IR could also potentially produce tremendous social inequality. This social disruption will be an added dimension that state cybersecurity governance needs to approach as a potential threat. This threat can also be extended to the conduct of corporate entities and their impact on other states. In regard to this, there is a real likelihood that an organization's plans for implementation of 4IR plans could cause "armed" international disputes (Smith and Pourdehnad 2018). In conclusion, the state remains at the core of the response to the disruption that the 4IR will bring. Consequently, it becomes critical to interrogate cybersecurity governance from the perspective of the state under a 4IR paradigm.

3 State Cybersecurity Governance and International Law

The 4IR developmental challenges of state cybersecurity governance produce fundamental obstacles to the creation of universally accepted norms and standards for states and their cybersecurity. The primary objective of state cybersecurity governance has been the maintenance of international peace and security in accordance with Article 1 (1) of the United Nations (UN) Charter. However, this objective is originally premised on two arcade notions of international law; (1) that the state is the exclusive subject of international law and (2) that the 'use of force' is limited to conventional acts of warfare and/or acts of aggression. In the absence of international consensus on the norms and principles, which are to steer global cybersecurity governance, states have resorted to a proliferation of their offensive cyber capabilities. This is a rational response under a peace and security, international law paradigm for state cybersecurity governance.

Figure 1a, is demonstrative of the global arms race for offensive cyber capabilities, even though some states might argue that their intentions are more defensive. This has divided the globe into two cyber centres of power. A developed world block spearheaded by the United States and are developing world block represented by Russia and China (Fang 2018). How should governance be comprehended by states in terms of executing their cybersecurity policy? The United Nations Development Programme (UNDP) records that:

> Governance is the system of values, policies and institutions by which a society manages its economic, political and social affairs through interactions within and among the state, civil society and private sector. It is the way a society organizes itself to make and implement decisions-achieving mutual understanding, agreement and action.

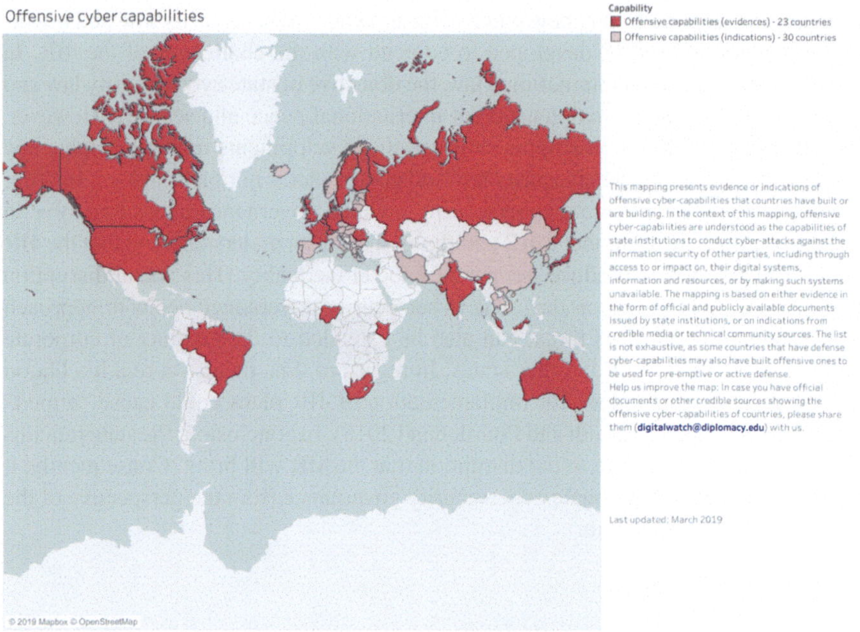

Fig. 1 **a** The global offensive capabilities of states (https://www.diplomacy.edu/cybersecurity)

State cybersecurity governance should have the capacity to accommodate a normative framework, which will be responsive to the challenges of the 4IR. Traditionally, governance is more concerned with the conduct of states rather than their citizenry, but the 4IR produces challenges both with conduct from the state and private citizens. Consequently, through a 4IR paradigm, it is not only states who pose a threat to international peace and security, but private citizens too. Cyber acts of aggression by private citizens of a particular state could easily fit the mould of 'use of force' in order for another state to proffer a justification for its own aggression. It has been argued that even though Article 2 (4) of the UN Charter prohibits 'use of force' by states and broadening the definition of 'use of force' to include private citizens could assist international law (Kulesza 2009), this is not the position in international law as yet. However, a cyber-crime or -attack, in essence, could be construed as the use of force against the political and economic interest of a state or its citizenry. Though states generally struggle with these definitions in developing domestic legal instruments. For example, that "a cyber-crime is a broad concept analytically distinct from cyber-attack. While, as with the concept of cyber-attack, there is no universally recognized definition of cyber-crime (Hathaway 2012)."

International law further lacks consensus where the cyber conduct of a private individual rises to an act of aggression against a state and such action benefits that person's own state. However, it would be possible to attach liability for the conduct of a private person or groups of people to a state via the international law concept of state responsibility. This will require proof that the state had ‘effective control’ over the perpetrator of a cyber-attack. So also, the traditional international law test of ‘effective control’ in order to a tribute state responsibility to the conduct of an individual will need rethinking. The International Court of Justice (ICJ) produced this test by stating:

> State’s responsibility can be incurred for acts committed by persons or groups of persons—neither state organs nor to be equated with such organs—only if, assuming those acts to be internationally wrongful, they are attributable to it under the rule of customary international law reflected in article 8 [of the ILC’s draft articles].[6]

Under this definition, international law does provide for a degree of accountability where a private person or group of people, but only at the direction and control of a state executes a cyber-attack. State cybersecurity law and governance need to develop in a manner that is more human-centred and allows for liability inclusive of a developed view of the subjects of international law. Especially because state cybersecurity law and governance resides at the heart of our 4IR future. The maintenance of the rule of law and efficiently articulating state responsibility under a 4IR paradigm will be the core functions of state cybersecurity law. However, international law would first have to resolve the primary issue of a definition of cybersecurity. Internationally cybersecurity remains beset with the challenge of its own definition (Kosseff 2017). Kosseff (2017) ventures a definition in that:

> [C]ybersecurity law promotes the confidentiality, integrity, and availability of public and private information, systems, and networks, through the use of forward-looking regulations and incentives, with the goal of protecting individual rights and privacy, economic interests, and national security.

This definition allows for a progressive perspective of cybersecurity law that embraces a 4IR paradigm. However, cybersecurity law and governance would need to develop legal norms and principles that also capture ever-evolving forms of cyber-attacks and still maintain the rule of law and the protection of human rights. In addition to this, cybersecurity needs to develop beyond the notion of protecting the infiltration of a secure computer network or critical infrastructure, to include for example 4IR cyber threats. From contemporary examples, these include the distributed denial of service attacks and planting inaccurate information (Hathaway 2012). The accessibility and technological equality that 4IR will produce allows for these normal threats. The underdeveloped nature of international cybersecurity law and policy presents a fundamental challenge to the attainment of a universally accepted state cybersecurity governance regime with accepted norms and principles.

In presenting the question, whether the international law of cybersecurity is in crisis Macak (2016) identifies three apparent crisis indicators. First, the area of cybersecurity appears resistant to the codification of the applicable rules in a comprehensive multilateral binding treaty. Secondly, Macak (2016) observes that “states have

[6]Case Concerning the Application of the Convention on the Prevention and Punishment of the Crime of Genocide (Bosnia and Herzegovina v. Serbia and Montenegro), ICJ Rep 2007, para. 406. See also Military and Paramilitary Activities in and Against Nicaragua (Nicaragua v. USA) ICJ Rep 1986, paras. 109 and 115.

shown extreme reluctance to contribute towards the development of cyber-specific customary international rules." Finally, Macak (2016) remarks that:

> … the third concerns their (*states*) actual conduct in relation to cyber governance. It would be inaccurate to claim that states have entirely given up on standard-setting. However, instead of interpreting or developing rules of international law, state representatives have sought refuge in the vacuous term 'norms'.

The crisis or criticism that Macak (2016) remarks on is the pluralization of international norm-making (Berman 2007; Aspremont 2012). Objective observation suggests that one has to side with Macak (2016) on the first two crisis indicators. However, what Macak (2016) does not observe is that the rationale for the third crisis indicator resides within the first two. It is due to the inability of states to produce a consolidated legislative and regulatory regime for state cybersecurity law and governance, that the approach of norm-setting is justifiable. Further, the challenge of codification is even more insidious on a domestic level with fears of government controlling the personal data of its citizens (Janssen and van den Hoven 2015). The development of norms seems to be the only manner of reaching some sort of international consensus for cybersecurity law and governance.

Where states have developed domestic legislation to regulate cybersecurity law and governance, one could discern customary international law from such State practice. However, international law requires that such State practice needs to be both extensive and uniform amongst states. In the North Sea Continental Shelf cases, the ICJ held that:

> An indispensable requirement would be that within the period question, short though it might be, state practice, including that of states whose interest are especially affected, should have been both extensive and virtually uniform in the sense of the provision invoked; and should moreover have occurred in such a way as to show a general recognition that a rule of law or legal obligation is involved…

Consequently, state practice has not been consistent in order to produce customary international law rules because the subject matter and challenges of such rules vary from state to state. It is critical to the integrity of international law that the distinction between current law (*lex lata)* and future law (*lex ferenda)* be maintained. The ICJ has been consistent in confirming that current law or positive law is the starting point for all judicial inquiry. Thus through the application of international law, the legal scholar needs first to establish what the law is before it can delft into what the law ought to be. The purpose of this is not to favour one norm over another, but to foster their relationships. In the *Islands of Palmas case,*[7] the arbitrator Max Huber remarked that international law "… has the object of assuring the co-existence of different interests which are worthy of legal protection." Consequently, a normative approach to state cybersecurity law and governance would not only serve to produce an international regime but also lead the way towards international state consensus.

[7] *Island of Palmas Case (or Miangas), United States v Netherlands, Award* (1928) II RIAA 829.

4 State Cybersecurity Governance Under a Disruptive Fourth Industrial Revolution 4IR Paradigm

4.1 The Future of Legislative and Regulatory Regimes

The process of law and policy-making on cyber-related matters is primarily legislative within domestic legal orders. At present international law does not have a legislator for all states. This has produced a plurality of competing and equivalent legal sources, both within domestic jurisdictions and in international law. International lawyers find solace in the prescripts of Article 38 of the ICJ Statute in order to have some remnants of legislative order for retrieving international law. Article 38 of the 1945 ICJ Statute provides for a hierarchy of legal sources starting with treaty law, customary law, and general principles. These are then complemented by other sources usually deemed as ancillary they are case law and the writings of eminent experts in the field of law (Teitel 2013; Chirwa and Chenwi 2016). In a 4IR world that is fast-paced; interconnected; autonomous and self-learning, the slow turning wheels of legislative bureaucracy stand in stark contrast. The production of laws and policies would be unable to keep pace with the rate of technological change. Where constant law-making becomes the response to the fast pace development in 4IR technologies, the result would be the fragmentation of an already fragile international state cybersecurity regime. Legislative intervention is a response, would also be further exacerbated through the horizontal integration of 4IR technologies both globally and municipally.

Consequently, what is needed is "a set of higher normative essentials that guides the governance and application of existing laws (Michelman 1995)." The challenge for the future of the legislative and regulatory regimes of state cybersecurity governance would be the development of higher norms, similar to constitutional principles under a constitutional democracy. The 4IR will challenge the "character of security threats while also influencing shifts of power, which are occurring both geographically, and from state to non-state actors (Schwab 2016)." This presents a further challenge to a legislative and regulatory approach as a solution. States would have to develop laws for technologies that they do not understand and which is operated from stables of power that they cannot identify. Schwab (2016) observes that the "critical danger is that a hyper-connected world of rising inequality may lead to increased fragmentation, segregation and social unrest, which in turn creates the conditions for violent extremism." Developing guiding and essential norms for all states will produce a safeguard against the rapid pace of technological development in the 4IR. International law would then not need to usurp the legislative functions of states through codification but will only be responsible for norm-setting. Such a normative approach will be consistent with contemporary international law in that it still premised on state consent and free will. In the *Lotus Case*[8] judgement the ICJ proclaimed the position of international law in that:

[8] *The Lotus Case* (France v. Turkey), PCIJ Reports, Series A, No. 10 1927, 18.

> The rules of law binding upon States, therefore, emanate from their own free will as expressed in conventions or by usages generally accepted as expressing principles of law and established in order to regulate the relations between these co-existing independent communities or with a view to the achievement of common aims.

A normative approach would also allow for legal certainty. Private actors would consequently follow the same set of higher norms that the state follows. What is certain about the future viewed through a 4IR lens is that only a multi-stakeholder, collective approach towards progress can ensure justice and the integrity of the state. The smart solutions and responses to the challenges of the 4IR will not be provided by a singular state or leader. However, by adopting novel leadership concepts, rather than responding via legislation, which can become arcade during the process of its enactment, stakeholders can adequately respond to the technological disruptions of the 4IR. Working together, both locally and globally states and production leaders from numerous businesses such as start-ups, large companies and labour can develop much more efficient solutions to these disruptions brought about by the 4IR. Kennedy aptly reminds one that:

> Any so-called ‘realism’ that attends only to the overt acts of national sovereigns is no longer realistic. In our world, power lies in the capillaries of social and economic life. Myriad networks of citizens, commercial interests, civic organisations and government officials are more significant than interstate diplomacy. Statesmen and stateswomen act against a background fabric of expectations—the legitimating or de-legitimating gaze of world public opinion—and they act in the shadow of all manner of public and private norms.

Where the state follows the path of norm development, one has to guard against supranational institutions following the path of legislating international law for state cybersecurity. The legitimacy of international law depends critically on the institutions of international law to be also legitimate. Where legislative or rulemaking functions are embedded in global governance institutions outside of the democratic system of the national state, this compromises the legitimacy of international law. What we have seen with global governance institutions such as the World Trade Organisation, the UN Security Council, the World Bank and even the European Union; is enduring patterns of coordinated, organised and persistent rulemaking governance. Global governance institutions, “though created and sustained through treaties made by states, are increasingly taking on lawmaking functions (Besson and Tasioulas 2010).” The sovereignty of states could potentially be compromised where we do not protect its integrity.

4.2 *The Ascend of Cyber Sovereignty*

State sovereignty is a keystone principle of international law (Benvenisti 2013; Alvik 2011). Cyber sovereignty can be interpreted in two ways. The first is cyber sovereignty as “a natural extension of national or state sovereignty in cyberspace (Fang 2018).” Cyber sovereignty is consequently directly linked with territorial integrity, as it is comprehended in international law. This view supports and has as its main content the state's authority to exercise jurisdiction in cyberspace (Fang 2018). The second is cyber sovereignty as an external quasi-geographical global cyberspace, which is boundary-less and encompasses an all-inclusive international jurisdiction of the global cyberspace. This second hypothesis is much more consistent with the notion of the 4IR. However, under the first hypothesis, cyber sovereignty follows the traditional tenants of state sovereignty. The principle of cyber sovereignty is lamented with traditional international law principles such as "equality among nations and the principle of non-interference in other nations' internal affairs and by implication also each other's cyberspace (Fang 2018)." Defining cyber sovereignty outside of the traditional notions of state sovereignty would mean that states will lose much of the exclusive authority and dominium, they currently enjoy over the domestic activities in cyberspace. It has been observed that for "superpowers such as the United States, China and Russia it is of critical importance to define cyber sovereignty to be consistent with the notion of state sovereignty (Fang 2018)."

Traditionally, international law perceives state sovereignty to be integrally linked to the territorial integrity and political independence of a state (Besson and Tasioulas 2010). However, in a 4IR world, states will become much more interconnected and dependent on each other both politically and economically. Is it then possible for sovereignty to accommodate both independence and interdependence of states? In the *Islands of Palmas-case*, it was noted that sovereignty means "[i]ndependence in regard to a portion of the globe is the right to exercise therein, to the exclusion of any other State, the functions of a State." From this position, the arbitrator Max Huber seems to suggest that the functioning of a state's sovereignty links critically to its independence. The normative order of international law is inherently dependent on this nature of statehood. The implications for state cybersecurity law and policy is paradoxical. The *Lotus Case* went on to confirm that "[n]ow the first and foremost restriction imposed by international law upon a State is that-failing the existence of a permissive rule to the contrary, it may not exercise its power in any form in the territory of another State." The very nature of state sovereignty seems to be challenged by the notion of cyber sovereignty.

The second hypothesis of cyber sovereignty seeks to disrupt the traditional notion of state sovereignty in cyberspace. This notion of sovereignty is not wholly inconsistent with the origins of sovereignty as it emanates from the transfer of an individual's autonomy to a collective, represented by the state. The legal theorist De Vattel (2011) commented that:

> … sovereignty is that public authority which commands in civil society, and orders and directs what each citizen is to perform, to obtain the end of its institution. This authority originally and essentially belonged to the body of the society, to which each member submitted, and ceded his natural right of conducting himself in everything as he pleased, according to the dictates of his own understanding, and of doing himself justice."

The notion of sovereignty is consequently linked with that of cyber justice. The concept of sovereignty presents a unique dichotomy for state cybersecurity governance under a disruptive 4IR paradigm. States in favour of the 4IR developments who seek to control systems and networks will use cyber sovereignty as a justification under the first hypothesis. On the other side, States who lag behind of fear the change will employ protectionism and nationalism to insulate their citizenry from 4IR disruptions. We have already observed this with states like the United States, with the rise of popular nationalism (Bonikowski and DiMaggio 2016). These States too will justify their decisions under the banner of cyber sovereignty. The concerns around the disruptions of the 4IR should not wholly be dismissed, because popular fear holds the potential of derailing the potential benefits of 4IR technologies. Goldring (1998) correctly warns that:

> "If a majority of voters in some countries feel that their political sovereignty is threatened by the free play of market forces, they are right. It is a characteristic of national sovereignty that nations can choose whether to submit to the interests of free trade and transnational business or not. Their choice need not be rational."

Cyber sovereignty needs to be defined and applied in a manner that produces justice for the individual citizen. Although, international law is not overly interested in the political tools of democratic control it is critical to incorporate notions of democratic legitimacy and its processes, in particular where it intrudes into national law. In the "probably most characteristic example, the ICJ did not decide, in its *Advisory Opinion on the Threat or Use of Nuclear Weapons*, on the question of whether the state interest or humanitarian principles prevail when in conflict with each other. The then ICJ President Bedjaoui created the category of 'neither allowed nor forbidden' for a clash between state and individual values."[9] However, international law would have to be much more decisive in terms of which hypothesis of cybersecurity prevails. What is critical is that whichever notion of cybersecurity international law adopts it needs to produce justice in order to maintain international peace and security in cyberspace.

4.3 *Justice Disrupted*

The ultimate normative order that international law proffers under a 4IR paradigm, would need to be conscientious of the advances in human rights in order to protect individual liberties in cyberspace and produce justice as an objective and as an outcome. As demonstrated above, following a regulatory or legislative approach domestically would also eventually just result in fragmentation of the international law regime. The objective of justice has been overlooked in the development of cybersecurity law and governance. The primary focus of states has been related to questions around adjudication and jurisdiction. This has usurped the objective of justice (Dekker and Werner 2004). State cybersecurity governance needs to develop policy objectives that link directly to established notions of justice. This should be open to an objective test such as, "suppose we say that, to justify directly a political

[9] *Legality of the Threat or Use of Nuclear Weapons*, Advisory Opinion, 1996 ICJ Rep, 226.

act is to show the correspondence of this very act (not just some higher-level act that authorized this one) to ideals of justice or conduciveness to general human goods (Michelman 1995)." In order to govern the disruptions of the 4IR in a stable manner, state cybersecurity governance needs to be evolve inclusive of justice. Mihr (2016) ventures a working definition of cyber justice that is:

> … based on good cyber governance and its human rights-based approach for (1) more accountability, (2) more transparency, and (3) more participation by multiple stakeholders and actors through the use of cyber tools, such as the Internet and other mobile devices.

The definition of cyber justice in itself holds normative characteristics to guide state cybersecurity governance. Consequently, state cybersecurity governance will achieve justice by adopting the normative tenants presented by cyber justice. These norms are accountability to the general populace, transparency, openness or flexibility to multi-stakeholder participation. These normative tenants will allow resilience in order to manage the disruptions of the 4IR and potentially avoid violent international conflicts between States. International law generally, suffers from the trauma of world wars as the fundamental rationale to its legal development. This is the rationale for the maintenance of peace and security as the primary objective in international security and consequently for state cybersecurity. An evolved objective for state cybersecurity under 4IR paradigms should be cyber justice.

5 The Fourth Industrial Revolution as Cyber Justice

States will in effect have two choices in determining how to respond to the challenges and potential international conflicts that the 4IR will produce for state cybersecurity governance. States can either follow a 'to war/law of war' (*jus in bello/ad bellum) a*pproach or a cyber justice approach. The notion of cyber justice presents a normative foundation for managing the degree of disruption that 4IR technologies will produce. Weil (1983) remarks that "the capacity of the international legal order to attain the objectives it was set up for will largely depend on the quality of its constituent norms." International law needs to be decisive and place at the centre of its normative order the foundational notion of cyber justice. However, these norms do not and cannot automatically offset the sovereignty of a state (Weil 1983). Jayasuriya (2001) argues that:

> As law and the territorial state are uncoupled, power of governance is becoming increasingly fragmented and diffused within the market and civil society; this poses an immense challenge to the traditional antinomies—between legality and legitimacy as well as between sovereignty and society—that underpinned the 'government' model of sovereignty.

The hypothesis of cyber sovereignty discussed above is consistent with a 4IR future. This view of cyber sovereignty will primarily be responsible for the uncoupling of the state power of governance in cyberspace. This is where the state leaders agree with this assertion not (Fang 2018). In order for the contemporary international order to hold, the governance of state cybersecurity needs to elevate itself to a set of higher legal norms. These higher legal norms need to find broad collective consensus amongst states, but they also need the recognition of all the players in cyberspace. The notion of cyber justice is but only one such normative foundation. However, it presents a valiant opportunity for the preservation of the state and its central function of protecting its citizenry against threats in cyberspace. Where states would want to impose the rule of law in cyberspace based on the traditional notions of sovereignty and international law, they would proceed at their own peril.

6 International Cooperation and Mutual Assistance as the Existing Normative Order for State Cybersecurity Governance

6.1 Background: The Objectives for State Cybersecurity Governance

Even though it can be argued that no consolidated or codified international legal order exists for state cybersecurity governance, an argument can be made for the presence of a normative order that is predicated by the objectives of international law. Kelsen's (1961) statement that "[t]he legal order is a system of norms", finds particular application for state cybersecurity governance under international law. International law is focused on three objectives in developing state cybersecurity governance protocols. The primary objective is the maintenance of international peace and security, the second is the harmonisation of legislative frameworks and thirdly cooperation amongst states. All three of these objectives face disruptive challenges under a 4IR paradigm. However, international cooperation and mutual assistance seem to have been elevated as a norm in state cybersecurity policy. It is problematic that the interpretation and application of the notion of international peace and security are premised on resolving cyber threats by applying the mechanisms and thinking of conventional warfare. The technological advances of the 4IR will produce novel surreptitious threats and clandestine perpetrators. As shown above the objective of the harmonisation of legislative frameworks, in itself would be challenged by the slow-pace of bureaucratic mechanisms relative to the swiftness of 4IR technologies.

However, international cooperation and mutual assistance seem to have filtered through in the majority of international state cybersecurity policy propositions. International cooperation and mutual assistance as an aspiration are challenged by the contemporary inequalities already existing between states. International law does not at the moment have a consolidated state cybersecurity governance regulatory framework or legal principles that are recognised by the majority of states. In the absence, of such governance infrastructure, individual states might be tempted to follow a 'to war/law of war' approach to resolve threats or attacks against their cybersecurity. Grotius said, on the idea of 'just war' when in the face of a multiplicity of sovereigns:

> I observed that men rush to arms for slight causes, or no cause at all, and that when arms have once been taken up there is no longer any respect for law, divine or human; it is as if, in accordance with a general decree, frenzy had openly been let loose for the committing of all crimes. (Grotius 1925)

Although it is accepted, that there is "no single path to war and peace but multiple possibilities (Marwala and Monica 2011)." The application of the 'to war/law of war' approach to incidents of breaches in states cybersecurity could have disastrous effects for the maintenance of international peace and security, especially where a private individual is involved. Currently, under the 'to war/law of war' approach, there are three incidents or categories of individuals which a state can lawfully target; these "are combatants, civilians directly participating in hostilities, and civilians acting in a continuous combat function (Iyi 2016; Gill et al. 2014; von Heinegg 2014)." However, it is clear how the law of war approach to cyber-attacks; cybercrime's or cyber-warfare could be apocalyptic in a world with nuclear capacity. The traditional notion of going to war or declaring war does not apply under a 4IR paradigm. Cyber-warfare is much more fluid, the participant more obscure and the interest or motives greatly variant.

The methodology followed below is to present the key international and regional policy documents that in essence have already produced an international normative order for state cybersecurity governance. The key norm that the majority of these policy interventions have produced is that of international cooperation and mutual assistance.

6.2 UNGGE Report

The United Nations Group of Governmental Experts on Developments in the Field of Information and Telecommunications in the Context of International Security (UNGGE)[10] released its third report in 2015. The Group emphasized "the importance of international law, the Charter of the United Nations and the principle of sovereignty as the basis for increased security in the use of Information Communication Technologies (ICTs) by States (Group of Governmental Experts on Developments in the Field of Information and Telecommunications in the Context of International Security 2015)." The UNGGE report did an analysis of existing and emerging threats and concluded that generally there is been a proliferation of malicious use of ICTs globally. These increased threats perpetrated by both non-state and state actors pose significant challenges to the maintenance of international peace and security. The advent of 4IR technologies will increase the capacity of the malicious use of ICTs. The UNGGE report identifies that "the most harmful attacks using ICTs are those directed against a state's critical infrastructure and the communication and information systems linked to them."

[10]The UNGGE was established through the UN General Assembly resolution 68/243.

In the threat analysis of the UNGGE report, the panel identifies a novel threat to international peace and security. The report recognises that there has been an increase in non-State actors perpetrating malicious attacks indirectly against a States. What makes the emergence of these new non-state actors particularly challenging is that they are; firstly, diverse in their make-up and motive;[11] secondly the speed with which they execute malicious attacks; thirdly the challenge of establishing the source or the origin of incidents. The focus of these attacks is producing destabilising misperceptions and inciting conflict and harm to the citizenry. One can imagine a future where AI potentially could be developed and deployed to execute such attacks. In effect, the citizenry of state can be manipulated into a weapon against that state itself. The UNGGE report improves "the security of critical ICT infrastructure; develop technical skills and appropriate legislation, strategies and regulatory frameworks to fulfil their responsibilities; and bridge the divide in the security of ICTs and their use (UNGGE 2015)."

Although the UNGGE failed to produce binding international law, it did show that the norm of international cooperation and mutual assistance was generally accepted by all states. The report proposes as a solution "[v]oluntary, non-binding norms of responsible State behaviour can reduce risks to international peace, security and stability." Unfortunately, the UNGGE could not produce consensus amongst member states and what followed was the development and adoption of two new resolutions. One was sponsored by the European Union, with the involvement of states such as Japan, the United States, Australia and Canada. In essence, creating a completely new group of government experts. The second resolution created an Open-Ended Working Group (OEWG) under the auspices of the UN. The second resolution was primarily sponsored by China and Russia together with African and Asian countries. The OEWG focuses on, the development of international law rules, norms and principles for how states should deal with cybersecurity. Although international cooperation and mutual assistance presented itself under two centres of power, it was the only surviving consensus from the UNGGE.

6.3 *Budapest Cybercrime Convention*

The Council of Europe produced the Convention on cybercrime or as it's properly known 'the Budapest Convention' in 2001 and it entered into force on 1 July 2004 (Council of Europe 2004). Although the focus of the convention is cybercrime, it does produce norms for international law that would be directly applicable to state cybersecurity governance under a 4IR paradigm. With five ratifications, the Budapest Convention is known as the only binding multilateral international law instrument on cybercrime (Clough 2014). The convention seeks to serve as a framework or model

[11]These non-state actors, consists of terrorists, criminal groups as well as individuals operating independently.

law for the enactment of domestic laws that holds international cooperation at its core.

The convention provides for international cooperation and mutual assistance through allowing for multi-stakeholder involvement in the combating of cyber-crimes, by recognising corporate entities together with natural persons. This is consistent with the notion of cyber justice that advocates for accountability, transparency and participation.

The staggered adoption of the Budapest Convention is evidence of the fallacy of international harmonisation of laws. This supports the argument for a normative approach to developing international law and norms for state cybersecurity, rather than enacting more legislation or regulations. There is been a debate around the manner in which international cooperation and mutual assistance should be developed, following the Budapest Convention. However, "the binary debate about the convention versus a UN convention in some way presents a false dichotomy (Clough 2014)." Where states are required to respond to their national security, the self-interest produced by state sovereignty will always move decision-making inwards. Consequently, "each country will determine what it considers necessary to effectively combat cybercrime, looking to national, regional and international standards in enacting laws that best suit its national circumstances (Clough 2014)."

6.4 NATO Tallinn Manual

International cooperation and mutual assistance as a norm probably find its greatest perversion through the Tallinn Manual. The 'to war/law of war' approach is probably best expressed through the perspective adopted by the North Atlantic Treaty Organisation (NATO) via its Tallinn Manual. The NATO Cooperative Cyber Defence Centre of Excellence (CCD COE), based in Tallinn, Estonia embarked on a research project in 2009 in order to produce international law's most comprehensive body of work on cyber warfare. The majority of cyber-attacks or cyber-crimes does not satisfy the international law requirements for 'the use of force' under the prescripts of *jus ad bellum* and *the jus in bello* (von Heinegg 2014). Although the Tallinn Manual does recognise international cooperation and mutual assistance as its core consensus, promising the development of a normative framework towards cyber justice is more consistent with contemporary international law than a 'to war/law of war' approach. The Tallinn Manual is focused on inter-state relations and approaches state cybersecurity from the perspective of public international law (von Heinegg 2014). Consequently, under a 4IR paradigm, the Tallinn Manual does not accurately keep pace with evolving realities of state cybersecurity law or governance. von Heinegg (2014) correctly criticises the manual for its almost exclusive analysis of the 'to war/law of war' approach.

6.5 SADC Model Law: Computer Crime and Cybercrime

In the African context, international cooperation and mutual assistance have found a direct application through the work of the Southern Africa Development Community (SADC). SADC has been consistent in its approach to see cybersecurity as a public security sector threat. Consequently, cybercrime is identified as a challenge under public security by SADC's Directorate Organ on Politics, Defence and Security. This is consistent with the notion of cyber justice. This approach produces a particular challenge for the development of normative standards for state cybersecurity governance in order to effectively and efficiently respond to disruptions brought about by the 4IR. However, SADC has traditionally followed the prescripts of developed international law. The approach in Southern Africa, through the endeavours of the SADC to achieve the three objectives of state cybersecurity governance, was the development of a model law. The norm, international cooperation and mutual assistance find a procedural application through SADC's model law. The SADC normative approach involves the harmonisation of state cybersecurity policy, through the 'establishment of harmonised policies for the ICT market in the Group of African, Caribbean and Pacific states (ACP) countries' (SADC 2013).

7 Normative Modelling for Fourth Industrial Revolution State Cybersecurity Governance

State cybersecurity governance for the 4IR needs to be developed through an evolved notion of cyber justice. Such 4IR normative modelling of state cybersecurity governance underpinned by cyber justice cannot be a closed list (*numerus clausus)* of policy suggestions but must be open to other governance norms and principles. These include governance principles viewed as good, cooperative, analytical in the outcome, future-forward and adaptive. The achievement of cyber justice would mean that these norms are reconcilable with norms such as the respect for human rights, accountability, transparency and participation. How these norms function and their relationships within a system of governance should be consistent with the legal system. In order for a norm to belong to a system, it needs to demonstrate connectivity or a rational validity, which traces that norm back to a base norm. The connectivity can manifest through an assumption of truth borne from objective experience or the immediately observable reality (Kelsen 1961). Various governance models have been developed in the past, each of which adopts characteristics that could produce cyber justice. Consequently, it is important to briefly discuss these models and their reconcilability with cyber justice.

7.1 The Continued Relevance of Good Governance?

The political, economic and social governance of a state is consequently an inclusive approach. What is envisioned for state cybersecurity governance is not only governance on a political, economic, social and cyber front, but also the cross interaction between these spheres. What follows is the question, how does one then measure the achievement of these objectives around in policy? Generally, the common policy yardstick towards which the international community aspire to is good governance. Consequently, what is good governance? The UN General Assembly confirmed through the famous Resolution 66/288 "The Future We Want" that good governance is a cornerstone for development. This would be inclusive of state cybersecurity governance for 4IR development. UN Resolution 66/288 states:

> Democracy, good governance and the rule of law at the national and international levels, as well as an enabling environment, are essential for sustainable development including sustained and inclusive economic growth, social development, environmental protection and the eradication of poverty and hunger.

Good governance is consequently founded on principles such as participation, populace voice, direction, performance, accountability and fairness. These principles postulates that good governance is where all people affected should be part of decision-making through credible institutions. Good governance focuses on broad consensus in order to achieve the best interest of peoples through policies and the procedures to effect such (Kriangsak 2017). State cybersecurity governance directed towards the strategic vision of good governance involving all affected stakeholders would be consistent with the theory of cyber justice. Good governance further holds a long-term perspective directed towards human development and drives government, the private sector and the public at large. This should be understood as inclusive of a collective understanding of what is needed to achieve good governance. Consequently, through its definition, good governance demands an awareness of the socio-economic, historical, cultural and political nodes that shape a community or state.

Performance as a principle of good governance is more focused on the institutional dynamics that underpin good governance. The principle demands institutional responsiveness through their processes. It further seeks the cultivation of a non-discriminatory institutional culture in servicing all affected parties. Together with this, the principle of performance would endeaviours to achieve the best utilisation of cybersecurity resources through being effective and efficient.

The principle of accountability premises good governance on the need for decision-making parties to be accountable to all stakeholders, especially the affected public. The nature of the decision would obviously affect the nature of accountability. Central to this principle is the institutionalisation of transparency. Transparency premised on the access to information, institutions and due process for concerned parties. This access should be strengthened through support that assists better comprehension of information and monitoring of it.

The last principle of good governance is fairness. Fairness is pillared by two sub-principles being equity and the rule of law. Good governance should consequently be interpreted as firstly, equal opportunity amongst states towards their development under a 4IR paradigm. Secondly, good governance postulates that under the rule of law all legal frameworks should be fair and their implementation should impartial. It is clear from the above that good governance is not only consistent with the notion of cyber justice but a necessary accompaniment for 4IR state cybersecurity governance.

7.2 Cooperative Governance

The prevailing contemporary international law norm of international cooperation and mutual assistance needs to be developed for state cybersecurity governance. International cooperation viewed under liberal theory presents a causal path between economic interdependence and interstate wars (Marwala and Lagazio 2011). The more states depend on each other economically it should reduce the risk of following a 'to war/law of war' approach and rather embrace cooperative state cybersecurity governance. This approach will also find support through the notion of cyber justice. The fundamental idea of cooperative governance in state cybersecurity needs to transform under a 4IR paradigm, from a state-centred idea to a broader inclusive governance model. Schwab (2016) notes that "it is, therefore, critical that we invest attention and energy in multi-stakeholder cooperation across academic, social, political, national and industry boundaries." Cooperative governance also played a critical role in developments within the Third Industrial Revolution. The recognition of multi-stakeholders by governments and involving them in policy development assisted the advances within the telecommunications software industries. Consequently, it is reasonable to argue that science policy, technology policy, and intellectual property rights policy have been crucial during the Third Industrial Revolution (Colli and Nicoletta 2013).

International law can learn from its past in the manner that international cooperation and mutual assistance consensus that led to binding international law. Article 11 of the Moon Treaty states "The moon and its natural resources are the common heritage of mankind, which finds its expression in the provisions of this Agreement, in particular in paragraph 5 of this article."[12] Paragraph 5 expresses a commitment by:

> States Parties to this Agreement hereby undertake to establish an international regime, including appropriate procedures, to govern the exploitation of the natural resources of the moon as such exploitation is about to become feasible. (United Nations 1979)

The paragraph further expresses that the common heritage of mankind is a doctrine clearly developed as an anticipatory measure to produce a new form of international

---

[12]Agreement Governing the Activities of States on the Moon and Other Celestial Bodies, Dec. 5, 1979, 1363 U.N.T.S. 3.

cooperation. This measure is born from a period of international conflict and developments in science, which necessitated an increased awareness of the environment and our relationship with it. So too in this period of technological advancement and with the knowledge of the disrupted future, international cooperation and mutual assistance would assist in not only managing the transition but producing cyber justice and legal certainty.

7.3 *Analytical Outcome Governance*

State cybersecurity governance has much to benefit from the discipline of policy statistical analytics in order to achieve cyber justice as an outcome. Analytical outcome governance will assist states to manage the degree of disruption pre-emptively rather than reactively. The introduction of analytical outcome governance into state cybersecurity brings with it the capacity for policy analysis which is more focused and involves competencies and capabilities as well as effective knowledge acquisition and utilisation of the processes of policies (Howlett 2018). A primary characteristic of the 4IR is the availability of enormous amounts of basic data. With cyber justice as the prescriptive objective, introducing analytical capacity into state cybersecurity governance will allow for interventions that are more precise. Decision-making outside of an analysis of these sets of data would naturally place a state at a disadvantaged position relative to other states. Consequently, analytical outcome governance is a crucial prescriptive characteristic for 4IR state cybersecurity governance. Howlett (2018) finds that "governments, as a whole, exhibit an uneven distribution of capacities, technical capabilities, and utilization practices across different organizational and thematic venues." In order to achieve success with the introduction of analytical outcome governance, safeguards need to be introduced. Analytical outcome governance needs to have transparency and depressed protection of the right to privacy at its core, consistent with the prescripts of cyber justice. This would involve the decentralisation of state cybersecurity governance mechanisms and retraining state cybersecurity governance officials (Vyas 2018). Such an approach would be consistent with both cyber justice and the accepted theory of cyber sovereignty above.

7.4 *Future-Forward Governance*

State cybersecurity governance in the 4IR would need to keep pace with the ever-fleeting and expanding set of technologies such as machine learning and artificial intelligence. As one is reading this chapter without a doubt high-velocity advancements in robotics, automation, digital transformation, artificial intelligence, and 3-D printing have considerably advanced (Groscurth 2018). In order to be equal to the challenges and disruptions brought about by the 4IR, state cybersecurity governance needs effective leadership. Such leadership needs to be embedded in the normative framework of state cybersecurity governance in order to produce future-forward thinking or smart, connected leadership (Groscurth 2018). Groscurth (2018) postulates that smart, connected leadership has—at its core—five fundamental demands: presence, agility, collaboration, development, and discernment. Without future-forward governance, indiscriminate 4IR technology will lead the world into the future and not the inverse. In terms of the values that future-forward governance underpins, it does hold the potential to be reconcilable with cyber justice in its outcomes.

7.5 *Adaptive Governance*

State cybersecurity governance, which is adaptive, would also need to be responsive in order to achieve cyber justice as an outcome. Including the potential for adaptation into the normative modelling of state cybersecurity governance will introduce flexibility and resilience into the regime. Noting the characteristics of 4IR technologies and especially the pace of change that these technologies are capable of, state cybersecurity governance needs to be flexible through adequately responding to their potential disruptions. Adaptive governance for the 4IR includes and requires adaptive leadership (Weiler 2001). This means, "with the high speed of change, the challenge for leaders is learning faster than the world around them changes. To ensure success, leaders may need to abandon old behaviours, habits and beliefs, only keeping those that best serve them and their people (World Economic Forum 2019)." In the case of adaptive governance, cyber justice would best serve as the conscience for states and their policymakers.

8 Conclusion

The research question was what policy approach should states undertake under international law to produce state cybersecurity governance that is just in the 4IR? This was underpinned by a hypothesis that accepted that the 4IR would disrupt state cybersecurity governance in two ways. The first is that the traditional mechanisms for intervention, which are legislative and/or regulative responses, will be rendered ineffective and redundant at worst. The second was that the 4IR would challenge traditional notions of sovereignty and by implication statehood, through what has been termed cyber sovereignty. The theory of cyber justice is advanced to answer the research question and to serve as an equitable and objective norm-setting approach to state cybersecurity governance. The presence of cyber justice in the development of state cybersecurity governance would mean good cyber governance that is premised on a human rights-based approach, with more accountability, greater transparency and broad multi-stakeholder participation.

This research has sought to make a broad argument for a developmental shift in international law's approach to state cybersecurity governance. The potential 4IR disruptions to state cybersecurity governance demand a change in focus for the adequate development of state cybersecurity. It is conceivable future, that for state cybersecurity governance these disruptions will hit at the core of the integrity of the state as well as render ineffective our customary legislative and regulatory interventions to legal challenges. This shift needs to happen from an international law perspective on the maintenance of peace and security towards the achievement of cyber justice. This precipitates and needs to adopt a normative approach to state cybersecurity governance, rather than a legal positivist approach. The consensus would be easier to reach around norms and principles for state cybersecurity governance in particular and international cybersecurity law in general. Once such consensus has already been achieved through the norm of international cooperation and mutual assistance. Following traditional international law, it can be said that through this development in state practice generally in future international cooperation and mutual assistance has risen to become a general principle of customary international law.

The 4IR is forging a path for the fast pace of technological development and the rapid disruptions to our traditional notions of how the world functions. The law and legal principles as the organizing structure to human development will not be spared. The global order is already experiencing a proliferation in protectionism and nationalism to guard against these disruptions. However, states cannot develop in isolation and an inwards retraction would produce domestic security instabilities. State cybersecurity governance for the 4IR should be pliable enough to accommodate normative standards such as international cooperation and mutual assistance as well as cyber justice.

Acknowledgements This paper is part of activities, assistance and incentives provided at the University of Johannesburg to be at the vanguard of research involving the 4IR. The assistance and patience of Mallissa Lenong have also been invaluable.

References

Alvik I (2011) Contracting with sovereignty: state contracts and international arbitration. Studies in international law, vol 31. Hart Pub, Oxford, Portland, Or

Aspremont J (2012) From a pluralization of international norm-making processes to a pluralization of the concept of international law. In: Informal international lawmaking. Oxford University Press, pp 185–199. https://hdl.handle.net/11245/1.378436

Benvenisti E (2013) Sovereigns as trustees of humanity: on the accountability of states to foreign stakeholders. Am J Int Law 107(2):295. https://doi.org/10.5305/amerjintelaw.107.2.0295

Berman PS (2007) A pluralist approach to international law. Yale J Int Law 32:301

Besson S, Tasioulas J (eds) (2010) The philosophy of international law. Oxford Univ. Press, Oxford

Bonikowski B, DiMaggio P (2016) Varieties of American popular nationalism. Am Sociol Rev 81(5):949–980. https://doi.org/10.1177/0003122416663683

Chirwa DM, Chenwi LM (eds) (2016) The protection of economic, social and cultural rights in Africa: international, regional and national perspectives. Cambridge University Press, Cambridge, United Kingdom, New York, NY

Clough J (2014) A world of difference: the Budapest convention on cybercrime and the challenges of harmonisation. Monash UL Rev 40(3): 698

Colli A, Nicoletta C (2013) The role of the state in the third industrial revolution. In: Giovanni D, Louis G (eds) The third industrial revolution in global business. Cambridge University Press, Cambridge, pp 229–251. https://doi.org/10.1017/CBO9781139236706.008

Council of Europe (2004) Convention on cybercrime. No. 185. vol. ETS No. 185. https://www.coe.int/en/web/conventions/full-list/-/conventions/treaty/185

De Vattel E (2011) The law of nations: or, principles of the law of nature, applied to the conduct and affairs of nations and sovereigns. In: Chitty J (ed). Cambridge University Press, Cambridge. https://doi.org/10.1017/CBO9781139

Dekker IF, Werner (eds) (2004) Governance and international legal theory. Nova et Vetera Iuris Gentium Series A, Modern International Law 23. Nijhoff, Leiden

Dugard J, Plessis MD, Katz A, Pronto AN (eds) (2011) International law: a South African perspective, 4th edn. Juta, Cape Town

European Union (2016) Regulation (EU) 2016/679 of the European parliament and of the council on the protection of natural persons with regard to the processing of personal data and on the free movement of such data, and repealing directive 95/46/EC (General Data Protection Regulation), vol 2016/679

Fang B (2018) Cyberspace sovereignty: reflections on building a community of common future in cyberspace. Springer, Singapore Beijing

Freeman DH, de Jongste H (1984) A new critique of theoretical thought: the necessary presuppositions of philosophy (Dooyeweerd, H. Works. Ser. A, V. 1). Paideia Press Ltd

Gill TD, Robin G, Robert H, Tim M, Christophe P, Jessica D (eds) (2014) Yearbook of international humanitarian law vol 15, 2012. T.M.C. Asser Press, The Hague. https://doi.org/10.1007/978-90-6704-924-5

Goldring J (1998) Globalisation, national sovereignty and the harmonisation of law. Unif Law Rev 3:435

Groscurth CR (2018) Future-ready leadership: strategies for the fourth industrial revolution. Praeger, Santa Barbara, California

Grotius H (1925) De Jure Belli Ac Pacis "On the Land of War and Peace." Carnegie ed. vol. chapter I

Group of Governmental Experts on Developments in the Field of Information and Telecommunications in the Context of International Security, and UN Secretary-General (2015) Report of the Group of Governmental Experts on Developments in the Field of Information and Telecommunications in the Context of International Security. A/70/150. New York, NY. https://digitallibrary.un.org/record/799853?ln=en

Hathaway, OA, Rebecca C (2012) The law of cyber-attack. Digital Commons Yale Law, Faculty Scholarship Series, no. Paper 3852

Howlett M (2018) Policy analytical capacity: the supply and demand for policy analysis in government. In: Xun W, Michael H, Ramesh M (eds) Policy capacity and governance: assessing governmental competencies and capabilities in theory and practice. Springer International Publishing, Cham, pp 49–66. https://doi.org/10.1007/978-3-319-54675-9_3

Island of Palmas Case (or Miangas), United States v the Netherlands, Award, (1928) II RIAA 829. n.d

Iyi J-M (2016) Humanitarian intervention and the AU-ECOWAS intervention treaties under international law. Springer International Publishing, Cham. https://doi.org/10.1007/978-3-319-23624-7

Janssen M, van den Hoven J (2015) Big and Open Linked Data (BOLD) in government: a challenge to transparency and privacy? Gov Inf Quart 32(4):363–368. https://doi.org/10.1016/j.giq.2015.11.007

Jayasuriya K (2001) Globalization, sovereignty, and the rule of law: from political to economic constitutionalism? Constellations 8(4):442–460. https://doi.org/10.1111/1467-8675.00252

Kelsen H (1961) General theory of law and state. Russell & Russell, New York

Kosseff J (2017) Defining cybersecurity law. Iowa Law Rev 103:985

Kriangsak K (2017) Public international law of cyberspace. Law, governance and technology series, vol 32. Springer, Cham

Kulesza J (2009) State responsibility for cyber-attacks on international peace and security. Polish Yearbook Int Law 29:139–151

Macak K (2016) Is the international law of cyber security in crisis? NATO CCD COE Publications 127, Tallinn, Estonia

Marwala T, Lagazio M (2011) Militarized conflict modeling using computational intelligence. Advanced information and knowledge processing. Springer London, London. https://doi.org/10.1007/978-0-85729-790-7

Michelman FI (1995) Always under law? Const Commentary 12:227

Mihr A (2016) Cyber justice: cyber governance through human rights and a rule of law in the internet. US-China Law Rev 13(4). https://doi.org/10.17265/1548-6605/2016.04.002

Pascariu C, Ionuţ-Daniel B, Ioan B (2017) Investigative analysis and technical overview of ransomware based attacks. Case Study: WannaCry. Int J Inf Secur Cybercrime 6(1): 57–62. https://doi.org/10.19107/IJISC.2017.01.06

SADC (2013) Harmonization of ICT policies in Sub-Saharan Africa—computer crime and cybercrime: South African development community model law

Schwab K (2016) The fourth industrial revolution. World Economic Forum, Cologny/Geneva

Sendler U (ed) (2018) The internet of things: industry 4.0 unleashed. Springer Vieweg, Berlin, Germany

Smith PAC, Pourdehnad J (2018) Organizational leadership for the fourth industrial revolution: emerging research and opportunities. Advances in Logistics, Operations, and Management Science (ALOMS) Book Series. IGI, Business Science Reference, Hershey PA, USA

Steyn L (2017) Data loss will make you wanna cry. The M&G Online. https://mg.co.za/article/2017-05-19-00-data-loss-will-make-you-wanna-cry/

Teitel R (2013) Author's response to Martti Koskenniemi's review of humanity's law. Ethics Int Affairs 27(02):233–234. https://doi.org/10.1017/S0892679413000154

The Lotus Case (France v. Turkey), PCIJ Reports, Series A, No. 10. 1927

United Nations (1979) Agreement governing the activities of states on the moon and other celestial bodies, vol 1363 U.N.T.S. 3

Venktess K (2017) Up to million computers hit in biggest cyber attack ever | Fin24. https://www.fin24.com/Tech/Cyber-Security/up-to-million-computers-hit-in-biggest-cyber-attack-ever-20170515

von Heinegg WH (2014) The Tallinn manual and international cyber security law. In: Terry DG, Robin G, Robert H, Tim M, Christophe P, Jessica D (eds) Yearbook of international humanitarian law, vol 15, 2012. T.M.C. Asser Press, The Hague, pp 3–18. https://doi.org/10.1007/978-90-6704-924-5_1

Vyas L (2018) Re-Invention of the Public Sector Training. In: Ali F (ed) Global encyclopedia of public administration, public policy, and governance. Springer International Publishing, Cham, pp 5411–5416. https://doi.org/10.1007/978-3-319-20928-9_2545

Weil P (1983) Towards relative normativity in international law? Am J Int Law 77(3):413. https://doi.org/10.2307/2201073

Weiler J (2001) The rule of lawyers and the ethos of diplomats : reflections on the internal and external legitimacy of WTO dispute settlement. Harvard Jean Monnet Working Paper 9. Harvard Law School, Cambridge, MA

World Economic Forum (2019) Leading through the fourth industrial revolution: putting people at the centre. World Economic Forum

Does the Global South Need to Decolonise the Fourth Industrial Revolution?

Naiefa Rashied and Muaaz Bhamjee

Abstract Colonialism and coloniality play a pivotal role in modern economies, particularly in the Global South. One of the most important tools of economic oppression, industrialisation, has often benefitted the Global North at the expense of the Global South through expropriation and exploitation of colony resources. The fourth industrial revolution is no exception, although it has the potential to empower the Global South through decolonisation. Conversely, it could perpetuate renewed coloniality at the hands of the more developed parts of the Global South. This chapter contributes to the literature by (a) providing a concise and comprehensive history of global colonisation; (b) describing the extent to which (economic) coloniality prevails and is being renewed in the Global South, and (c) subsequently arguing for ways in which the Global South needs to break coloniality in order to reap the benefits of the fourth industrial revolution. South-South cooperation, particularly in less-developed parts of the Global South, using post-developmentalist and other strategies, are recommended in order for the Global South to reap the benefits of the fourth industrial revolution.

Keywords Fourth industrial revolution · Decolonisation · Global south · Africa rising

---

N. Rashied
School of Economics, College of Business and Economics, University of Johannesburg, Johannesburg, South Africa
e-mail: naiefar@uj.ac.za

M. Bhamjee (✉)
Department of Mechanical Engineering Science, Faculty of Engineering and the Built Environment, University of Johannesburg, Johannesburg, South Africa
e-mail: muaazb@uj.ac.za

W. Doorsamy et al. (eds.), *The Disruptive Fourth Industrial Revolution*, Lecture Notes in Electrical Engineering 674,
https://doi.org/10.1007/978-3-030-48230-5_5

1 Introduction

Industrial revolutions are characterised by intense technological innovations, resulting in accelerated commercialisation. This commercialisation often translates into increased country-level growth and improved standards of living. Industrial revolutions tend to influence every aspect of human life and are meant to improve productivity through technological convenience. Despite its benefits, industrialisation initially flourished as a result of deeply rooted colonialism, implemented and propagated by Europe, North America and even parts of Asia. Although they were in the early stages of industrialisation, these nations exploited indigenous people and resources from their less-developed colonies to further their own industrial development agenda (Alam 1998). As a result, there seems to be a degree of coloniality present in these post-colonies, who seem inclined to be at the forefront of the fourth industrial revolution (4IR) (Naudé 2017). The developed world typically leads an industrial revolution, with the Global South following suit decades later, a ramification of historical underdevelopment. The 4IR is probably the first exception to this, with China, Brazil and Singapore taking the lead with respect to digitisation alongside the usual technology leaders from the Global North namely, the United States, the United Kingdom, Germany, Sweden and Holland, among others.

Decolonisation and industrialisation are somewhat conflicting notions, in principle. Industrialisation is a Western notion, rooted in colonialism, while decolonisation represents political freedom from the coloniser. In literal terms, decolonisation refers to "the process of a state withdrawing from a former colony, leaving it independent" (Oxford English Dictionary 2002). While the literal definition captures political withdrawal, it does not encompass the economic, epistemic and political aftermath of colonisation; in other words, this definition does not capture coloniality. Coloniality, according to Maldonado-Torres (2007), is everything that survives colonialism, including colonial forms of domination, culture, structure and knowledge. Coloniality also represents the economic disadvantage that former colonies face today as a consequence of colonialism.

In this chapter, a critical inquiry is used to establish whether the Global South needs to decolonise the 4IR. The chapter begins with a definition of the 4IR and a brief history of colonisation. Thereafter, the chapter explains the economics of the colony and how it manifests in the Global South presently. Economic growth in the South and the disruptive 4IR, with respect to Africa, are also explored. The chapter concludes by arguing that the Global South needs to decolonise the 4IR through several cooperative post-developmentalist initiatives within the South.

2 What Is the Industrial Revolution?

The simplest definition of the industrial revolution is the transition of labour from human-sourced to machines (Liao et al. 2018). Subsequently, an industrial revolution is characterised by intense technological innovations, resulting in accelerated commercialisation. The explicit link between scientific and technological progress indicates that the major factor that signals the transition from one industrial revolution to the next is technological advancement.

The first industrial revolution spanned 1760–1840, with the onset of mechanisation over muscle power, particularly with the development of steam engines and the construction of railroads, which led to population growth and urbanisation (Lasi et al. 2014; Schwab 2016; Bartodziej 2017; Xu et al. 2018). Between the late 19th century and early 20th century, the use of electricity and the birth of the assembly line made mass production possible, thus establishing the second industrial revolution (2IR) (Lasi et al. 2014; Schwab 2016; Bartodziej 2017; Xu et al. 2018). The third industrial revolution (3IR) began in 1960, and the driving technologies were semiconductors, mainframe and personal computing and the internet, also known as the computer/digital revolution (Lasi et al. 2014; Schwab 2016; Bartodziej 2017; Xu et al. 2018).

Germany coined the term Industry 4.0 with the launch of its own strategic initiative in 2011 to bolster and cement Germany's leadership in global manufacturing (Bartodziej 2017). The term became popular in other parts of the developed world with Klaus Schwab, German engineer, economist and World Economic Forum (WEF) chairman declaring its arrival in his book 'The fourth industrial revolution' (Schwab 2016).

The 4IR builds on the 3IR through greater integration of physical, mechanical and digital systems (Lasi et al. 2014; Schwab 2016; Bartodziej 2017). Schwab (2016) argues that the 4IR is broader than just artificial intelligence, machine learning and bio-digital-mechanical integration. He contends that the scope includes renewable energy, gene sequencing, nanotechnology and quantum computing, among others, and highlights that the emergence of the 4IR is the greatest time for gain or risk (Schwab 2016).

Contrary to Schwab (2016), many academics and professionals regard some of the developments of 4IR as simply part of the 3IR. However, Schwab (2016) insists that there are three reasons why these elements are part of a new industrial revolution, citing velocity, breadth and depth in addition to the systems' impact of the 4IR technological change exceeding the 3IR.

It is therefore clear that the transition from one industrial revolution to the next is signalled by technological advancement, particularly technology that causes social and economic disruption. No one formally decides the stage or status of an industrial revolution. Rather, societies' acceptance (voluntary or coercively) rests on economic transition and the status or stage of the industrial revolution in their respective context.

Schwab (2016) acknowledges that 17% of the world (approximately 1.3 billion people) are yet to experience the 2IR due to a lack of access to electricity. Also,

52% (approximately 4 billion people) are yet to experience the 3IR due to a lack of internet access (Schwab 2016). What Schwab (2016) fails to acknowledge is the role that slavery and colonialism played in previous industrial revolutions, and how this laid the foundation for the inequality in access to key resources observed today.

Initially, slavery, accompanied particularly by inhumane working conditions, was seen as the height of economic efficiency (Gray 1930). Economic philosophy 'giant' Karl Marx argued that colonialism—which includes slave labour—underpinned industrial and commercial supremacy (Williams 1994). Similarly, another 'giant', Adam Smith, who opposed slavery on moral grounds (Pack 1996), argued for his negative position on slavery from the perspective of economic inefficiency—"A person who can acquire no property, can have no interest by to eat as much, and to labour as little as possible" (Williams 1994: v). The abolishment of slavery on economic grounds would later be termed "lucrative humanity" (Williams 1994: 170).

Colonial empires are among the earliest beneficiaries of global resources and industrial development. Most colonies seldom benefitted directly and only reaped indirect benefits after World War II, following generations of deprivation. Subsequently, the poorest parts of global society absorb the largest part of the disruption, while the wealthy reap the benefits due to their resource and developmental advantages as a result of colonisation. The often ignored yet well-known historical reality is that colonisation fuelled the first industrial revolution (van Neuss 2015), as colonies provided cheap black slave labour which replaced white labour (Gray 1930). Furthermore, colonisation and the colonies were essential to the success of the industrial revolution as they were the source of necessary natural resources (van Neuss 2015). Subsequently, former colonies are the countries that have the highest level of inequality in terms of access to the key resources to take them into the second, third and 4IR. Thus, while Schwab (2016) mentions the statistics related to lack of access to such resources, he ignores the factors that have caused it—the inherent colonial nature of industrialisation. The next section provides a brief history of colonisation to illustrate how economic and political power was distributed historically and how coloniality is linked to underdevelopment in the Global South.

3 A Brief History of Colonisation and Coloniality in the 4IR

Colonisation is a practice that entails controlling another country, occupying its land with settlers, usurping its resources and exploiting it economically, and otherwise (Stam and Spence 1983; Oxford Living Dictionary 2019). In addition, colonisers often imposed their belief systems and altered the native identities of the colonised—who were typically indigenous people of the colony.

Colonialism (synonymous with 'expansion' and settlement), the belief system underlying the practice of colonisation, is rooted in Greece, eight centuries B.C. (Cawkwell 1992; Figueira 2008). Greece and Phoenicia were known to colonise local tribes, initially solely for agricultural purposes. Two centuries B.C., Roman colonisation took place across parts of Western Europe, North Africa and Western

Asia (Boardman et al. 2001; Tweedie 2011). The Chinese also colonised Northern Vietnam between the first and sixth centuries A.D. (Holmgren 1980). As a result, the Greeks, Romans and Chinese were the main drivers of colonisation during the classical period of colonisation, approximately between eight centuries B.C. and 475 A.D. (Hanson 1972).

The fall of the Roman Empire gave rise to colonisation in the middle ages. There was a rise in the migration of people from the Roman Empire's Eastern Europe to Western Europe and from Asia to Yemen and Oman (Koebner 2017). Between 632 and 750 A.D., Arabs began their expansion into North Africa, Mesopotamia and the Levant (Richards 1974; Nebel et al. 2002; Nicolle 2012). Somewhat parallel to Arab expansion, the Viking expansion took place between the eighth and mid-eleventh century A.D. Otherwise known as the Viking Age, this era is synonymous with exploration, conquest, settlement and piracy across parts of modern-day Russia, Ukraine, England, Ireland, and Scotland, along with most of Europe, Western Asia and the eastern seaboard of Northern North America (Barrett 2008; Brink and Price 2008).

Modern colonialism, otherwise termed the colonial era was initiated in the 15th century. Western Europe, particularly Spain, Portugal, France, Italy, Britain, Netherlands and Prussia went on to colonise in North America, South America, Asia, Africa and the Pacific (Fieldhouse 1965). During this era, Japan colonised Korea (Cumings 1987), China colonised Tibet (Sautman 2006) and Russia invaded neighbouring Siberia (Armstrong 2010). The end of the modern colonial era was brought about by World War I as a result of colonial rivalry—Europe had colonised about 85% of the Earth by this time. Decolonisation gradually took place after World War II, purely as a result of the financial collapse of colonial empires making structural domination of colonies unaffordable and financially unsustainable (Stam and Spence 1983; Yazzie 2000).

While colonisation may have formally ended, the factor endowments and many other associated economic benefits derived from colonialism were never returned to the colonised. Complementing this is coloniality in the Global South, and particularly in Africa. Coloniality is defined as a "global power structure" emerging from colonialism and affecting "culture, labour, knowledge production and intersubjective relations" (Ndlovu-Gatsheni 2015: 15). In other words, coloniality is the modern-day legacy of colonialism. Ndlovu-Gatsheni (2015) argues that coloniality is perpetuated by six core elements characterised mainly by a Eurocentric, techno-scientific, capitalist and imperialist world order. While the Global North has a significant part to play in this global order, more developed parts of the Global South are at the forefront of executing a similar world order under the guise of '4IR partnerships with and investment in Africa'.

China, India and Brazil are perfect examples of this. Africa experienced immense growth during the 1990s, partially attributed to some of the developed South's 'new scramble' for Africa (Southall 2008). The scramble has since intensified with Japan and Russia joining the race for Africa's land, mineral, oil, gas and coal resources, in addition to labour supply (Carmody 2011).

The new scramble for Africa comes with renewed coloniality under the guise of cooperation, development, economic growth and the 4IR. A case in point is the role China has played in Africa over the last decade; it has taken over from the US as Africa's major financial and trade partner (Van Mead 2018; McKinsey and Company 2017). With Chinese loans, investment and trade in Africa have risen exponentially over the last 10 years. A closer look at the price that Africa is paying for this 'partnership' will be explored in the next section. The discussion starts by exploring the economics of the colony. More specifically, the section highlights how colonies were subjected to an array of deprivations in order to accelerate the growth and development of the colonising nation. Thereafter, the section describes how similar behaviour manifests in less-developed parts of the Global South, sometimes perpetuated by developed parts of the Global South.

4 Economics of the Colony

Fieldhouse (1965: 381) argues that imperial profit-making is a myth and that modern colonial empires are "not artificially constructed economic machines". On the contrary, Brown (1974) convincingly claims that imperialism is underpinned by a strong marriage between capitalism and colonialism, resulting in an artificial division of labour and the subsequent creation of a dual economy. Rodney (1982) also presents strong economic arguments for how Europe used oppression to further itself economically in Africa, Asia and Latin America. The independent state typically had its own colonial economic policies, rooted in capitalist industrialisation, which economically disadvantaged its colonies (Amin 1972). Fieldhouse himself acknowledges this by highlighting six ways in which Europe might have "complicitly" benefitted economically from its colonies (1965: 382):

1. Looting of the colony, whether the resources were of use to the colonial empire or not. The colony took control of these resources and colonies were no longer allowed to benefit or profit from them directly.
2. Extraction of profit from transnational companies established in the colonies. While these companies made use of labourers who were native inhabitants of the colony, it was the company and the colonising nation who benefitted from the profits and not the indigenous people. Returns for labourers was low and stagnant (Amin 1972).
3. Transfer of money or goods from the colony to the metropolis.
4. Unfair terms of trade by monopolising mercantilism. Colonies were unable to protect their own industries resulting in hindered industrial progress.
5. Exploiting the natural endowments of colonies.
6. Higher returns on investments in colonies.
7. Furthermore, colonies were forced to speak the language of the coloniser and disregard their native language and culture. The language, heritage and culture

of the colonised were often assigned less-prestige and importance (Migge and Léglise 2007).
8. Inhumane working and living conditions for colony labourers (Pack 1996).

The ramifications of colonial policies manifest in the economic stature of the present day third world. Africa, parts of Asia and Latin America continue to experience high levels of poverty, limited industrial development and are largely reliant on less technologically intensive sectors of the economy, such as agriculture, to drive economic growth—a testament to the aftermath of the economics of the colony. The abovementioned behaviour continues to manifest itself in less-developed regions of the Global South.

1. Exploiting the natural endowment of colonies continues in parts of the Global South today. China's illegal extraction of natural resources, such as timber and fish in parts of Africa, are a case in point (McKinsey and Company 2017). While the extent may not be comparable to that of European exploitation during colonisation, there is no guarantee that this trend will not grow in the years to come. Similarly, in Zambia, the Konkola Copper Mine, responsible for 65% of mineral production in Zambia, was sold to Indian-based Vedanta Resources. Subsequently, Zambia only received $2 million of the $2 billion revenue generated. Once costs are taken into account, the Zambian government made no profit while the environmental damage in one of the province's rivers was irreversible (Carmody 2011). In addition, Indian-owned, ArcelorMittal established a 'quasi-sovereignty' in Liberia through an iron-ore contract which was later re-negotiated (Carmody 2011). Carmody (2011) further argues that China replicated similar unfavourable deals in Africa in specific Special Economic Zones.
2. The extraction of profit from transnational companies is a common trend in less-developed parts of the South. China's role in various sectors of Kenya, Uganda, South Africa, Tanzania and Angola is an example of this. It is estimated that 90% of Chinese firms operating in Africa are private, profit-driven and not state-owned Chinese enterprises (McKinsey and Company 2017). This is an indication that profit generated across various sectors in parts of Africa is channelled back to China's economy.
3. Developed parts of the Global South are known to collude with respect to resources in parts of Africa. A case in point is the cooperative agreement between India and China concerning African oil and gas companies. In 2006, Indian and Chinese firms outbid each other for oil and gas. To avoid spiralling prices of oil and gas, the two governments signed a cooperative agreement (Carmody 2011).
4. Purporting inhumane working conditions remains prevalent in parts of Africa. For example, Chinese-owned businesses in Africa are known for their violation of labour and environmental laws (McKinsey and Company 2017). Moreover, China is known for violating human rights in its own country by oppressing minorities and forcing them to embrace the Chinese value system (Lin 1997; Chuah 2004). The Indian government also has a reputation for imposing oppressive policies onto certain religious groups (Jeffrelot and Gayer 2012). Again,

there is no guarantee that this trend will not spread to economic partners of China and India, particularly in the South.

5. Linguistic and cultural oppression is also on the rise. In parts of Africa, China's official language of Mandarin is made compulsory in schools and universities. For example, in Kenya and parts of South Africa, it is compulsory for children to learn Mandarin in their foundational years of schooling (Adeoye and Mukhtar 2019). In Kenya, children are also made to recite the Chinese national anthem in their classrooms, an indication of loyalty to China's value system (Adeoye and Mukhtar 2019). Similar trends are observable in parts of South America (Forero 2006). In Nigeria, Chinese investors are awarded tribal chief titles as a token of appreciation for their investment in the region (Van Mead 2018). Mandarin in the classroom and tribal chiefship are therefore signs of African leadership's willingness to sacrifice the language, culture and identity of their people in exchange for a 'quick buck', otherwise termed 'Chinese investment'.

Governments in the Global South, particularly in Africa and Latin America, use the 4IR and 'Africa rising' narratives to perpetuate the kinds of exploitation outlined above. Using statistics on economic growth and the East Asian Crisis as a case in point, the succeeding section argues that the Global South, Africa and Latin America in particular, are in no economic position to maintain the status quo outlined in this section.

5 Economic Growth in the Global South

Despite contradictory theories suggesting that life satisfaction reaches a threshold before it begins to diminish, such as the Easterlin Paradox, countries strive for improved economic growth because of its associations with improved subjective well-being at the individual level (Stevenson and Wolfers 2008). Orthodox theories measure well-being more rigidly by examining improved standards of living and increased income, among other factors (Obeng-Odoom 2015). Regardless of the measures used, the general developmentalist consensus is that economic growth improves lives.

Economic growth in the Global South has been mixed. Africa is often seen as a 'lion on the move', boasting doubled GDP growth between 2002 and 2007 at 6.2 and 5% between 2007 and 2017 in Sub-Saharan Africa alone (Obeng-Odoom 2015). Moreover, the International Monetary Fund (IMF) forecasts average growth on the continent to range between 3 and 10% over the next five years (IMF 2019). However, critics argue that African people remain poor and jobless, and that growth statistics for Africa are misleading (Devarajan 2013; Obeng-Odoom 2015). The truth may lie somewhere in between considering that the largest sector in Africa is agriculture and not technology-intensive industries, such as ICTs, which are typically associated with higher levels of growth (McKinsey Global Institute 2010).

Latin America's growth has lagged behind other developing economies, at around 3% between 2007 and 2017 (McKinsey Global Institute 2017). Despite its current developments in green energy production, technology, textiles and social infrastructure, Latin America remains at the centre of looming disruptions in the form of declining fertility, the end of the commodity supercycle and increased protectionism (McKinsey Global Institute 2017).

Like Latin America, Asia's major sectors include natural resources (Southwest Asia), agriculture (central and Southeast Asia) and manufacturing (East Asia). At 5.6%, Asia has experienced significant growth over the past 15 years (McKinsey Global Institute 2017). Despite the Asian crisis and the 'clash of the capitalisms' (as a result of American Imperialism in Asia) in the late 1990s (Johnson 1998), Asia continued to specialise and later regained control of its economic policies. As a result, Asia is in a better position to develop and control its own technology to (a) keep up with the developed world and (b) mitigate the disruptive elements of the 4IR. In addition, Asia remains focused on building stronger trade ties with the developed world, which could also place the continent at the forefront of economic development and growth in the Global South.

While Africa and Latin America have improved their trade ties with economic giants such as China and the United States, they have less to contribute with respect to technology and innovation, given the large financial investments required. It is often the Global North that leads the way because they are almost always financially better off, relative to the majority of the Global South.

The aim of the next section is to examine the kinds of 4IR technologies developed in the first and third world, and the sectors that benefit from them.

6 Global Leaders in Key 4IR Technologies

The technologies of the 4IR are built on the exponential increase in computing power and digital advancements (Schwab 2016). Thus, information and communication technology, as well as ubiquitous computing, play a central role in all the technologies of the 4IR (Schwab 2016; Bartodziej 2017). Ubiquitous computing is the pervasive availability of computing in society through widely distributed devices as opposed to centrally located computing availability (Schwab 2016; Bartodziej 2017). The driving technologies of the 4IR are wide in scope and go beyond artificial intelligence, machine learning and bio-physical-digital integration. Table 1 illustrates global leaders in some key technologies of the 4IR and the sectors in which those technologies are primarily used.

It is depicted (Table 1) that the Global North is the global leader in key 4IR technologies. While parts of the Global South, such as China, Japan and South Korea, are penetrating the market and challenging the Global North in many of these key areas, these countries often remain second to the Global North. Unlike their Asian counterparts in the Global South, it appears as though African and Latin American counties do not lead or show any significant signs of development in these areas.

Table 1 Global leaders in key 4IR technologies. [Sourced from popular literature: (The Climate Council 2019; Davenport 2019; Phillippidis 2018; StatNano 2018; The Economist 2018; Phillips 2019; Strumpf 2019; Technavio 2019; Chapman 2019)]

Technology	Global leader		Sector
	Global north	Global south	
Renewable energy	Germany		Energy
Artificial intelligence and machine learning	USA[b]	China[c]	Pervasive[a]
Gene sequencing	USA		Genetic Engineering and medicine
Nanotechnology	USA[b]	South Korea[c]	Materials, manufacturing, energy, electronics, medical/biomedical etc.
Quantum computing	USA		Computing
Biotechnology	USA[b]	Singapore[c]	Medicine agriculture, food production etc.
5G		China	Information and communication technologies
Robotics	Germany[c]	Japan[b]	Manufacturing, agriculture, automotive, medical, military/defence etc.
Additive manufacturing	USA[b], Germany[c]		Manufacturing

[a]Pervasive is used to indicate that a technology that is across numerous sectors to the point where it is growing to be all-encompassing
[b]Global leader
[c]Second place global leader

Manufacturing forecasts alone suggest that Africa and Latin America will remain less developed and less competitive (Deloitte Touche Tohmatsu Limited and US Council on Competitiveness 2016).

Table 1 clearly shows that former colonisers lead the 4IR in the Global North while greater developed parts of Asia, particularly China, lead the 4IR in the Global South. This suggests that if lesser developed parts of the South, such as the majority of Africa and Latin America, continue unfavourable cooperation with Asia, they run the risk of continued or renewed coloniality and adverse economic conditions, which the next section discusses.

7 Disruptive 4IR: The Case of Africa

The 4IR has far-reaching consequences for Africa, particularly with respect to disruption. On the one hand, as described in previous sections, coloniality has kept Africa marginalised and dependent, despite the global 'Africa rising' narrative. On the other hand, the 4IR will have adverse consequences for the continent should the Global North continue to accelerate its development of 4IR technologies, as highlighted in Table 1. These consequences will further disadvantage the continent and perpetuate coloniality.

The Global South usually mobilises capital and obtains technology transfers through the provision of cheap labour. However, the Global North, through its use of 4IR technologies, will move away from using cheap labour and move towards (cheaper) automation. This will create excess labour supply and increase unemployment for impoverished masses in Africa. Moreover, this will adversely affect capital accumulation and hinder technology development and technology transfer. To make matters worse, Africa has engaged in premature de-industrialisation, a movement from manufacturing to the services sector, in an attempt to boost its own economic growth (Naudé 2017; Harvey 2017).

The Global South runs the risk of losing its own sovereignty and becoming economically dependent on 4IR giants due to its lack of development in 4IR technologies. The presence of China in Africa is a case in point. In 2008, China surpassed the United States and the European Union as Africa's largest trading partner (Van Dijk 2009). In addition, China also provides aid to Africa, foreign direct investments, and leases or purchases land, mainly for agricultural purposes. There is also an increased Chinese military presence in Africa (Van Dijk 2009). Yet other countries, from the North and even Asia, such as Russia and India, are also trying to increase their presence in Africa (Winters and Yusuf 2007). This raises questions about Africa's ability to thrive on its own. If China, or any other nation, maintains a dominant presence in Africa, the country is likely to lose its sovereignty because of its economic underdevelopment and dependence on other countries for major facets of its own growth.

8 Way Forward for the Global South

Throughout this chapter, we highlighted that colonialism and coloniality are the reasons for the Global South remaining underdeveloped and largely dependent on external drivers of economic growth. In this section, we argue how the Global South can decolonise the 4IR to mobilise its own development.

Many argue that the Global South is 'catching-up' with Western (and capitalist) narratives of development. Alam (1998) argues that in the past, Latin American countries found themselves more integrated into the world economy due to greater control over their social and economic policies, relative to their Asian and African counterparts. Similarly, as Asian countries exercise greater control over their own economic policies in the 21st century, they find themselves experiencing greater integration in the world economy, which includes higher levels of industrialisation and economic development. As a result, Africa is often pushed to focus on industrialisation (commonly via 4IR more recently) under the guise that it stands to benefit from industrialisation as it will decrease dependency on production and exports of commodities (Naudé 2017).

The reality is that the Global South needs to take control of their economic policies by moving away from developmentalism and towards post-developmentalist ideology. Developmentalism refers to rising standards of living as a result of improved economic growth (Rapley 2004). Here, improved standards of living refer to improved nutrition, education, health and personal autonomy. Developmentalism is a capitalist mechanism that the first world uses to impose individualist and pro-Western worldviews and policies. It also reinforces the Western ideals of industrialisation and disregards cultural and other contexts (Escobar 1995). On the other hand, post-developmentalism places more value on third-world voices and value systems (Ferguson 1994). While post-developmentalism is not without its weakness (Ramírez-Cendrero 2018), it certainly presents an alternative. Some argue that post-developmentalist notions are not really accompanied by innovative designs for social change. Instead, they offer alternatives to accommodate capital (Ramírez-Cendrero 2018). The next few sections highlight possible designs for social change within the post-developmentalist framework.

The Global South needs to redefine its own growth path, without sacrificing its value system (which includes economic values) or culture. One way to do this is to use decentralised systems of local development otherwise termed local economic development (LED). LED is defined as a development strategy that is territorially-based, locally owned and managed and primarily aimed at improving economic growth and employment (Rodríguez-Pose and Tijmstra 2005). While 'development strategy', 'economic growth' and 'employment' are developmentalist notions, one cannot escape the fact that most individuals desire the simplest of conveniences, such as electricity, clean water and a comfortable yet basic standard of living. Africans and Latin Americans already have strong self-developed informal economies, used to fill the gaps of the failing formal economy. This suggests that with the right kind of support, these informal markets can be improved further for the benefit of the communities they serve.

Less-developed countries in the Global South, particularly in Africa and Latin America, need to cooperatively support each other in their quest for a new growth path. For example, Latin America needs to remove obstacles to competitiveness (McKinsey Global Institute 2017). One way to do this would be to draw on Africa's expertise in agriculture and manufacturing to expand its own high value-added activities across important value chains within the continent. In a similar light, Africa has a lot to learn from Latin America with respect to technological development. A partnership between the two continents could break the cycle of coloniality in the third world and lead to third-world solutions for third-world problems. As it stands, trade alone between the two continents has declined significantly since 2011 due to economic instability across both regions (Trade Law Centre (Tralac) 2016). Despite the immense potential, a possible obstacle to such a partnership would be a lack of financing due to the relatively poor financial positions of both continents, on average.

One way for Africa and Latin America to enhance their financing is through reparations. Financial reparations are a suitable way to ensure redistributive justice in former colonies (Wenar 2006). If the Global North is serious about redressing the wrongs of the past, it should place significant emphasis on remunerating former colonies for centuries of economic oppression. German reparations to Israel are examples of how modern states take responsibility for past colonisation and human rights violations. However, backwards-looking reparations—as in the case of Germany and Israel—cultivates victimhood (Torpey 2001). We recommend a more forward-looking approach to reparations (Torpey 2001), which aims to equalise imbalances between rich and poor countries while providing mutual economic benefits to both regions. Funds from forward-looking reparational programmes could be used to accelerate technological developments in the Global South and could help Africa and Latin America depend less on former colonisers.

Concerning the 4IR, the Global South should pursue a post-developmentalist strategy, as discussed in this section, by increasing South-South cooperation, particularly between 'lesser developed' parts of the South. Asia can also provide assistance through expertise, capital and technological developments via these cooperative mechanisms. However, renewed coloniality perpetuated from South-South cooperation needs to be addressed. It would therefore be necessary for the respective governments to establish firm requirements that ensure cooperation without a sacrifice of value systems, sovereignty and diversity. Cooperation that focuses on the concept of *Ubuntu* and diversity as opposed to individual gain and uniformity may lead to post-developmental success and meaningful self-reliance without coloniality.

References

Adeoye A, Mukhtar I (2019) China's influence in Africa grows as more young people learn to speak Mandarin. Cent News Netw (CNN), 11 Apr. Available at https://edition.cnn.com/travel/article/mandarin-language-courses-africa-intl/index.html. Accessed 17 Jan 2020

Alam MS (1998) Colonialism and industrialization: empirical results. MPRA paper 37866, Munich, Germany

Amin S (1972) Underdevelopment and dependence in Black Africa-origins and contemporary forms. J Mod Afr Stud 10(4):503–524

Armstrong, T. (2010). Russian settlement in the north. Cambridge University Press

Barrett JH (2008) What caused the Viking age? Antiquity 82(317):671–685

Bartodziej CJ (2017) The concept Industry 4.0: an empirical analysis of technologies and applications in production logistics. Springer Gabler

Boardman J, Griffin J, Murray O (eds) (2001) The Oxford history of the Roman world. Oxford Paperbacks

Brink S, Price N (eds) (2008) The viking world. Routledge

Brown MB (1974) The economics of imperialism. Penguin Books

Carmody P (2011) The new scramble for Africa. Polity Press

Cawkwell GL (1992) Early colonisation. Class Q 42(2):289–303

Chapman S (2019) HP deems the US as the leader of additive manufacturing. Global Manuf, viewed 06 June 2019. https://www.manufacturingglobal.com/leadership/hp-deems-us-leader-additive-manufacturing

Chuah O (2004) Muslims in China: the social and economic situation of the Hui Chinese. J Muslim Minority Affairs 24(1):155–162

Cumings B (1987) The legacy of Japanese colonialism in Korea. In: The Japanese colonial empire 1895–1945, pp 487. Princeton University Press
Devarajan S (2013) Africa's statistical tragedy. Rev Income Wealth 59:9–15
Davenport TH (2019) Opinion: China is overtaking the U.S. as the leader in artificial intelligence. Market Watch, viewed 06 June 2019. https://www.marketwatch.com/story/china-is-overtaking-the-us-as-the-leader-in-artificial-intelligence-2019-02-27
Deloitte Touche Tohmatsu Limited and US Council on Competitiveness (2016) 2016 global manufacturing competitiveness index, viewed 30 May 2019. https://www2.deloitte.com/global/en/pages/manufacturing/articles/global-manufacturing-competitiveness-index.html
Escobar A (1995) Encountering development: the making and unmaking of the Third World. Princeton University Press
Fieldhouse DK (1965) The colonial empires: a comparative survey from the Eighteenth Century. Macmillan Education
Ferguson J (1994) The anti-politics machine: "development," depoliticization, and bureaucratic power in Lesotho. University of Minnesota Press
Figueira T (2008) Colonisation in the classical period. In: Greek colonisation, pp 427–523. BRILL
Forero J (2006) Across Latin America, Mandarin is in the air, Washington Post, 22 Sept 2006. Available at https://www.globalpolicy.org/component/content/article/162-general/27647.html. Accessed: 17 Jan 2020
Gray LC (1930) Economic efficiency and competitive advantages of slavery under the plantation system. Agric Hist 4(2):31–47
Harvey R (2017) The 'fourth industrial revolution': potential and risks for Africa. Nat Sci Technol Innov Inf Portal (NACI), viewed 06 June 2019. http://www.naci.org.za/nstiip/index.php/analytical-contributions/technological-progress/40-the-'fourth-industrial-revolution'-potential-and-risks-for-africa
Hanson RPC (1972) The reaction of the church to the collapse of the Western Roman Empire in the Fifth Century. Vigiliae Christ 26(4):272–287
Holmgren J (1980) Chinese colonisation of Northern Vietnam: administrative geography and political development in the Tonking Delta, first to sixth centuries AD. Faculty of Asian Studies in association with Australian National University Press
International Monetary Fund (2019) World economic outlook 2019—gross domestic product, viewed 03 June 2019. https://www.imf.org/external/datamapper/NGDP_RPCH@WEO/OEMDC/ADVEC/WEOWORLD
Jaffrelot C, Gayer L (2012) Muslims in Indian cities: trajectories of marginalisation. Columbia University Press
Johnson C (1998) Economic crisis in East Asia: the clash of capitalisms. Camb J Econ 22:653–666
Koebner R (2017) The settlement and colonisation of Europe. In: The expansion of Central Europe in the middle ages, pp 39–59. Routledge
Lasi H, Fettke P, Kemper HG, Feld T, Hoffmann M (2014) Industry 4.0. Bus Inf Syst Eng 6:239–242
Liao Y, Loures ER, Deschamps F, Brezinski G, Venâncio A (2018) The impact of the fourth industrial revolution: a cross-country/region comparison. Production 28:e20180061
Lin J (1997) Policies and practices of bilingual education for the minorities in China. J Multiling Multicult Dev 18(3):193–205
Maldonado-Torres N (2007) On the coloniality of being. Cult Stud 21(2–3):240–270
Migge B, Léglise I (2007) Language and colonialism. Applied linguistics in the context of creole communities. In: Hellinger M, Pauwels A (eds) Language and communication: diversity and change. Handbook of applied linguistics. Mouton de Gruyter, Berlin, pp 297–338
Mckinsey and Company (2017) Dance of the lions and dragons. How are Africa and China engaging, and how will the partnership evolve?, viewed 17 Jan 2020. https://www.mckinsey.com/~/media/McKinsey/Featured%20Insights/Middle%20East%20and%20Africa/The%20closest%20look%20yet%20at%20Chinese%20economic%20engagement%20in%20Africa/Dance-of-the-lions-and-dragons.ashx

Mckinsey Global Institute (2017) Where will Latin America's growth come from?, viewed 03 June 2019. https://www.mckinsey.com/~/media/mckinsey/featured%20insights/employment%20and%20growth/how%20to%20counter%20three%20threats%20to%20growth%20in%20latin%20america/mgi-discussion-paper-where-will-latin-americas-growth-come-from-april-2017.ashx
Mckinsey Global Institute (2010) Africa's path to growth: sector by sector, viewed 03 June 2019. https://www.mckinsey.com/featured-insights/middle-east-and-africa/africas-path-to-growth-sector-by-sector
Naudé W (2017) Entrepreneurship, education and the fourth industrial revolution in Africa. IZA discussion papers, No. 10085. Institute of Labour Economics (IZA)
Ndlovu-Gatsheni SJ (2015) Genealogies of coloniality and implications for Africa's development. AfrDev XL(3):13–40
Nebel A, Landau-Tasseron E, Filon D, Oppenheim A, Faerman M (2002) Genetic evidence for the expansion of Arabian tribes into the Southern Levant and North Africa. Am J Hum Genet 70(6):1594–1596
Nicolle D (2012) The great Islamic conquests AD 632–750. Bloomsbury Publishing
Oxford Living Dictionary (2019) viewed 10 May 2019. https://en.oxforddictionaries.com/definition/colonization
Obeng-Odoom F (2015) Africa: on the rise, but to where? Forum Soc Econ 44(3):234–250
Oxford English Dictionary. (2002). Concise oxford english dictionary. Oxford University Press
Pack SJ (1996) Slavery, Adam Smith's economic vision and the invisible hand. Hist Econ Ideas 4(1/2):253–269
Phillippidis A (2018) Top 10 sequencing companies. Genetic Eng Biotechnol News (GEN), viewed 06 June 2019. https://www.genengnews.com/a-lists/top-10-sequencing-companies-2/
Phillips T (2019) The top countries for biotech firms and research. Balance, viewed 06 June 2019. https://www.thebalance.com/ranking-the-top-biotech-countries-3973287
Rapley J (2004) Development studies and the post-development critique. Prog Dev Stud 4(4):350–354
Ramírez-Cendrero JM (2018) Limits and contradictions of post-developmentalism as a heterodox approach to capitalist development. Capital Nat Social 29(4):68–87
Richards JF (1974) The Islamic frontier in the east: expansion into South Asia. South Asia: J South Asian Stud 4(1):91–109
Rodney, W. (1982). How Europe underdeveloped Africa. Bogle-L'Ouverture Publications
Rodriquez-Pose A, Tijmstra S (2005) Local Economic Development as an alternative approach to economic development in Sub-Saharan Africa (An abbreviated report for the World Bank). Evaluating and disseminating experiences in Local Economic Development (LED). The World Bank, Washington, D.C
Sautman B (2006) Colonialism, genocide, and Tibet. Asian Ethn 7(3):243–265
Schwab K (2016) The fourth industrial revolution. Penguin Random House
Southall R (2008) The 'New scramble' and labour in Africa. Labour Cap Soc 41(2):128–155
Stam R, Spence L (1983) Colonialism, racism and representation. Screen 24(2):2–20
StatNano (2018) Top ten countries in nanotechnology patents in 2017. StatNano, viewed 06 June 2019. https://statnano.com/news/62082
Stevenson B, Wolfers J (2008) Economic growth and subjective well-being: reassessing the Easterlin paradox, No. w14282. National Bureau of Economic Research
Strumpf D (2019) Where China dominates in 5G technology. Wall Street J, viewed 06 June 2019. https://www.wsj.com/articles/where-china-dominates-in-5g-technology-11551236701
Technavio (2019) Top 21 industrial robotics companies in the world 2019. Technavio, viewed 06 June 2019. https://blog.technavio.com/blog/top-21-companies-in-the-industrial-robotics-market
The Climate Council (2019) 11 countries leading the charge on renewable energy. Climate Counc, viewed 06 June 2019. https://www.climatecouncil.org.au/11-countries-leading-the-charge-on-renewable-energy/

The Economist (2018) Quantum spring: the race is on to dominate quantum computing. The Economist, viewed 06 June 2019. https://www.economist.com/business/2018/08/18/the-race-is-on-to-dominate-quantum-computing

Torpey J (2001) Making whole what has been smashed: reflections on reparations. J Mod History 73(2):333–358

Trade Law Centre (Tralac) (2016) Latin America–Africa trade relationship, viewed 07 June 2019. https://www.tralac.org/resources/our-resources/10493-latin-america-africa-trading-relationship.html#moredownloads

Tweedie FC (2011) The case of the missing veterans: Roman colonisation and veteran settlement in the Second Century BC. Historia: Zeitschrift für Alte Geschichte 4:458–473

Van Dijk MP (ed) (2009) The new presence of China in Africa. Amsterdam University Press

Van Mead N (2018) China in Africa: win-win development, or a new colonialism? *The Guardian* 31 July. Available at https://www.theguardian.com/cities/2018/jul/31/china-in-africa-win-win-development-or-a-new-colonialism. Accessed 17 Jan 2020

Van Neuss L (2015) Why did the industrial revolution start in Britain? In: University of Liège working paper series. University of Liège, Economics Department. Available at https://orbi.uliege.be/handle/2268/188592. Accessed: 20 Jan 2020

Wenar L (2006) Reparations for the future. J Soc Philos 37(3):396–405

Williams E (1994) Capitalism and slavery. UNISA Press

Winters AL, Yusuf S (2007) Dancing with the giants: China, India, and the global economy. World Bank

Xu M, David JM, Kim SH (2018) The fourth industrial revolution: opportunities and challenges. Int J Financ Res 9(2):90–95

Yazzie R (2000) Indigenous peoples and postcolonial colonialism. In: Reclaiming indigenous voice and vision. UBC Press

Decentralizing Emerging Markets to Prepare for Industry 4.0: Modernizing Policies and the Role of Higher Education

Wynand Lambrechts, Saurabh Sinha, and Tshilidzi Marwala

Abstract This chapter investigates socioeconomic policies, technical focus, and academic necessities in core competencies and skills development to prepare emerging markets for the technological disruption of the Fourth Industrial Revolution (Industry 4.0 or 4IR). This chapter reviews the potential that Industry 4.0 has on creating sustainable work opportunities in emerging markets that have typically not benefitted equally (compared to developed markets) in earlier industrial revolutions. Industry 4.0 is unique in its ubiquity through the internet and allows remote participation in Industry 4.0. Emerging markets are urged to address challenges in adapting Industry 4.0 early and prepare for its maturation. Policy gaps in broadband connectivity and equal gender distribution in the information and communications industry are reviewed and presented in this chapter. Following this review, the role that higher education must play in skills development for Industry 4.0. Brazil, Russia, India, China, and South Africa have been identified as significant role-players in addressing policy gaps that could hinder Industry 4.0 development in emerging markets. Through Industry 4.0, a global shift towards a decentralized industry is occurring and this chapter additionally reviews the factors that influence this shift.

1 Introduction

Industry 4.0 will provide early adopters with a competitive advantage in numerous sectors; however, the primary areas where successful implementation could lead to a clear advantage are within

W. Lambrechts (✉) · T. Marwala
Faculty of Engineering and the Built Environment (FEBE), University of Johannesburg (UJ), Johannesburg, South Africa
e-mail: wynand.lambrechts@vodamail.co.za

S. Sinha
Research and Internationalisation, University of Johannesburg (UJ), Johannesburg, South Africa

W. Doorsamy et al. (eds.), *The Disruptive Fourth Industrial Revolution*, Lecture Notes in Electrical Engineering 674,
https://doi.org/10.1007/978-3-030-48230-5_6

- data-driven enterprises relying on cloud technology,
- the internet of things (IoT),
- advanced analytics,
- enhanced customization, and
- improved communications.

Industry 4.0 is proving to be a disruptive force within the manufacturing and service sectors and companies are realizing the importance and significance of preparing infrastructures towards this shift. Due to Industry 4.0 not yet in a mature state, emerging markets have an opportunity to boost economic growth and upskill their workforce to participate in Industry 4.0 and this chapter focuses on identifying challenges and opportunities to achieve this. Preparing for Industry 4.0 is however, necessary in various sectors, and should be included in governmental policies being modernized to facilitate and incentivize Industry 4.0. Furthermore, higher education needs to play a significant role in preparing the workforce by teaching skills needed for the modern industry. From the five primary areas that could lead to clear advantages for businesses (and countries) within Industry 4.0, it is clear that these technology-related areas require skilled and educated populations. The nature of Industry 4.0 also allows that upskilling populations can be done by taking advantage of technologies that drive Industry 4.0, enabling anyone with an internet connection to be educated, and eliminating various forms of geographical limitations especially in rural areas. This chapter also reviews the role that the BRICS nations (Brazil, Russia, India, China, and South Africa) has in preparing themselves, and assisting other emerging markets, for Industry 4.0. Examples of efforts put in place by several BRICS nations to spearhead Industry 4.0 are presented in this chapter.

The following section reviews how countries can become competitive and maintain their competitiveness in Industry 4.0 through a process of reindustrialization. Reindustrialization refers to a process that is economically, socially and politically driven in which economies re-establish their industries through a belief that manufacturing and other industrial jobs are more desirable than jobs in the service sector or in the financial sector. In Industry 4.0 and the reindustrialization of manufacturing, jobs for skilled workers will offer innovative and challenging environments that are stimulating and rewarding for workers, as opposed to traditional service-oriented careers. Industry 4.0 also requires a vital service-oriented approach, but new technologies and digitization will offer additional benefits for workers to apply their skills and innovate new and customized products and services to customers. As a result, historical industrial powerhouses such as China, India and South-East Asia are pushing hard to realize reindustrialization in time for when Industry 4.0 becomes an indispensable part of business globally. Reviewing the importance of reindustrialization also serves as an introduction to identifying policy gaps that exist in developing nations within BRICS (and a wider group of countries in the Global South) that must be addressed to reflect self-sufficient economies and be the go-to markets for Europe and the USA towards Industry 4.0 mass industrialization.

2 Industry 4.0 Competitiveness Through Reindustrialization

In the wake of the 18th to 20th century colonization, historical industrial driving forces such as China, India and South-East Asia suffered a great loss of industrial production. Reindustrialization is an economic, social and political shift towards consolidating national resources for the primary goal of re-establishing industries that were once flourishing. Globally, countries have realized that a reinvigorated national economy lies at the heart of sustainability, growth and economic desirability. For a company, and a country, to succeed in entering Industry 4.0, maintaining competitiveness, and taking advantage of an industrial revolution in its infancy, some forms of reindustrialization are required.

In the USA, Germany, and China, reindustrialization reform programs in AI and environmentally friendly (green) technologies have become the modern growth drivers in the manufacturing industries. Through these drivers, new service-oriented products are developed, essentially through the pervasive use of information technology (IT) and the IoT. All countries aiming to reindustrialize their economies in accordance with Industry 4.0 should adopt fitting and modular policies to support and invigorate these drivers in their current economies. Policies should help develop high value-added and modern manufacturing industries to support a modern industrial revolution, where the shift towards autonomous cyber-physical systems (CPS) is seen as the dominating space for future prosperity. Modern manufacturing is not only about producing and distributing products, but has an equally important obligation to provide online service platforms for after-sales support, advertising and personal relationships between an industrialist and the client. These services require accessible communication infrastructure penetration, a high degree of freedom of information and accessibility throughout the product-service continuum. Accessible and cost-effective energy, low labor cost and a general habitual attractiveness are likewise important to drive reindustrialization and bring the operations of local businesses and manufacturing houses back to a country. Favorable energy and labor trends lead to rebounding strengths for manufacturing and industrial production within the borders of a country, limiting the requirement to outsource services or import materials and positioning them to dominate this space. A firming economy leads to relaxation of policies such as quantitative easing, where a central bank electronically introduces new money to purchase financial assets. This cash boost drops the cost of borrowing and increases private sector expenditure in the economy to stabilize inflation rates to the targeted values. If inflation values overshoot, the bank can sell these assets and reduce the amount of money and spending in the economy, therefore aiming to a reach a pre-defined equilibrium.

Manufacturing in its traditional sense, producing physical items to satisfy the growing demands for goods as living standards increase, has a significant impact on the environment. The depletion of raw materials and energy demands in the manufacturing space are constantly accumulating and constitute one of the major causes of harmful emissions and waste production. However, it is still recognized that manufacturing is a critical factor that determines the longevity and sustainability of emerging markets, driven primarily by progress in science and technology. Manufacturing of modern green technologies allows emerging and developed markets to penetrate traditional as well as new industries. Adapted processes, policies and business models are required to dominate in industries and to provide social and economic benefits. There is also a close affiliation between a manufacturing industry and disciplines in research, education and innovation. This is an often-overlooked synergy that has long-lasting beneficial effects in a socio-economic setting. A high level of industrialization, or reindustrialization, results in an excessive number of patent submissions, requirements for tertiary education in engineering, industrialization, production, science, and other technology-oriented disciplines, and often leads to an increase in learners pursuing post-graduate qualifications.

A considerable extent of *deindustrialization* is however noticeable in developed countries such as the USA, Japan, and Europe, whereas a need for products and consumption has shifted to emerging regions such as the BRICS nations. To its advantage, the population of the combined BRICS nations is comparable to that of the rest of the world, allowing for large investments in human capital. Developing countries also typically have larger population growth rates compared to developed countries (Lambrechts and Sinha 2016), which further solidifies future investment opportunities in a developing youth. Figure 1 presents a population comparison between the BRICS nations and the rest of the world.

As seen in Fig. 1, the total population of the BRICS nations is significant with respect to the total world population. China and India have the biggest populaces, 1.42 billion and 1.35 billion, respectively (making up 87% of the population of all

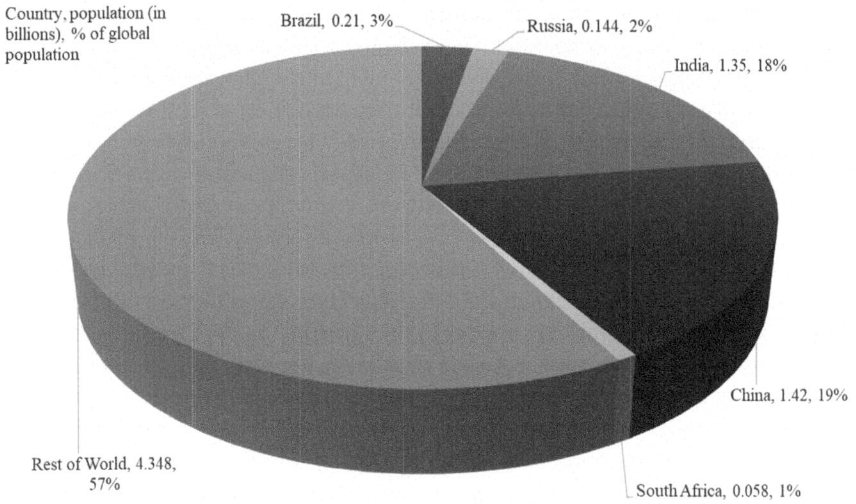

Fig. 1 Summary of the population of BRICS nations related to the rest of the world (data from 2018 and 2019 World Bank estimates), indicated by country, population (in billions) and percentage of global population

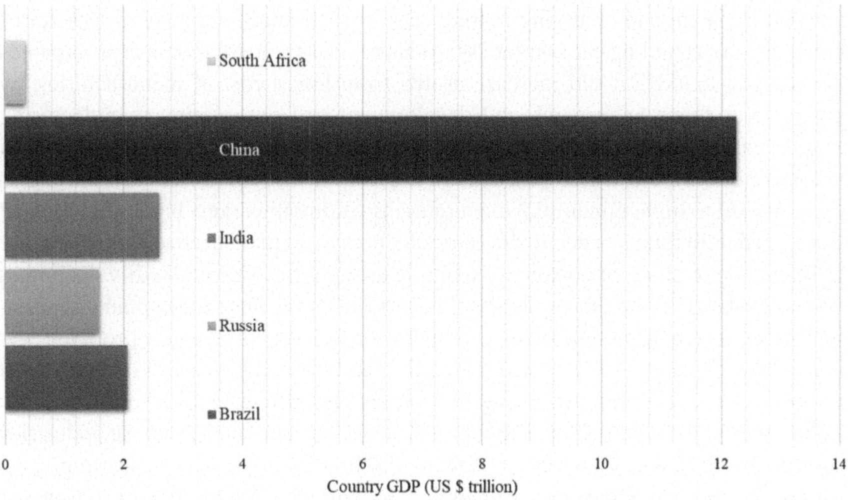

Fig. 2 Summary of the absolute GDP of BRICS nations (data from 2018 and 2019 World Bank estimates) in US $ trillion. The gross world product (GWP) (not shown in this figure) was estimated in 2017 at approximately US $78 trillion

BRICS nations), followed by Brazil and Russia, 210 million and 144 million, respectively, and South Africa, which has the smallest population of the BRICS nations, at approximately 58 million people. Investments in the workforce, manufacturing and ultimately Industry 4.0 are therefore crucial not only for these nations, but for global economy adoption of Industry 4.0. At approximately 43% of the world population, BRICS nations offer vast workforce advantages if skills development is prioritized. In terms of their contribution to the global GDP, Fig. 2 lists the BRICS nations with their respective GDP in US $ trillion.

As shown in Fig. 2, China, as well as having the largest population, also has the highest contributing GDP of the BRICS nations, much higher than that of India, which has a comparable population. The estimated global GDP is US $78 trillion; therefore, China contributes approximately 15% of the worldwide GDP, a significant percentage considering it is a single country. India only contributes roughly 3% of the global GDP, which shows that industrialization and manufacturing in this country can be increased significantly, since the workforce in India is large. India is expected to lead growth in the available workforce in the Asia-Pacific region by 2050, with an estimated 1 billion people qualified to come into the job market at that time, according to the UNDP. Brazil, Russia and South Africa all make relatively low global GDP contributions, but all these nations have the potential and capabilities to improve upon their global dominance and become industry leaders in the emerging market domain. In certain sectors (especially natural resource availability and prospective skilled workforces), they are also able to compete with developed nations such as Europe and the USA.

Primarily, a shift to emerging markets can be attributed to a general transfer of knowledge in expanding economies (Westkämper 2013). To their benefit, companies in developed markets could increase profits from lower cost of manufacturing by shifting their focus to research and development and outsourcing manufacturing capabilities. Consequently, deindustrialization started occurring in these markets and the focus on service-oriented practices in a quest for more innovative technologies eventually led to higher unemployment rates, as common workers had to be replaced by educated individuals to add to the value chain. In most industrialized countries, the traditional value-share of customary manufacturing industries is steadily declining, whereas in the BRICS nations the trend is opposite. This increase in manufacturing capabilities in emerging countries is imperative to expand research proficiencies, improve the education of the local population and drive a new wave of innovation to improve products, manufacturing techniques and efficiency, leading to greener technologies. For the developed countries, especially in Europe, these industries cause a continual loss of national GDP, as the process typically causes migration of products and consumption to expanding economies. The figure of merit of the economic well-being of a country can indirectly be measured by its manufacturing cycles. Economic cycles moreover accelerate the process of deindustrialization from lower investments in research and development and a decline in available resources (Westkämper 2013). Historically, such events are noticeable in, for example, the introduction of flexible manufacturing systems between 1970 and 1980, as well as the implementation of lean production systems during the 1990-to-2000 period. The global recession around 2008 also showed a significant decline in economies; however, this event can be localized to a financial crisis that had adverse effects on most economies worldwide.

As listed in Aulbur et al. (2016), the South African public sector has made investments in research and development for Industry 4.0, especially in the 3D printing industry. The Department of Science and Technology, the Council for Scientific and Industrial Research, and collaborations with Aerosud have invested the equivalent of approximately US $26.5 million in 3D printing, but this is not enough to establish the country as a dominant player in Industry 4.0. Furthermore, the Technology Innovation Agency (TIA), a parastatal South African agency, has established an advanced manufacturing unit with the main objective to promote investment to enrich the knowledge concentration of industrialization in South Africa. This unit also aims to build technological and industrial capabilities, as well as information systems. Additionally, TIA has advocated making investments in the regions of

- additive manufacturing,
- mechanization and robotics,
- cutting-edge integrated circuit technology,
- photoelectric engineering, and
- aero-structures,

while persistent to endorse expansion of the cutting-edge manufacturing sectors (Aulbur et al. 2016). The South African government is heading the development of establishing distinctive economic zones to endorse the competitiveness of South

African firms by supporting venture capital in export-focused engineering industries and through the export of value-added manufactured products. A key objective of this initiative is to entice advanced overseas fabrication and technology procedures to learn from and gain familiarity nationally, with the intention to transfer knowledge to the local workforce. In 2019, the major theme at the WEF annual meeting held in Davos, Switzerland, focused on Globalization 4.0 and influencing the global architecture in the age of Industry 4.0. A pertinent question to be addressed at such a meeting is how Africa, in the development and (economic) climate challenges it faces, takes on a prominent role in Globalization 4.0, especially considering that Africa has the world's youngest population. One opinion, as proposed by Ndabeni-Abrahams (2019), is to import Industry 4.0 technologies and combine these with low-cost manufacturing, essentially leapfrogging within the value chain and boosting economic growth through creating new jobs.

However, there is little to no alignment between governmental policies and those proposed by industry leaders. Investments in core competencies such as research, development and innovation remain low. Africa/South Africa has traditionally adopted a philosophy that shies away from local manufacturing to save cost, outsourcing many of these tasks to international organizations. Political and economic uncertainty in African countries have led to many international affiliates withdrawing their operations from Africa/South Africa, leaving behind a severe lack of manufacturing capabilities. Further concerns of information security, inconsistent energy availability from power utilities and lack of digital infrastructures and the cost involved to establish them have necessitated the implementation of new policies to be implemented if African countries are to succeed.

Successful policies require strong political leadership and the full commitment of all levels of government (AEO 2017). Government should adopt key innovative policies to facilitate inclusive growth in African countries, and implement legislative amendments to consolidate sluggish growth into the fourth industrial revolution, as is done by its European and American counterparts. Innovations and policies currently driven particularly in Germany, such as relaxed taxation, assisting start-ups, helping new business founders, and decreasing arithmetical and information duties are some steps that could be taken in this regard. In response to the many challenges faced by Industry 4.0, EU governments have made Industry 4.0 a primacy, embracing large-scale policies to surge efficiency and competitiveness, as well as improving the technical skills of their workforce (Klitou et al. 2017). Again, in Africa/South Africa, sufficient drive towards such developments is severely lacking.

To create and overview the policies, strategies and qualities of a country required to adopt and develop Industry 4.0 rapidly, it is firstly required to review the policy gaps that prevent this from happening. The following section specifically addresses the gaps that are evident in African countries. These policy gaps can also be used to identify issues where educational institutions, with help from the government, can assist in building nations that address the gaps and implement policies that aim not only to educate a population for future generations, but also to sustain this development and safeguard a growing economy.

3 Policy Gaps of the African Continent Preventing Industry 4.0

Published works such as Carroll et al. (2016) reference that compared to the rest of the world, Industry 4.0 adoption in the African continent is still low. Although this statement should be defended by comparing various market segments in various emerging markets and comparing it to the same areas in developed countries, in a general sense, this statement is considered acceptable. Since the mid-1990s, economic policies in African nations have fixated mostly on cultivating the business environment, aligning with the *Washington Consensus*. The term *Washington Consensus* was initially used in 1989 to designate 10 economic policies considered to constitute standard restructuring for developing countries experiencing economic crises. As a holistic approach, this consensus encouraged macroeconomic equilibrium, economic projections in trade and investments, and the enlargement of market forces within national economies. It was highly criticized and many argued that the underlying policies were too inflexible to succeed. Some critics maintained that these guidelines were constructed for a period of swift economic growth, and not for developing economies that were experiencing crises. Although these policies have had positive impacts on African governments, their recorded progress has been slow and the generic policy recommendations typically overlooked country-specific provisions and often required capabilities outside the scope of African administrations. The manufacturing sector in Africa faced a number of challenges as a result (adapted from AEO 2017). These challenges included

- infrastructure bottlenecks as policies called for capacities outside the competences of national governments,
- inadequate productive competences,
- a scarcity of trained employees,
- immature fiscal markets, and
- high levels of income inequity.

Fortunately, inventive industrialization policies on the African continent are progressively being considered by business leaders and governments. The reason for this is primarily the impact that smart technologies can have on the socio-economic level. New opportunities are also arising owing to external factors, such as

- enhanced automation and from digitization,
- the slow-down of global growth after the commodity super cycle, and
- rising labor cost in East Asia.

3.1 Broadband Connectivity

The greatest challenge that Africa/South Africa must mitigate is its undersupplied access to connectivity. To alleviate this, more private and public incentives are required in an arena on which policy makers should focus their attention. To its advantage, Africa/South Africa is not weighed down by legacy infrastructure issues that could thwart necessary change. To prepare for the rapid change that Industry 4.0 introduces on a technological and economic level, African and South African manufacturers will need to upgrade or replace traditional ways and means of conducting business. The two largest deterrents to Industry 4.0 in African countries are the cost implications of upgrading or replacing current infrastructure and electricity constraints in rural as well as in many urban areas. Sufficient government funding and legislative backing, alongside embracing renewable energy and energy harvesting, will make it possible to implement self-sufficient and autonomous systems regardless of financial constraints, especially when considering expected future benefits. To unlock the potential of digital transformation towards Industry 4.0, new policies, practices and actions are required on a national and international level. Unfortunately, in Africa/South Africa, individual foundations of Industry 4.0 have been realized by organizations without substantial backing from the government.

Connectivity, specifically broadband, has the potential to offer a common podium for global partnerships for development, ensure sustainability, improve health services, promote gender equality and improve education. The WEF estimates that a 10% surge in broadband infiltration in low- and middle-income countries could potentially effect in as much as 1.38% surge in economic growth (Barnes 2015). However, statistically, the number of internet users in African countries (not accounting for the quality and cost of their internet connection), remains low. The Internet Telecommunications Union (ITU) annually posts ICT data, and according to their 2016 findings, a digital divide remains in developing countries, with Africa still showing the lowest number of internet users. Table 1 lists the number of individuals in the regions of the world who do *not* have access (or are not using) the internet.

According to the data provided in Table 1, in 2018 Europe was the most connected region, with only 20.4% of individuals not accessing the internet, therefore 79.6% of the population actively used the internet for business and pleasure. The Commonwealth of Independent States (CIS, also called the Russian Commonwealth) has 28.7% of non-active internet users, followed by the Americas, where the relatively low number of internet users is primarily attributed to user numbers in South America

Table 1 The number of individuals in the regions of the world who do not use the internet (adapted from the ITU Facts and Figures 2016 and 2018 annual reports)

Region	Percentage of individuals *not* using the internet (2016)	Percentage of individuals *not* using the internet (2018)	Difference
Africa	80.2	75.6	−4.6
Arab States	57.3	45.3	−12
Asia and Pacific	58.8	53	−5.8
The Americas	35.0	30.4	−4.6
CIS	33.4	28.7	−4.7
Europe	20.9	20.4	−0.5

and Mexico. The average percentage of internet users in North America, including Canada, is typically above 75%. In Asia and the Pacific, as well as the Arab states, respectively 53% and 45.3% of individuals are not using the internet. In Africa, however, only approximately 25% of the population is using the internet for daily activities, and this is a significant problem with low connectivity penetration in African countries.

Broadband penetration and ICT skills in a country have far-reaching effects, on not only productivity gains, but also extending to economic and social transformation. ICT transforms the way people communicate and interact and creates new and innovative business opportunities in most sectors. The technology divide in developing countries between urban centers and rural areas has a significant effect on the potential to increase income levels through enhanced communication, ubiquitous computing and capabilities to manage essential tasks. Developing countries, especially in the rural areas, are urged to develop their ICT ecosystems to remain competitive in an evolving marketplace. Again, governments are partly responsible to enable environments that promote competition through regulation of policies in the ICT market. Such contributions, strategies, and innovative policies can create economic opportunities and social and political inclusion, inevitably leading to shared prosperity of the local population. Economically, Industry 4.0 and more specifically enabling ICT services can

- increase productivity,
- reduce transaction costs,
- improve accessibility to high-quality and relevant information (knowledge),
- reduce the cost of acquiring information,
- spearhead new models of collaboration, and
- increase the efficiency and flexibility of workers.

Statistically, mobile broadband and fixed broadband penetration in developing countries remains low. Mobile broadband typically works through a connection to a mobile network, using a SIM card, whereas fixed broadband is usually delivered through a physical line such as copper of optic fibers. In developing countries, mobile broadband subscriptions have rapidly increased, a good sign showing increased uptake by users capable of accessing the internet. Investments in Africa in telecommunication infrastructure from local and international investors have effected far-reaching improvements of intercontinental connectivity and terrestrial fiber networks. Although these investments have brought connectivity to many African inhabitants, continuous and sustainable investments are needed to improve the connectivity experience for users in Africa, specifically referring to infrastructures that enable Industry 4.0 development. Changes to existing policies are needed to maximize internet service providers' ability to distribute internet services to all areas, either mobile of fixed line. Table 2 summarizes some mobile and fixed line broadband penetration statistics obtained and adapted from the ITU Facts and Figures 2016 and 2018.

As shown in the data presented in Table 2, there is a large discrepancy between the number of mobile broadband subscription and fixed line subscriptions (such as

Table 2 The mobile broadband and fixed line broadband penetration in 2016 and 2018 in developing countries such as Africa and the least developed countries (LDCs), compared to Europe and the Americas (adapted from the ITU Facts and Figures 2016 and 2018 annual reports)

Region	Mobile broadband subscriptions (2016/2018 per 100 inhabitants)	Fixed line subscriptions (2016/2018 per 100 inhabitants)
Developed countries	97.9/111.2	30.6/32.7
Developing countries	42.9/61.0	8.6/10.4
World average	52.2/69.3	12.3/14.1
Least developed countries	19.6/28.4	1.1/1.4

DSL and fiber). In addition, the number of subscriptions of both these broadband alternatives is significantly lower in developing countries when compared to developed countries. LDCs had a mobile broadband subscription base of only 19.6 per 100 inhabitants in 2016, up to 28.4 in 2018. Although these numbers are low, they are significantly higher than fixed-line subscriptions and the most viable alternative for internet penetration in these countries.

Although the mobile broadband uptake in developing countries shows much promise, it will not directly contribute to developing Industry 4.0. Mobile broadband remains expensive when considering the amount of data used. Mobile broadband penetration is more affordable compared to fixed line broadband, but when data usage becomes high, the cost exponentially increases too far beyond that of fixed line connections. Competitive pricing of internet services for the end user is a strong measure of successful policies set by governments. Low prices for users indicate a fair and competitive market eliminating (usually state-owned) monopolies. It also leads to a *snowball* effect; lower prices lead to more users, further leading to a reduction in unit costs (running cost per person), thus increasing the capabilities of service providers to reach more users. In 2015, mobile broadband prices per person, in US dollars, were US $30.8 per person, compared to US $67.3 per person for fixed line broadband. This however refers mainly to its potential to penetrate the market as a service, typically below about one gigabyte, and mobile broadband prices above this point become expensive. Because of the easier and less costly market penetration of mobile broadband, fixed line broadband (which is necessary for many Industry 4.0 applications) remains very low globally, especially in developing countries. In Africa and the least developed countries, fixed line broadband subscriptions are below one person per 100 inhabitants. This is compared to also relatively low figures of 30 and 18.9 subscriptions per 100 inhabitants in Europe and the Americas, respectively. Policy makers in developing countries should consider certain solutions to improve internet access in the countries. European and American policies are good examples showing that these strategies have the desired effect, ultimately offering broadband internet to as many of its population as possible, at low cost. At the most basic level, these solutions include:

- Policy makers ought to not make it more challenging and costly for investors to partake in evolving terrestrial fiber networks by unnecessary taxes or high cost for licenses and limiting access to rights over borders.
- The policy makers should serve as regulatory bodies but still encourage investments from the private sector. If private sector investments are limited, budgetary allocations from government should command infrastructure development.
- Overseeing agencies should be appointed to act as intermediaries between governments and organizations investing in infrastructure development. These agencies should be technologically educated and have the authority to serve as consultants in all aspects of the value chain.
- Emerging markets must ensure that they are part of the important global discussion pertaining to Industry 4.0.

Broadband internet or at least a high-speed internet connection is a necessity in the 21st century for conducting business remotely around the world. Countries that are able to harness the power of technology in communications are typically more successful in relaying economic benefits and shared prosperity to the local population. Economic benefits of ICT are also evident in the education sector, with online open courses allowing students to access course material from anywhere, if for example commuting or financial constraints hinder these students from getting to class.

Demographic internet penetration between men and women is also divided, with typically higher internet usage by men, compared to women. Table 3, adapted from the ITU annual post of their findings in 2016 and in 2018, summarizes the divide between men and women internet users.

Table 3 The percentage of men and women across the regions of the world that use the internet (adapted from the ITU Facts and Figures 2016 and 2018 annual reports—the 2018 report published the data gathered during 2017)

Region	Internet penetration for men (%) 2016/2017	Internet penetration for women (%) 2016/2017
Africa	28.4/24.9	21.9/18.6
Arab States	46.1/47.7	36.9/39.4
Asia and Pacific	47.5/47.9	39.5/39.7
The Americas	65.6/65.1	64.4/66.7
CIS	68.5/69.8	65.0/65.8
Europe	82.0/82.9	76.3/76.3
Combined regions		
Developed world	82.3/82.2	80.0/79.9
World average	51.1/50.9	44.9/44.9
Developing world	45.0/44.7	37.4/37.5
Least developed countries	18.0/21.0	12.5/14.1

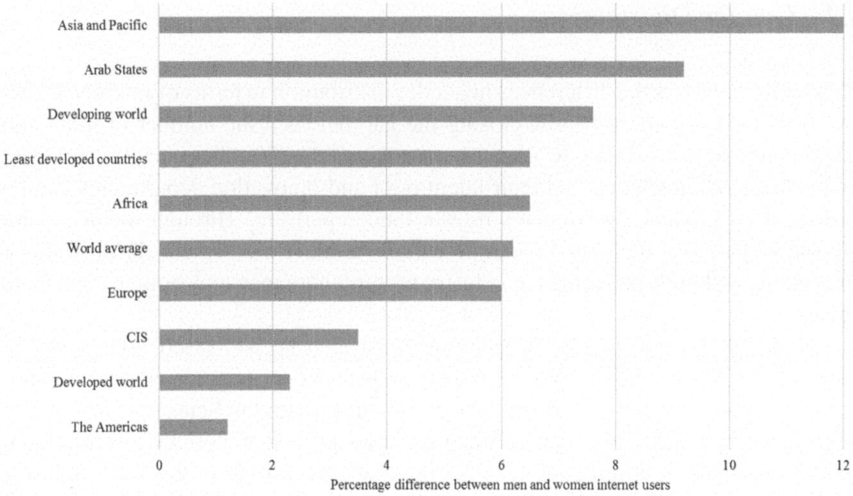

Fig. 3 The percentage difference between men and women internet users across the regions of the world

From Table 3, it is evident that internet usage among men is typically higher than among women in all regions, therefore this is not specific to developing countries. However, it is noticeable that the difference in internet penetration between men and women is larger for countries with an overall smaller community using the internet. Figure 3 represents the differences in the percentage of men and women internet users across the regions of the world.

As seen in Fig. 3, there is a definite discrepancy between the percentage of men and women internet users worldwide, with a significant difference in the developed world. In Africa, the difference is 6.5%, in the Arab States it is almost 10%, with Asia and the Pacific regions having a difference of 12%. In America and the CIS, the differences are only 1.2% and 3.5%, respectively. In Europe, the difference is somewhat larger, approximately 6%, but with a much higher overall internet penetration compared to the other regions and still below the world average of 6.2%. Industry 4.0 has the potential to increase women's participation in the workforce.

The following section addresses the consequences of a digital divide between men and women—a form of gender inequality. Demographics in technology gives an indication of the distribution of the workforce. It can be used to identify policy gaps in gender equality. In many nations across the world, especially in developing countries, some form of gender discrimination still exists. In terms of higher education and its place in Industry 4.0, the digital divide in all forms must be mitigated and ideally be eliminated to offer fair and inclusive opportunities for innovation, industrialization, manufacturing and consequently economic growth.

3.2 Gender Distribution

This section reviews the differences in gender distribution in terms of internet access and if there is any merit to 4IR closing the gap between the number of male and female internet users. In the ICT sector, approximately 37% of companies surveyed by the WEF acknowledge that their talent pool and innovation departments can be expanded by broadening women's role in their repertoire. Through inclusive and innovative policies, Industry 4.0 can potentially achieve a complete shift towards a sustainable and high-performance industry sector where men and women contribute equally.

Policies that equal gender distribution must also be addressed holistically in Industry 4.0 development. National beliefs and rules can shape equal participation of men and women in the workforce. Sectoral cultures and policies play an equally important role. Even while many companies are adapting their policies towards equal gender distribution, eliminating bias is important. According to the WEF industry gender gap report[1] from 2016, the distribution of women in the workforce is still unbalanced, with

- junior level staff made up of approximately 33% women,
- mid-level staff represented by 24% women,
- senior staff only made up of 15% women, and
- alarmingly, only 9% of chief executives being represented by women.

The lowering percentage from junior level to higher levels in the workforce is not only attributed to appointments in those positions, but also to companies' inability to retain women's talent. In the 2018 WEF global gender gap report,[2] further key findings in terms of unequal gender distribution in the workplace were presented and some of these include:

- There is still (by the end of 2018) a 32% average gender gap globally.
- The largest gender disparities are in political empowerment, and secondly in economic participation and opportunity.
- Only 17 out of 149 countries assessed have female heads of state.
- Access to financial services and land ownership differs by 60 and 42% between men and women.
- Towards Industry 4.0, AI professionals are made up of only 21% women and 79% men.

Industry 4.0 can be seen as a new beginning to a more inclusive work environment that broadens gender distribution owing to the changes brought about by the shift towards automation, robotics, machine learning and AI. Industry 4.0 is driven in many ways by the internet, allowing workers to perform their duties at home. This is beneficial to both men and women having responsibilities of, for example, nurturing a newborn child. It remains a priority for policy makers to acknowledge,

[1] Available from www3.weforum.org.

[2] Available from www3.weforum.org.

understand and address the benefits of Industry 4.0 for all genders. They have to ensure equal growth for both genders in the IT and engineering fields, which could increase gender diversity in the future. The WEF lists several elementary policies serving as guidelines to promote gender diversity in the workplace. Each country, sector and industry must however adapt these strategies to its specific issues, limitations and cultural background; several recommendations offered by the WEF are adapted and listed below:

- Identifying and evaluating causes of gender inequality within the organization and tracking the progress of the strategies to eliminate (or at least alleviate) the gap. This includes being transparent in highlighting any dissimilarities in remuneration and compensation within the organization.
- Investing in training programs that educate the employees within the organization about the cultural background of the workforce, how to treat a varied workforce and how to entice, preserve and indorse talent.
- Implementing transparent management policies that do not encourage gender discrimination and making senior management accountable for these policies.
- Tailoring the policies for men and women with respect to a healthy and sustainable work-life balance.

Furthermore, diversity guidelines include that companies fully support any gender in management, respect their authority, lead efforts in diversifying culture and maintaining an inclusive culture outside the office.

Gender equality should be promoted globally; however in Africa there are various additional factors that policy makers must account for in addressing policy gaps that lead to low levels of gender diversity in the workplace. The African gender equality index developed by the African Development Bank (AfDB) provides evidence on gender inequality for 52 out of 54 African countries. The AfDB gender equality index not only identifies gender inequality in Africa, it also aims to promote development in industry to address these problems, tailored to African nations. Policy makers can refer to this index to identify barriers that prevent gender equality in Africa from engaging in the workforce and to promote equal rights. According to the African gender equality index of 2015[3] (which has not been updated since then), the top 10 countries in Africa with respect to performance in gender equality, ranked from best to worst, are given. These results are listed in Table 4.

The performance index adapted from the 2015 AfDB African gender equality index and presented in Table 4 is essentially a score taking into account equality in economic opportunities, human development, and laws and institutions in African countries. As shown in Table 4, the top 10 countries in Africa in terms of performance of gender equality achieve relatively high performance scores (a complete breakdown of the performance criteria is available in the 2015 AfDB African gender equality index report). South Africa was the leading country in gender equality in 2015 (74.5%). With one of the highest gender parity rates in Africa for wage employment (not accounting for agriculture). Rwanda, second on the list at a performance

[3] Available from http://www.afdb.org.

Table 4 African gender equality index overall performance of the top 10 countries out of 52 provided by the AfDB 2015 gender equality index

Rank #	Country	Performance index	Rank #	Country	Performance index
1	South Africa	74.5	6	Lesotho	70.0
2	Rwanda	74.3	7	Botswana	69.4
3	Namibia	73.3	8	Zimbabwe	69.1
4	Mauritius	73.2	9	Cape Verde	66.8
5	Malawi	72.8	10	Madagascar	65.0

Table 5 African gender equality index overall performance of the bottom 10 countries out of 52 provided by the AfDB 2015 gender equality index

Rank #	Country	Performance index	Rank #	Country	Performance index
43	Cote d'Ivoire	43.7	48	Guinea	39.5
44	Chad	42.2	49	Libya	37.9
45	Niger	42.2	50	Mali	33.4
46	Mauritania	41.9	51	Sudan	31.9
47	Djibouti	41.1	52	Somalia	15.8

index of 74.3%, was the first and only country in the world where more than half of the parliamentarians were female, partly contributing to its high gender equality performance index according to the 2015 AfDB African gender equality index report. The third highest performance in 2015 was achieved by Namibia, with a constitution that guarantees gender equality before the law. Namibia has a gender equality index of 73.3%. In Table 5, the 10 countries with the lowest gender equality index are listed.

According to Table 5, Somalia performed worst in 2015 with respect to gender equality, having a performance index of only 15.8% compared to South Africa's 74.5% (to reiterate—equality in economic opportunities, human development, and laws and institutions in African countries). Somali women experience a higher unemployment rate than men (74% compared to 61% for men); most women have to undertake roles as providers for basic needs in their families and most Somali women are excluded from decision-making and asset ownership. Sudan had the second lowest gender equality index, 31.9%. The difference in education between young men and women is one of the most prominent inequalities in Sudan. A large religious influence from a predominantly Islam nation also structures the gender roles in Sudan, further contributing to gender inequality. Each country listed in the 2015 AfDB African gender equality index report, especially the lower ranking countries, experiences significant difficulty to overcome policy gaps with respect to gender equality in order to create fair, sustainable and empowering conditions. Industry 4.0 is not necessarily one of the prioritized reasons to encourage transformation in African countries with vast discriminatory policies, but it could serve as an enabler for change. Apart from religious concerns, the quality of and equality in education must be addressed at all levels. Both men and women should be given equal opportunities to be educated to a level where a higher education is obtainable. Industry 4.0 can provide the means to participate in higher education through remote access to institutional content through the internet, for example.

The AfDB promotes gender equality, inclusive growth and resilient societies. A strategy to promote gender equality, formulated by the AfDB, is provided in the 2015 AfDB African gender equality index report. This strategy aims to contribute to social and economic development in African countries through human development. The strategy outlines non-country-specific guidelines to identify strategic opportunities to stimulate gender equality. These strategies are adapted from the 2015 AfDB African gender equality index report and given in Fig. 4.

The strategies in Fig. 4 address specific barriers against the equal participation of men and women in the development of African countries. The strategy envisions all Africans as equal producers for their families, in human development and equal as citizens and leaders. To accomplish this, policies are designed to enhance access for women to land, education and equality before the law, among other policies

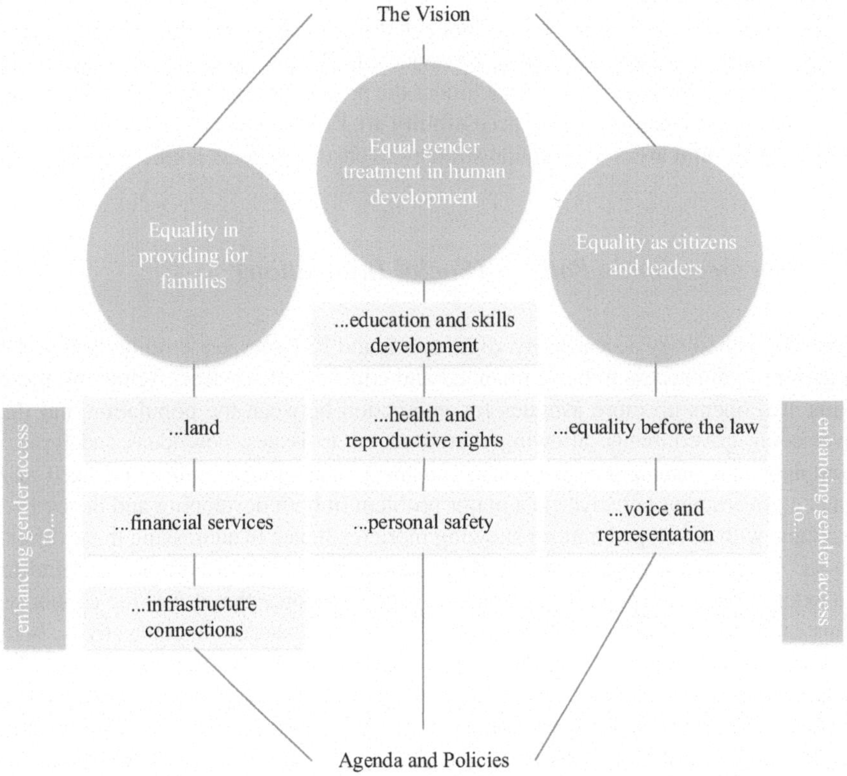

Fig. 4 The agenda for action on empowering African women adapted from the 2015 AfDB African gender equality index

outlined in Fig. 4. In addition, as stated in the UNDP Africa Human Development Report 2016 and reiterated in the UNDP National Human Development Report 2018, political commitments from government to promote gender equality can be achieved through reinforcing options, such as

- macroeconomic policies to showcase the commitment of governments to provide access to resources and the efficacy of their use,
- encouraging institutions aimed at the advancement of the minority gender,
- effective and targeted interventions that are specifically aimed at identifying and mitigating any forms of gender inequality,
- legislative policies that hold companies and individuals accountable for gender-based prejudice combined with consistent enforcing of these policies, and
- universally agreed, equal human rights and social norms.

The UNDP Africa Human Development Report 2016 and the UNDP National Human Development Report 2018 detail how these strategies should be implemented. The reports explore the methods that governments can adapt to respond to gender inequality and accelerate equal growth and empowerment of men and women in African and other nations.

The following section reviews socioeconomic policies that allow shared access and information sharing through ICT services. In Africa, specifically, connectivity remains one of the largest issues that hinder the progress of Industry 4.0. Addressing these issues and gaining mass accessibility to ICT services in African nations, especially in rural areas, is a crucial strategy to ensure socioeconomic growth.

3.3 Socioeconomic Policies (Social Innovation)

Socially, technology accessibility, digitization and ICT services promote and make it easier to gain access to basic financial and educational services. Achieving these aims also opens up more avenues for interaction between the population and the overseeing governments, allowing shared access to issues, new ideas and typical complaints. A discrepancy between what local populations require and need and what a government perceives is a major problem in both developing and developed markets, with developed nations showing more resilience in addressing these differences. The general population should be an essential cog in participating and to an extent administering policies, since it is the workforce that drives the economy. This can largely be achieved by improving the public's access to connectivity, both in rural and urban areas. This increase in connectivity leads to big data, if administered correctly these data can be used to generate better policies that are more inclusive. Real-time data sharing between ubiquitous sources (for instance in rural areas) will give policy-makers first-hand access to what is required, where potential risks are identified and what the community is lacking. This therefore eliminates the need for constant human interface, since machines are able to relay important information. Human interaction is still a critical part of developing policies, but the regularity of this can decrease significantly and the quality of the interaction will increase when all the parties involved have the same data. Digital platforms such as social media can also be used to provide an approximation of how *digitally literate* a population is. Industry 4.0 promotes social entrepreneurship that develops potential solutions, through collaboration using the internet, to a variety of social, cultural and environmental issues. A digitally literate and socially accessible population provides an ecosystem where innovations offered by Industry 4.0 can flourish, yielding socioeconomic benefits.

Social innovation is driven by technical innovations; however, the opposite is also true. Innovations in technology can arise when organizational policies are reinvented or adapted for Industry 4.0 (Buhr 2017). Buhr (2017) lists opinions that aim to coordinate Industry 4.0-related strategy implementations in organizations such that technical innovations also become societal innovations, contributing to social progress and socioeconomic benefits. Among the points listed by Buhr (2017), the following are highlighted as it has some relevance in this chapter,

- a shift in organizational thinking towards considering Industry 4.0 an integral part of social innovation,
- embedding Industry 4.0 in a coordinated market economy,
- the interplay between social and technological innovations, which requires policy coordination, and
- stimulating Industry 4.0 as a driver of social progress.

Buhr (2017) further indicates that Industry 4.0 should not be viewed as just being about machines, but also as being about how human development will benefit from a paradigm shift in skills development. Since the design of integrated systems that spearhead Industry 4.0 is viewed as a central cog in digitization, further training, higher education and additional qualifications of people are critical in its success.

The role of higher education in Industry 4.0, specifically regarding accessibility of technology and adapting learning methods, is discussed in the following section. Higher education has become more prevalent in recent years to provide learners with the knowledge and skills required to function as contributing members of society. A surge in higher education enrollments in developing countries also poses several challenges, infrastructure and maintainable growth being among the most serious issues. Industry 4.0 and ICT enable workers to perform tasks and activities remotely and these advantages of a new form of industrialization must be addressed, especially in higher education institutions. Schools are still tasked to transfer knowledge of first principles in mathematics, science and languages, whereas institutions of higher education are responsible for developing individuals further towards readiness for skills they require in a changing industry.

4 The Role of Higher Education in Industry 4.0

In the provision of the fourth industrial revolution (Industry 4.0 or 4IR), innovation (evolutionary and revolutionary) should be a topmost priority in the higher education system of any country (Xing and Marwala 2017). At the core of modernization, economic growth and long-term sustainability is technology (Xing and Marwala 2017). Industry 4.0 is a very broad domain that includes

- improvement and new innovations of production (manufacturing) processes,
- increased efficiency of new processes, as well enhanced efficiency of existing processes,
- the management of large volumes of data,
- a more transparent relationship with clients of products and services, and
- higher levels of competitiveness through ubiquitous availability of technology—with many more possibilities existing (Piccarozzi et al. 2018).

There is a discrepancy between the education pursued by students at school level and at tertiary education institutions, and the qualifications sought by employers priming for Industry 4.0. Industry 4.0 has become an important theme for management scholars and business economic disciplines and research publications in these fields are increasing dramatically (Piccarozzi et al. 2018). Several forecasts suggest that a significant shortfall in information and communication technology (ICT) professionals will be experienced within the next decade if these issues are not addressed straightaway. Lack of clarity on where Industry 4.0 is heading, with rapid changes in market development, internationalization and growing competitiveness, are key challenges that could cause this shortfall.

Educational institutions aim to address two important aspects to mitigate these challenges,

- firstly categorizing what to instill in students to meet future (near and long-term) demands, and secondly,
- making education responsive enough to match the need for specific skills before they become obsolete or substituted.

It is acknowledged globally that the heightened use of ICT and the inclusion of new technologies and digitization in manufacturing, brought about by Industry 4.0, will require widespread retraining and additional education of students and staff. Industry 4.0 requires transformation of the business industry and a digital culture through properly training the workforce in ICT skills is needed (Fitsilis et al. 2018). A problem in an emerging market such as South Africa is that the capacity to adapt to future jobs is below average, which is attributed to the poor quality (Naicker 2018) of education. A large obstacle in South Africa is the fact that it has a lot to lose, resulting from dwindling education standards and above-average exposure to future disruption in the workplace, as well as a stagnating economy. Fortunately, Industry 4.0 is on the radar of businesses, government and the labor sector in South Africa, providing them with valuable resources (and a closing window of time) to rectify the current situation.

In Africa, implementation of digital education and training, addressing topics such as the internet of things (IoT), artificial intelligence, big data and preparing smart artisanship, still has a long way to go. Vocational training and competence development must be supported by governmental policies. Europe and North America have started addressing this approach, but there are still significant policy gaps on the African continent that are preventing Industry 4.0 from rapid growth. In a sense, vocational education and training for industry have become a key political issue. Educational establishments must apply for governmental support to obtain digital training equipment and pilot projects must be in place to display the progress of new digital learning methods. Policies that support this transformation are in no way generic and must be tailored for individual countries or groups of countries. Emerging markets, with the focus on the BRICS bloc (Brazil, Russia, India, China and South Africa), has its own set of limitations to overcome before these policies can take off; these limitations are typically different from those in developed countries in Europe and in the United States of America (USA). Identifying these key limitations is paramount and this section highlights and analyzes some of the unique characteristics that are holding back the emerging market. This section aims to identify policy gaps in developing countries, specifically on the African continent, that prevent similar results.

Through identifying the policy gaps that continue to plague developing countries, typically visible from the highest political offices, the importance of higher education to develop skills for Industry 4.0 must also be addressed. This section therefore stipulates the significance of higher education with the goal to *commercialize* Industry 4.0 for quicker government adoption. These higher education institutions need to address the challenges ahead to propel an immediate transformation into digitization, typically with reduced financial support. This forces these institutions to implement strategic methods to utilize their own identities and credentials to offer and support new and innovative services. Leadership also needs to be adapted to be less risk averse as disruptive technologies are introduced that could potentially steer Industry 4.0 in a different direction. However, collaboration between private companies, government and institutions of higher education is essential to scale education efforts and ensure the sustainability and adaptability of changing vocational requirements.

4.1 Sustainable Partnerships in Higher Education

The world's leading economies are increasingly dependent on innovation for long-term prosperity and development depends on technologically based intellectual property (Panarina 2015) developed through university-industry partnership sustainability (UIPS) (Kaklauskas et al. 2018). Developing countries are however facing many challenges in driving technology-based innovation; as outlined in Panarina (2015), these challenges (specifically in Russia but also relevant for emerging markets) are

- underdeveloped infrastructure and institutions,
- subdued competition,
- a decline in the quality of education offered to pupils,
- underdeveloped financial markets, and
- low levels of business sophistication (Panarina 2015).

According to Panarina (2015) and Kaklauskas et al. (2018), the fostering of effective and inclusive university-industry partnerships is essential to enhance commercialization efforts and sustainable goals for developing countries. In Panarina (2015), a center of competence is proposed to serve as an ecosystem for innovation development. Such a center is defined as a tool to integrate knowledge and expertise and to support entrepreneurial activity. Adapted from Panarina (2015), Fig. 5 represents an ecosystem for innovation growth.

According to the system proposed in Fig. 5, such an ecosystem is poised to interlink industry and institutions of higher education and assess resources from public and private investors to benefit socioeconomic growth. The work presented in Panarina (2015) researches such efforts in detail. Kaklauskas et al. (2018) pursued a neural evaluation scheme for a multiple-criteria scrutiny of UIPS. The evaluation scheme is warranted at three levels:

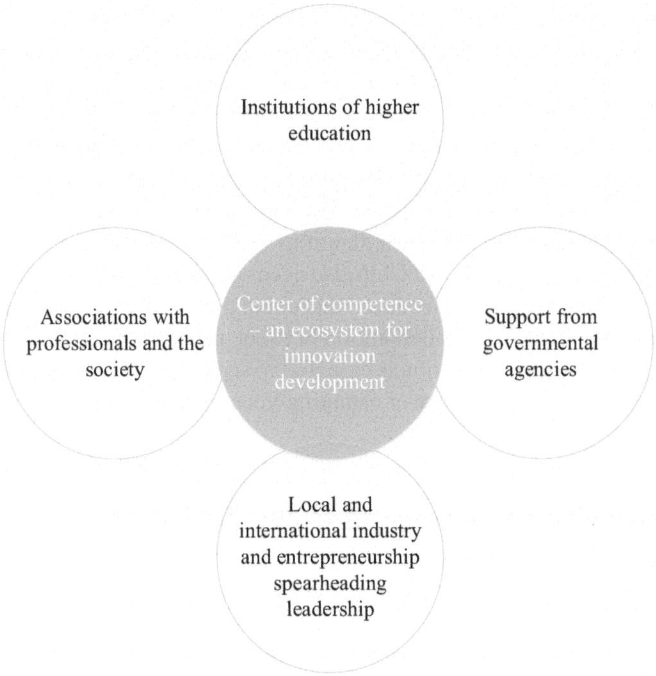

Fig. 5 Adapted from Panarina (2015)—a proposed system that acts as an ecosystem for innovation development

1. on a micro-level encompassing study and modernization performance, transferal and absorptive competence and technology progression,
2. on a meso-level encompassing formal measures, communication networks and native and aboriginal guidelines, and
3. on a macro-level encompassing supply and demand, rules, financing, duties, culture, ethnicities, markets, economic and political climate, demographics and available technology(ies).

The research presented in Kaklauskas et al. (2018) resulted in an evaluation system that makes it possible to assess alternative UIPS life phases and analyze their market and fair value, as well as conducting virtual discussions to select the most effective and proficient options.

Higher education and combinations of UIPS have become much more prevalent in the global space and have given emerging markets an opportunity to play a role in Industry 4.0 developments. Higher education enrollments have undergone unprecedented growth in the past decades, and continue to do so. Maturity in this growth is expected in the near future, but institutions can still capitalize on a surge of willing learners hoping to further their education and play their part in growing economies.

4.2 Global Growth in Higher Education

Worldwide, "*between 2012 and 2025 the number of students enrolled in higher education is expected to double, to about 262 million*" (Jelinek and Fomerand 2013). The relevance of this fact is that most of the growth in higher education enrollment will be from developing countries in the BRICS group; with India and China responsible for more than half of this number (Brazil also accounting for a large portion). In addition, students will be seeking to study abroad almost three times more often compared to the number in 2012. For developing countries, the stellar growth in higher education will provide many benefits, but accommodating the rising number of prospective students could become problematic with current infrastructure limitations (Jelinek and Fomerand 2013). In Africa, the Middle East, Asia, Central America and South America, the average annual growth in the demand for higher education is approximately 3–5% up to 2025, an estimation obtained inform the research of Jelinek and Fomerand (2013). Leading universities have already started, and are continuing to expand their infrastructure to accommodate foreign students and capitalize on the shifting dynamics of internationalization and globalization. Apart from additional sources of income through international students, universities must use globalization to take advantage of new avenues for collaboration and networking among higher education institutions.

KNBS (2016) reported that in Kenya the total number of university student enrollments increased by 22.8% in 2015, with a substantial increase in the number of female students (higher compared to the enrollment of male students). The overall increase in enrollments was primarily driven by three factors, namely

- large governmental support for infrastructure development in Kenya,
- new disciplines and courses introduced in Kenyan universities that are in line with international standards, and
- an increased number of satellite campuses opened across the country.

These three factors constitute a collective shift in the right direction for a developing country aiming to improve socioeconomic benefits and inevitably enter Industry 4.0. These factors improve the demographic distribution of students, with more females studying to contribute to the workforce, aiming to increase the quality of graduates by introducing new courses and providing accessibility for students from rural areas with transport limitations. However, with high growth comes high investments in its sustainability. The Kenyan government has been criticized for providing inadequate funding to sustain this growth and to ensure that infrastructure and the quality of education for the large number of students are maintained. Globally, the fast pace of higher education enrollment is also predicted to slow down in the coming decades as markets mature and demographic trends become less favorable as a result of declining birth rates since the beginning of the 1990s.

4.3 Education and Socioeconomic Status

According to the American Psychological Association (APA), "*socioeconomic status (SES) is an embodiment of not only the gross income of individuals, but also a measure of their financial security and educational attainment, and a generalized perception of social status and class*". Roksa and Kinsley (2019) examined two dissimilar methods of family sustenance (emotional and financial) among low-income students and found that emotional sustenance plays a large role in nurturing encouraging academic results. Roksa and Kinsley (2019) reported that financial support plays a limited role for students from low-income families (that are already enrolled) in attaining a tertiary degree. SES is a quintessence of opportunities and privileges afforded to a society and approximately quantifies quality of life. According to a study led in 2009 by Morgan et al. (2009), youth in households or communities with low SES develop slower academically compared to youth in higher SES conditions. The attributes in low SES communities that contribute to slower educational development, as given in Morgan et al. (2009), include

- sub-standard psychological development of cognitive skills,
- poor language development,
- difficulty storing and remembering knowledge and information, and
- inability to process socioemotional relationships.

As a direct result of these developmental concerns, youth in low SES conditions are statistically biased to be associated with a lower income and health issues during adulthood. The APA acknowledges that improvements at the early stages of a child's educational development and early intervention programs can reduce some of the risk factors identified with the correlation between SES and education. The APA also lists several research outcomes where the correlation between low SES and relatively lower academic achievement has been identified. These cases, listed by the APA and referred to on its website, conclude that

- the average literary skills of children entering high school from a low SES family is typically five years behind those of students from higher income families,
- a higher rate of students from lower income families drop out of high school compared to students from higher income families,
- low SES students tend to fare worse in mathematics, science, engineering and technology related subjects compared to high SES students, and
- a bachelor's degree is more likely to be obtained by students hailing from high income families than by students from low income families.

Furthermore, the vocational development of humans is more likely to be hindered if they are associated with minority classes, typically related to racial, ethnic and socioeconomic factors. These factors are more evident in the career aspirations of people in minority classes, but can also be attributed to slower and less pronounced development of an early education. The United States National Center for Education Statistics[4] have presented research that relates a higher average family income to higher aptitude test scores, specifically in mathematics. Although this research is based on statistics from a developing country, the results could potentially be reproduced for emerging markets and determined if a similar trends is witnessed. If so, higher education plays an imperative role in cultivating the socioeconomic status of a country.

4.4 Socioeconomic Status in Higher Education

During undergraduate studies in higher education institutions, research has shown that low SES students are less likely to complete a four-year degree or similar qualification at a university or college (Mompremier 2009). In these institutions, the socioeconomic characteristics of families, such as household income and parental education, have also been linked to attrition of graduates. Declercq and Verboven (2015) investigated the SES and enrollment in higher education from the perspective of costs related to attaining a higher education degree. In this work, Declercq and Verboven (2015) distinguish between three scenarios that might lead to perceived low SES students being less likely to attend institutions of higher education, or complete their graduation program. These three scenarios include:

- The cost of applying and studying at higher education institutions could be a higher deterrent for students from low income and disadvantages families, compared to students that have more funds and advantages.

[4]http://nces.ed.gov.

- Students from low SES background might be less concerned with pursuing a higher education and have lower expectations of what is required to form part of the workforce.
- Low SES students might already be academically behind their peers because of slower development during their schooling years and would therefore be less likely to participate in higher education.

Declercq and Verboven (2015) aim to provide research and statistics to highlight the effect of socioeconomic status and financial aid on enrollment in higher education. The paper not only studies the reasons for lower participation in education from some predispositions, but also investigates possible governmental incentives to alleviate the gap between social classes within education. An investigation on the impact of alternative tuition fee policies is presented. In their findings, Declercq and Verboven (2015) state that students are more sensitive to costs involved in pursuing a higher education. They find that students from low SES backgrounds are more likely to apply for disciplines and fields that are less academically driven—if they can afford to apply. In its conclusion, the paper states that there is potential from non-financial incentives and policies to increase enrollment across all social classes, when compared to financial incentives. This means that governmental policies that address the enrollment gap at school level and ensure that disadvantaged pupils acquire the same academic skills as higher SES students are more likely to succeed compared to policies targeted at adjusting tuition fees (subsidies and lower tuition for low SES pupils). Furthermore, the expansion in global higher education has been criticized by the Times Higher Education for increasing inequality. Universities that have been categorized as *elite* give graduates a stronger chance of being employed compared to universities that do not fit into the *elite* group. As a result, students with starting resources high enough to enter these *elite* universities inevitably have a higher probability of being guaranteed a job and career after graduation. In this sense, the higher education institutions must be averse to increasing social inequality, especially through a high increase in foreign students from previously disadvantaged backgrounds or developing countries.

4.5 Higher Education in Emerging Markets

Community engagement is an essential part of university activity in Africa and in many other emerging markets. Universities and higher education institutions in developing countries typically have high proportions of students from disadvantaged backgrounds and low SES. In countries such as South Africa, a strong governmental equity agenda can be used to accommodate students from disadvantaged backgrounds. However, in many instances, governmental financial backing and interventions from the early developmental stages in a pupil's life are limited and policies must be adjusted to accommodate the consequent backlog. Universities that have a large proportion of disadvantaged students can adjust their internal policies to build social communities among students, since these students are not in the minority in the traditional sense. Students must be given fair opportunities and institutions must understand and improve the culture and societal bonds within communities.

Socioeconomic advances can only be accomplished in developing countries by acknowledging the importance of higher education. Through Industry 4.0, educated youth employment will change the understanding of work and employment. Education and upskilling of the youth is imperative to prepare them for the changes that accompany work and employment in a technological-driven era. The skills and competences that the educated youth require include

- critical thinking,
- problem-solving,
- creativity,
- innovation,
- collaboration,
- communication and information literacy, and
- social justice.

Basic education and higher education must therefore be refocused towards a competence-based approach and integrate modern skill requirements into modules. The changing job landscape would require the youth to apply these skills and competencies in everyday tasks and to use technology as a tool to find solutions for problems. An argument that technology, specifically automation, will replace jobs performed by certain groups of unskilled workers is valid; however, through education this prospect could be mitigated. The youth, and upskilled workers, will have the ability to develop new jobs and roles in such a landscape and apply their skills beyond what only automation can achieve. In WEF (2016), a report titled "*The Future of Jobs*", presents an in-depth analysis on employment, skills, and the workforce strategy for Industry 4.0 and the report is publicly available. The detail of this report is not repeated in this chapter and the reader can consult WEF (2016) for an in-depth analysis on this research (and accompanied data).

Developmental agendas can only be driven forward if the local population is informed and educated, especially in science and technology. In the emerging markets, the role of universities to transfer knowledge and create sustainable social welfare is critical. Referring back to Fig. 6, the findings that the education of parents are linked to the abilities and wellbeing of children further solidify the importance of a formal education to promote global wellbeing. As Industry 4.0 looms, skills in science and technology (specifically ICT) can only be obtained at higher education levels, with a strong background in fundamental principles taught at school level. Universities are also tasked to uphold the integrity and standards of knowledge being transferred to students, an increasingly important issue since the global *commercialization of education.* Privately owned institutions of higher learning are typically not as strictly standardized as universities, which could lead to inferior and irrelevant knowledge being passed down to students. In developing countries, various inadequacies in the higher education systems could lead to more private institutions of higher learning and this could inadvertently have detrimental effects on education.

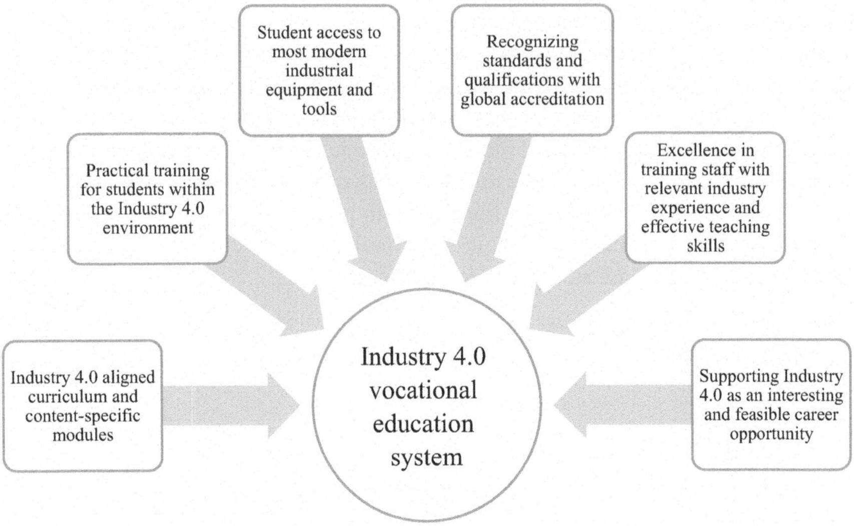

Fig. 6 The characteristics of an effective vocational educational system (adapted from Aulbur et al. 2016)

Emerging markets essentially requires broad-based and high-quality universities to develop the foundations to evolve into sustainable societies. Industry 4.0 not only requires knowledge of mathematics, science and engineering, but also needs to empower people to strive for innovative solutions to complex and pressing challenges. Adjustments in higher education should encourage curiosity and collaboration with international partners to promote willingness to learn and challenge issues that hinder emerging markets. Importantly, institutions offering higher education must be capable of adapting to the needs of Industry 4.0. Knowledge and skills required by the fourth industrial revolution are mutable, in contrast to earlier industrial revolutions, and these institutions are encouraged to acknowledge and change teaching methods to accompany the changing industry.

4.6 Adaptive Higher Education for Industry 4.0

Universities aim to generate and disseminate state-of-the-art knowledge, a theoretical background and the ability to learn in students, all combined with effective and efficient teaching methods. An organization of research must ideally provide opportunities that complement traditional structures with multidisciplinary institutes, focused on addressing societal challenges. Industry, especially Industry 4.0, requires graduates that are capable of solving complex problems in several sciences, with a respect for general interest and the needs (versus wants) of the population. In conjunction, teachers and staff are also required to teach skills that are not necessarily completely defined to learners, since there is still a general lack of understanding of where exactly Industry 4.0 is heading. Apart from relearning skills, teachers also have to adapt to changing learning programs, an enhanced learning experience and importantly, instill an attitude of lifelong learning (Xing and Marwala 2017). The fourth industrial revolution urges creative thinking around the manufacturing procedure, value chain, and dissemination and consumer service by strategically implementing technologies such as the IoT and CPS.

Comparing all the sectors that have been disrupted by rapid advances in technology, education (particularly higher education) has remained relatively untouched. From a unpublished statement made by Mezied (2016), universities are shaping this creative thinking by "*being the testbeds for innovation and the education of future generations*". However, the delivery of education must in many aspects be re-imagined, or at the very least adapted, to deliver the right sets of skills and knowledge to learners. Universities (and schools) are responsible for preparing the youth for skills and jobs that do not yet exist in anticipation of a completely different work environment by 2020–2025. It is a likely scenario that future jobs will rely less on the mathematical and engineering skills obtained through traditional engineering degrees, but rather rely heavily on problem-solving skills and critical thinking.

Teaching must become an immersive experience of technologies of the future, assisted by the most modern and recent implementations such as wearables (Xing and Marwala 2017), AI, robotics, and advanced materials. Learners are increasingly gaining more access to knowledge through the internet, cloud computing, and mobile devices. Massive open online courses (MOOCs) are a theoretically unsettling innovation with participants in MOOCs growing exponentially. As a result, many universities are also digitizing various modules and courses, while individualized face-to-face interaction remains crucial as a part of high-quality education. MOOCs are in fact embraced in many higher education institutions, since they lessen many constraints in respect of learners having access to lecture halls. An increase in applications to higher learning institutions leads to higher cost for universities to increase their teaching capacity, and this trickles down to the cost of an education that must be covered by the students themselves. MOOCs mitigate these issues to some extent in view of the increased popularity in mixing traditional education with MOOCs to benefit from the advantages offered by both. However, despite the initial promises and expectations of MOOCs since their initial mention in 2011, they have not proven to be as transformative in emerging markets as promised (Ziady 2017). According to Ziady (2017), it was found in 2013 that few students that were undertaking MOOC were persisting to the end of the course, although a majority already had undergraduate degrees in the traditional sense. Nevertheless, MOOCs are still in the running to transform education and as with any emerging paradigm, they require a significant period and sustained backing to succeed. In 2018, the best MOOC platforms of 2018 were published by Reviews.com; these results are listed in Table 6.

According to Table 6, there are still numerous (active) MOOC platforms available in 2019 and the user base of these platforms has increased significantly, typically in the millions of users. A certificate from a MOOC in 2019 might not be the same as a college or university degree, but they provide access to excellent education to

Table 6 The best MOOC platforms of 2018 as determined by Reviews.com in December 2018

Rank	MOOC	Score	Founded (by)	Users/Alexa rank
1	Coursera	8.0	Stanford	33 million (June 2018) Alexa Rank: 642 (April 2018)
2	edX	7.4	Harvard/MIT	14 million (December 2017) Alexa Rank: 983 (April 2018)
3	FutureLearn	6.4	The Open University	Alexa Rank: 6709 (September 2018)
4	Cognitive class	5.6	IBM	Undisclosed
5	University	3.4	Privately founded by Liepmann and Klöpper	>1 million
6	Udacity	0.4	Stanford	1.6 million Alexa rank: 1781

anyone with an internet connection. In addition, workers who cannot afford to leave their current workplace to pursue a tertiary education can further their career path and educate themselves through MOOCs.

Adapting delivery of knowledge has become more necessary, and in certain instances critical because of the progression of higher education from when it was only available to the elite, through *massification* (education for the masses), and in its current morphology termed post-*massification* (Xing and Marwala 2017). The accessibility enabling students to gain an education from virtually any university in the world requires the readiness of higher education to accommodate diversity and a digitalized classroom. Behavioral analysis, identification and verification of students, teachers, and content are part of adjusting for Industry 4.0 and individuals gaining a *global identity*. This has some major benefits for individuals from emerging markets that are unable to attend costly higher education institutions. There is however still a concern of inequality within digital higher education, since a large proportion of the population has limited or no access to connectivity.

The following section specifically addresses the role of BRICS (which can be extended to other emerging markets) in developing tailored policies for Industry 4.0. BRICS nations have several unique characteristics that many developed countries do not, for example, large populations and high rates of underprivileged people, and these issues must be addressed in collaboration with other developing nations as well as in accordance with successful policies in countries such as Germany. BRICS therefore plays an important role as the champion of new policies that encompass these unique challenges.

5 The Role of BRICS in Industry 4.0 Policies

In recent history, an upward GDP forecast and the accessibility of inexpensive labor have incentivized organizations from developed countries to create resident industrial plants in BRICS countries. China and India have provided critical capacities for these plants, primarily because of their enormous populations and large unpenetrated domestic markets. Although India has been dawdling in embracing Industry 4.0 expertise, with a low level of automation countrywide, it is facing a threat from other countries, both in BRICS and in developed nations, which aim to pass it and implement advanced manufacturing solutions. India has a robust background in IT; it has witnessed a number of startups in the digital space, especially in 3D printing, as well as in the automotive industry. However, it faces a threat emanating from its escalating labor cost compared to neighboring countries such as Vietnam and Indonesia. India has historically been an attractive destination, offering a cost advantage and manufacturing opportunities, but foreign investors have started looking elsewhere for such prospects. Both the government and private companies have observed these trends and are launching several local initiatives to mitigate the effects. National manufacturing policies and supporting initiatives such as Make in India have been successful governmental mediations to boost Industry 4.0 adoption and ensure that manufacturing in India remains a commodity. Private companies are also spearheading automation locally, implementing Industry 4.0 in manufacturing plants, establishing startups in the smart manufacturing space and providing funding for the first smart factory in India (Aulbur et al. 2016).

China, a historical world frontrunner in low-cost trades and a hub for the industrial sector, still has manufacturing advantages due to the availability of its large (albeit shrinking) labor force (of approximately 776 million)—twice the size of the labor force of India, and many times greater than that of most other economies. Despite China being considered a low-cost nucleus of the world, it is rapidly changing its profile to provide technologically advanced and customized solutions of automation within the Industry 4.0 framework. China is reportedly registering a vast number of patents annually, even more than the USA and Germany, especially in robotics. The Chinese government has acknowledged the integration of industry and the IT sector, and the Made in China 2025 initiative is a governmental aim to level the playing field between China and Western industrial nations with respect to Industry 4.0 (Aulbur et al. 2016). The government in China has also unveiled a plan to "*integrate mobile internet, cloud computing, big data, and the IoT with modern manufacturing, to encourage the development of e-commerce, industrial networks and internet banking. It also aims to grow internet-based companies to increase their presence in the international market*" (Aulbur et al. 2016). China's private sector is not dragging its feet either when it comes to Industry 4.0. It is developing and funding the first unmanned factory through robotic arms to produce cell phones; it produces robots locally and funds robotic research for students. There are multiple bids to acquire Industry 4.0-focused companies in the country, and hardware as well as software companies are taking part in many of these initiatives (Aulbur et al. 2016).

Brazil has a labor force analogous to that of other industrial-based countries, with approximately 70% of its salaried populace contributing in the labor force. Approximately 62% of this labor force has been working in the amenities sector and in response to an overall scarcity of labor in industrialization; the hourly recompense cost has climbed significantly.

Russia, although a large geographical area, has a relatively small population density. The amount of persons active in its labor force is also low associated to developing countries such as China and India, with about 74% of the population of working age being within the labor force, and roughly 13.5% of the labor force working in manufacturing. Russia is a minerals-driven climate and, therefore, manufactured trades subsidized a small fraction of total exports in 2014; only around 5.5% of total trades were contributed by manufactured exports. With its relatively lesser population concentration, Russia has been concentrating on evolving technologies such as drones to conduct reconnaissance of its large geographical space. Although the embracing of Industry 4.0 in Russia is currently low, it presents itself as a significant opportunity for Russia to confront the challenges it faces in terms of low productivity, an elderly population and low population concentration. Advanced digital technologies are seen as not only restricted to industry, but as playing a large role in overcoming the challenges posed by such a large geographical area. As a BRICS nation, Russia is employing government intervention to fuel the transition towards Industry 4.0 with several initiatives, the principal motivation being initiating research and development projects within the country. These initiatives include

- development of the manufacturing industry by improving the labor force with modernization of the manufacturing process using IT,
- robotics and 3D printing, which are beginning to play a significant role to focus adoption of individual Industry 4.0 components and promote widespread adoption of manufacturing within the next decade,
- initiatives that produce circumstances for the rise of companies that aim to be viable in new markets, especially within Industry 4.0 and digitization, creating *factories of the future* within the country,
- robotics and mechatronics, which have been added as a part of school and university curricula, and
- educational complexes that have been established by agencies aiming to promote research and development.

Furthermore, the private sector has been motivating the expansion of Industry 4.0 skills and some private leaders in the economy are getting ready for Industry 4.0 in Russia by several initiatives. These include establishing robotics centers, forming partnerships to promote industrial internet, medium-sized firms proposing engineering resolutions for industrial robotics, consulting services and fabrication management, and an inflow of global companies opening research laboratories in Russia. As a BRICS nation, however, Russia still has several challenges to deal with in terms of Industry 4.0.

Russian innovation is principally evident in the national market, and prospective global clients and clients have limited access to these innovations. In addition, there is significant deficiency of documentation of manufacturing procedures at manufacturing plants, which makes it difficult to promote and expand these practices. Russia also suffers from lack of contemporary technological infrastructure, inadequate mandate for innovation amid large economic firms, lack of faith of international partners and restricted flexibility of foreign software and computer-aided design tools. There are also signs of decaying investments, affluent credit and reserved access to global capital markets, which are several monetary complications that hinder the adoption of Industry 4.0 (Aulbur et al. 2016).

5.1 Focus on South Africa (BRICS)—Commercializing Industry 4.0

South Africa has also experienced an upsurge in the labor force over the last two decades, notwithstanding its 27% unemployment rate in 2018. Compared to other emerging countries, labor involvement is relatively low and the services segment has been a main proprietor of the labor force, inevitably lowering labor productivity in manufacturing per employee. This can in part be ascribed to the recent political changes and challenges the country has faced since apartheid was abolished. The historically large labor force, which used to be underpaid and worked long and tough hours, has only recently been able to leave behind these perils and given an opportunity to create its own wealth. As a result, a large percentage of the workforce is now moving towards the services industry where an education and relatively low capital investment can quickly lead to a sense of success and achievement, something that the workforce is not used to. The WEF estimates that by 2025, South Africa could create 462,000 extra professions by focusing on "*clean energy generation, energy efficiency, pollution control and natural resource management*" (Ziady 2017).

Fortunately, more than two decades after South Africa attained democracy, most of the newly discovered wealth and a rise in skilled workers are starting to bear fruit, as these entrepreneurs are discovering ways to increase their wealth further, to the advantage of the country and its people. However, another politically unstable period has been entered, with increased bureaucracy and cultural tension. In addition, South Africa is facing several political and socio-economic challenges with respect to Industry 4.0 adoption in the form of high input and high labor cost with low proficiency and lack of application and execution of reassuring policies defined by government. Infrastructure, especially poor connectivity and inaccessibility of digital elements are, as mentioned, among the most worrying of Industry 4.0 technical challenges. In South Africa, the adoption of Industry 4.0 is currently low when compared to other BRICS nations, with interest in the fourth industrial revolution within the country perceived mainly on a company-specific level or from large international organizations, with little governmental support. In November 2017, Sipho Maseko, the CEO of the largest telecommunication company in South Africa, Telkom, addressed the country's issues of extreme inequality where people with the means to seize opportunities in technology are obtaining more wealth compared to many more who are excluded from the digital economy. Maseko stressed the fact that it is not a sustainable evolution of society when a specific group of people is advancing at a rapid pace (through technology) and other groups are left further and further behind. Maseko also specifically referred to the fact that for many years, developing countries such as South Africa have had to play catch-up with developed markets in Europe and America. In his address, Maseko outlined a strategy to mitigate the increases in inequality, through

- implementing technology to be as inclusive as possible,
- improving education to ensure young people have the skills to apply technology in their socioeconomic growth, and
- investing in specific technologies to ultimately become leaders in that field.

A comprehensive Industry 4.0 government strategy and a countrywide central strategy to drive Industry 4.0 are clearly absent in South Africa. Several tertiary institutions have realized this and are implementing strategies from the ground up to address the lackluster adoption. This could prove crucial in the near future to boost the economy of the country to become the economic hub of Africa, a position it once held proudly. Sackey and Bester (2016) explore the potential of Industry 4.0 to change the knowledge and skill sets requirements for industrial engineers significantly with respect to assessing the role of such engineers to avoid a similar identity crisis as that experienced by IT during the 1990s. Sackey and Bester (2016) therefore examine the likely impact of Industry 4.0 on industrial engineering and propose enhancements to the engineering curricula in South Africa. In their findings, Sackey and Bester indicate that emphasis has already shifted from traditional industrial engineering towards data-driven functions and CPS. Importantly, mismatches between required and available workforce skills (such as big data and novel human-machine interfaces) are to be addressed by educational institutions, to prepare the next generation of the workforce to contribute to Industry 4.0 and establish South Africa as a dominant player in this sector. Sackey and Bester (2016) indicate that urgent attention is required in a number of areas, applied to industrial engineering (and other engineering disciplines) for Industry 4.0. The areas include

- big data which relays to data science and advanced analytics,
- progressive simulation and simulated plant modeling,
- data networking,
- overall mechanization and automation,
- human-machine interfaces (HMIs),
- digital-to-physical transference (3D printing),
- process excellence,
- real-time catalogue and logistics digitization, and
- current and future digital infrastructure development (Sackey and Bester 2016).

The areas that require enhancements in current curricula in South Africa are all categorically analogous to Industry 4.0 and should be taken into account by South African schools and traditional universities. Modules such as analytics, at least at the basic level, are already being offered by universities, and this is a step in the right direction of *commercializing* Industry 4.0, a necessary initiative. These modules should aim to

- provide broader skill sets to students in industrialization fields,
- offer job-specific capabilities, and
- narrow the IT skills gap.

To prepare for the changing job requirements of Industry 4.0, regulatory and program accreditation bodies should work with universities and industrial engineering professional bodies to develop a set of specific capabilities for industrial engineering, and design ways to assess these capabilities and competencies against set requirements (Sackey and Bester 2016). Bilateral and multilateral collaboration for skills development, as is done among BRICS nations such as China and India, allows countries to announce, after adjustment as per local requirements, best practices and sharing of skills and knowledge.

China, together with Germany, has introduced strong collaboration policies in higher education and vocational training. India and the UK also share a partnership to support the *Skills for Jobs* program in India, an initiative that promotes partaking of the private sector in cultivating the standing of skills development in the country. India has collaborated with the EU on a skills development project in 2012 as well, where the EU assisted in defining and realizing strategies to skill Indian men and women to advance their employability. Since there are numerous models of job-related schooling across the world, with similar goals and primary objectives for developed and emerging markets, it is reasonable to categorize these models or policies as educational, employment-based, or hybrid programs that address both these approaches. BRICS nations can however adopt various strategies, tailored to their own specific needs, but with essential characteristics that ensure an effective vocational education system. In Aulbur et al. (2016), a summary of the characteristics of an effective vocational educational system are presented; it is adapted here and given in Fig. 6.

As shown in Fig. 6, adapted from Aulbur et al. (2016), an inclusive Industry 4.0 development policy iterates several important and key aspects, most noteworthy those of vocational training of students and prospective engineers targeting Industry 4.0 and being recognized globally, with accredited learning and tertiary education offers being of crucial importance. Such a system focuses on practical training of skilled workers within enabling environments and ensuring that skills are developed in line with modern technologies that enable the fourth industrial revolution. Comparing this system to, for example, strategic objectives (policies) of South African universities, specifically the University of Johannesburg, several parallels can be drawn. The University of Johannesburg, South Africa, lists six strategic intents that provide a dedicated means of realizing global excellence and stature (these strategic objectives have been directly adapted from the website of the university):

- "Excellence in research and innovation by offering a range of innovative programs characterized by the highest level of scholarship that are incorporated into the curriculum, therefore attracting high-quality graduates from diverse backgrounds. The institution promotes a culture of innovation and entrepreneurship by providing stimulating and supportive environments, especially technology stations leading to potential patents and technology transfer."

- "Excellence in teaching and learning through maintaining an appropriate and rigorous enrolment profile and providing challenging and applicable curricula that respond innovatively to the challenges of the 21st century. The institution aims to achieve high student success rates, improved retention and increased graduate output to enter industry as experienced and skilled workers."
- "An international profile for global excellence and stature through recruitment of international students and staff, study-abroad programs, smart institutional partnerships and bilateral and multilateral collaboration. Its aim is to have an increased proportion of undergraduate and postgraduate international students through focused recruitment from Africa, BRICS nations, Asia, the Americas, Australasia and Europe, as well as increasing the stature of relationships with global universities, United Nations agencies, and the African Union."
- "Enriching a student-friendly learning and living experience through high-standard teaching and learning facilities, support through the student life cycle, learning and living communities, and a responsible and respectful student culture and philosophy. The learning and living objectives are to provide students with a nurturing and supportive environment to foster all-round development."
- "National and global reputation management by being a pan-African center of critical intellectual inquiry through extensive scholarship and participation in the knowledge networks of the continent."
- "Fitness for global excellence and stature with world-class financial systems to ensure good investment returns and overall financial stability and sustainability."

In comparing the strategic initiatives of a renowned South African university listed above with the vocational education system targeted at spearheading Industry 4.0 by Aulbur et al. (2016), it is noticeable that there are many similarities in these approaches. Context-specific content, attracting high-quality learners (analogous to skilled workers and apprentices) and teachers (analogous to training staff), access to modern and state-of-the art equipment and tools, as well as international recognition of study programs and qualifications, are vital aspects of multiple sectors. These aspects ensure successful integration of multiple disciplines, including, for example, a commanding presence in Industry 4.0 through policies and strategies invoked by educational institutions.

The Department of Higher Education and Training in South Africa has established three national skill development approaches for the period 2001–2016. Under these strategies, its objective is to intensify access to superior and applicable education as well as training for the advancement of new expertise. The strategies achieve this by providing bearing for sector expertise planning and execution in the "*Skill Education and Training Authority*", and additionally providing a basis on how to use the funds generated by imposing a skills development tax on businesses. To its credit, and according to the White Paper of Aulbur et al. (2016), the South African government has acknowledged five key skills development primacies to lead Industry 4.0 in the country. These five key priorities are listed in Aulbur et al. (2016) and have been adapted and listed here as

- the development and ongoing support of a sector labor market intelligence system,
- continued support as well as increased focus on artisan development,
- facilitating strategic partnerships (both locally and internationally) to improve the readiness of funds and the value of skills development,
- intensifying the stream of incoming skilled workers, and
- developing the skills of the present labor force.

Demographically, there has been a change in the trade industry, where intercontinental trade of digital information is becoming more lucrative. Decentralization is another defining characteristic of Industry 4.0, where sematic machine-to-machine communication through ubiquitous microcontrollers and internet protocols is revolutionizing decentralized and service-oriented control. Decentralized, autonomous and real-time control of factories and machinery enables smarter and more adaptable systems, spearheading the fourth industrial revolution. Intelligence in object networking can be created through decentralized process management. The interaction between the real world and virtual (simulated) environments is another defining aspect of Industry 4.0. Again, education and skills development are essential in a decentralized approach. Workers are now required to innovate, design and commission (and maintain) complex systems capable of generating big data. Analyzing and intuitively presenting these large volumes of data require a specific skills set from workers, enabled through effective higher education and vocational training.

6 The Move Towards a Decentralized Industry

6.1 How Demographics Influence Industry 4.0

Sustainable development and economic growth can only be achieved if all-encompassing growth throughout the demographics of the working populace is encouraged. "*Countries with the greatest demographic opportunity for development are those entering a period in which the working-age population has good health, quality education, decent employment opportunities and a lower proportion of young dependents*" (UNFPA 2016). A decrease in the dependents per domiciliary typically leads to bigger investments per youth, more independence for females to go into the professional workforce and larger availability of personal funds for retirement. This economic payout, if sustained, is broadly termed *demographic dividend*. To achieve a demographic dividend, a country needs to "*understand the size and distribution of the population, its current and projected age structure and the pace of population growth*" (UNFPA 2016).

The fourth industrial revolution is merging with socio-economic as well as various demographic rudiments to drive significant vicissitudes in business prototypes across many doings, resulting in significant disturbances to labor markets. New programs of occupations and vocations are emerging, in certain instances displacing or adapting traditional jobs. The skill sets essential to both traditional and novel occupations are changing in most industries and are shifting the dynamics of the industry demographics, especially the gender gaps. There is however general lack of consensus about the impact of Industry 4.0 on the labor landscape, where some predictions are focused on large-scale unemployment due to automation and AI, whereas others predict a major shift to highly skilled and educated workforces replacing the traditional roles condemned by automation and AI. In both circumstances, the labor market and quality of the workforce will look considerably different as Industry 4.0 gains traction and becomes the total industrial revolution it is designed to be. Emerging technologies, socio-demographic shifts, and political and economic reservations in both developed and emerging countries are already dramatically remaking the workforce. Unfortunately, technological changes are increasingly focused on substituting human workers as opposed to enhancing them, to drive down labor cost. People and organizations must therefore have adaptability at their core to safeguard the jobs and livelihoods of the workforce. In line with technological changes, demographic and economic factors are reshaping the labor market, with suggestions by the UK government that the workforce will be older, multigenerational, increasingly international, and with a higher proportion of female employees. Employers are also urged to change their approach to workforce planning to diminish scenarios of severe skills shortages in the next two decades. Technology and flexibility of employment create the opportunity for wealth creation across international boundaries, with the potential for local tax revenues to fall despite employment increasing. Therefore, traditional economic and social models no longer satisfy the need to plan effectively as Industry 4.0 technologies become mainstream and both political and new social factors can drive change.

Demographic evolution is linked with a variation in age arrangement resulting from a decrease in both mortality and fertility. Many industrialized nations are at an advanced stage of demographic evolution, while a high number of African countries (most in fact), and countries within the BRICS group are entering the initial stages of demographic transition. Fortunately, for developing countries entering into demographic transition, associated advantages of demographic dividends are evident, offering a window of opportunity for rapid human development and associated economic growth if effective and decent policies are executed. An increase in labor supply and human capital are among the demographic dividends associated with smaller numbers of dependents in households, with fewer economic burdens related to large families and households.

Furthermore, several studies, particularly on demographics, have been conducted, especially with regard to the observation during the Great Recession, where countries with large populations such as India and China were least affected by this economic crisis. Two alternative theories were developed, and economists are aiming to determine if this is primarily due to the vast populations of the two nations or to less dependence on international trade (particularly in India) (Basu et al. 2013).

BRICS countries differ considerably in terms of their growth forecasts, where demographic tendencies and labor supply dynamics are particularly favorable for Brazil, India and South Africa. In these countries, the working age population will remain to develop into the middle of the 21st century, and possibilities of obtaining demographic dividends are well within reach, especially with Industry 4.0 in its infancy. With fertility rates hurtling in Asia and Latin America, half of the proliferation in population over the next 40 years will be in Africa. Russia is already experiencing a shrinking workforce, while the one-child policy in China introduced in 1979 and only formally phased in 2015 will inevitably lead to a decline in the working age population. Although the labor force of Russia is well educated and skilled, it is undergoing remarkable change, largely mismatched to the rapidly changing needs of the Russian, and global, economy. Unemployment among women and young people in Russia is high, and after the breakup of the Soviet Union and economic dislocations, the standard of living has fallen dramatically. In China, the labor force participation rate (people who can work) fell from 77% in 2000 to 68.7% in 2018, and is still declining, further shrinking the labor market. There are many labor shortages in China, and economic growth has slowed down significantly to lower than approximately 7% in 2016. China is struggling to uphold manufacturing competitiveness, as swift aging causes its workforce to decrease and wages to rise faster than productivity gains. This creates complications and challenges for leaders in supporting growth while trying to reduce economic reliance on credit-fueled investments. Figure 7 presents the expected population aging and shrinking of the participation rate in China, comparing data from 2010 and estimated levels for 2030.

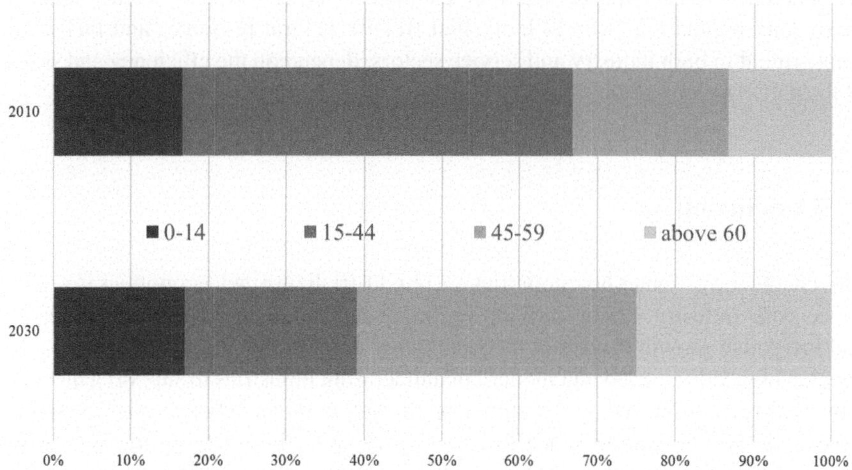

Fig. 7 The expected population aging and shrinking of the participation rate in China, comparing data from 2010 and estimated levels for 2030

As shown in Fig. 7, a significant decline in the Chinese labor force is expected by 2030, represented by a decline in most participation sectors. The young demographic group from birth to 14 years of age is expected to remain relatively constant at about 17% of the total population, with the largest decline in ages between 15 and 44 (from 50 to 22%). The older demographic group, nearing retirement age, will increase from approximately 20 to 36% of the population by 2030. The percentage of people older than 60, not contributing actively to the labor force, will according to the 2010 data increase from 13 to 25%, one quarter of the entire population.

Professor Wang Feng from the University of California proposes that a declining labor force is also inherent to various positive factors. These include

- rapid expansion of tertiary education,
- earlier withdrawal from the labor market as wealth prospects increase, and
- a general reduction in participation for age groups that are traditionally too young to work,

all aspects pointing to an overall increased quality of life. With more people capitalizing on retirement benefits, a potential boost in consumption is also perceived in many economies.

Fast growth of productivity in workers is in many instances linked to growth in GDP, where contribution of the service sector contributes most, followed by industry and agriculture. As Industry 4.0 is being introduced in many countries, the shift towards labor productivity in industry becomes more crucial, and demographics play a critical role in its early adoption. An imbalance in the workforce and trade opportunities, as will be the case in China in the near future, could potentially stunt growth of the fourth industrial revolution if applicable strategies and policies, such as international collaboration, bilateral and multilateral exchanges and re-training of the current workforce are not implemented rigorously. The effects of demographic dividends in BRICS nations, with high estimated GDP growth due to factors such as a young population (Africa, India, and Brazil), natural resources and increased employment in both industry and service sectors, depend on the efficiency and skills of the participation sector.

7 Conclusion

This chapter investigates how countries are reindustrializing their economies in accordance with Industry 4.0 by developing fitting and modular policies that support and invigorate growth drivers in the technology domain. Such policies should help develop high value-added and modern manufacturing industries to support a modern industrial revolution, where the shift towards autonomous cyber-physical systems is seen as the dominating space for future prosperity. In emerging markets, with specific focus placed on the African continent, policy gaps exist that prevent Industry

4.0 uptake. The chapter also identifies the policy gaps that are less evident in developed countries, but stifles growth in developing markets. Technical and infrastructure issues (for example limited and expensive broadband availability) and social, economic and political challenges are identified and reviewed in the chapter.

Talent management and skills development for Industry 4.0 remain challenging, especially in emerging BRICS countries, where political and economic instability is high. Furthermore, organizations should know exactly which skills to develop that would be beneficial and sustainable during the early adoption of Industry 4.0. BRICS countries must embrace collaboration in research by universities to reduce dependence on the West.

Digital transformation is presenting new challenges for many employers, and their current and future workforces should be adaptable in this regard. New business models and strategies of cooperation and collaboration constitute added value to Industry 4.0, but space for creativity and innovation must be created. The emphasis should be placed on developing new technical skills, especially in operating activities and mechanical processes, purchasing and warehouse logistics. These new demands are becoming a major challenge for existing employees, often requiring retraining or further skills development.

Education, specifically higher education, plays a crucial role in developing youth to be able to contribute to growing economies in both developing and developed nations. In developing countries, there are however several unique challenges that must be overcome. This chapter addresses the importance of higher education, with specific focus on Industry 4.0, to develop young and innovative minds with a solid background in theoretical fundamentals learned at school level, but with skills and relevant knowledge to take forward a changing industrial milieu.

Innovation and technology-based intellectual property have been identified as pillars in the world's leading economies to sustain long-term prosperity throughout Industry 4.0. Higher education, therefore, plays a significant role in educating workers to participate in a changing workplace. This chapter reviews the adaptive strategies and policies of higher education in emerging markets and the role that tertiary education plays in uplifting the socioeconomic status of developed countries. Focus is also placed on the role that BRICS countries can play in Industry 4.0 policies, specifically in the decentralization of technology development throughout the demographics of the working population in these emerging markets.

BRICS countries can cooperate mutually to effect expansion in the fourth industrial revolution, and these nations can learn from one another, and from developed nations, to implement best practices and effective infrastructure. The changing political, economic and social fortunes of these countries are creating space for innovative urban environmental practices in cities. Collaboration among BRICS nations, also in higher education, will strengthen ties between countries and improve appreciation of different cultures, labor practices and socio-economic improvements. Governments must make skills development a top priority and engage with partners, such as the private sector and universities, to widen the skills sets of young individuals

References

AEO (2017) African economic outlook 2017. Special theme: entrepreneurship and industrialization. ISBN 978-92-64-27425-9. AfDB, OECD, UNDP 2017

Aulbur W, Arvind CJ, Bigghe R (2016) White paper: skill development for industry 4.0: BRICS skill development working group. http://www.globalskillsummit.com/. Accessed 13 Oct 2017

Barnes S (2015) How better connectivity can transform Africa's economies. http://www.weforum.org. Accessed 12 Nov 2017

Basu T, Barik D, Arokiasamy P (2013) Demographic determinants of economic growth in BRICS and selected developed countries. In: XXVII IUSSP international population conference (IUSSP 2013), 26–31 August 2013 at BEXCO, Busan, Republic of Korea

Buhr D (2017) Social innovation policy for Industry 4.0. The Friedrich-Ebert-Stiftung foundation. http://www.fesdc.org. Accessed 20 Nov 2017

Carroll M, Fear U, Kediutwile E, Oelofse J, Rees G, Smith E (2016) Industry 4.0—is Africa ready for digital transformation? A study by Deloitte. http://www.deloitte.com. Accessed 18 Oct 2017

Declercq K, Verboven F (2015) Socio-economic status and enrollment in higher education: do costs matter? Educ Econo 23(5):532–556

Fitsilis P, Tsoutsa P, Gerogiannis V (2018) Industry 4.0: required personnel competences. Industry 4.0 3(3):130–133

Jelinek V, Fomerand J (2013) Higher learning institutions and global citizen education. United Nations Chronicle L(4)

Kaklauskas A, Banaitis A, Ferreira F, Ferreira J, Amaratunga D, Lepkova N, Ubartė I, Banaitienė N (2018) An evaluation system for university-industry partnership sustainability: enhancing options for entrepreneurial universities. Sustainability 10(1):119–136

Klitou D, Conrads J, Rasmussen M (2017) Key lessons from national industry 4.0 policy initiatives in Europe. Digital Transformation Monitor, May

KNBS (2016) Economic Survey 2016. Kenya National Bureau of Statistics

Lambrechts JW, Sinha S (2016) Microsensing networks for sustainable cities. Springer International Publishing Switzerland. ISBN 978-3-319-28358-6

Mezied AA (2016) What role will education play in the fourth industrial revolution? http://www.weforum.org. Accessed 7 Nov 2017

Mompremier L (2009) Socioeconomic status and higher education adjustment. SES and familial stress may work against adjustment to higher education. http://www.apa.org. Accessed 17 Nov 2019

Morgan PL, Farkas G, Hillemeier MM, Maczuga S (2009) Risk factors for learning-related behavior problems at 24 months of age: population-based estimates. J Abnorm Child Psychol 37:401–413. https://doi.org/10.1007/s10802-008-9279-8

Naicker SM (2018) Inclusive education in South Africa and the developing world: the search for an inclusive pedagogy. Emerald Group Publishing. ISBN 1787439623

Ndabeni-Abrahams S (2019) Industry 4.0: big opportunities, big risks for Africa. http://www.businesslive.co.za. Accessed 22 May 2019

Panarina E (2015) University-industry partnership as a key strategy for innovative sustainable economic growth. J Int Bus Res Market 1(1):24–27

Piccarozzi M, Aquilani B, Gatti C (2018) Industry 4.0 in management studies: a systematic literature review. Sustainability 10(10):3821, 1–24

Roksa J, Kinsley P (2019) The role of family and support in facilitating academic success of low-income students. Res High Educ 60:415–436

Sackey SM, Bester A (2016) Industrial engineering curriculum in Industry 4.0 in a South African context. South Afri J Indust Eng 27(4):101–114

UNFPA (2016) Demographic dividend. http://www.unfpa.org. Accessed 28 Oct 2017

WEF (2016) The future of jobs. Employment, skills and workforce strategy for the fourth industrial revolution. Global challenge insight report. http://www3.weforum.org. Accessed 17 Nov 2019

Westkämper E (2013) Towards the re-industrialization of Europe: a concept for manufacturing for 2030. Springer Science and Business Media, 4 August

Xing B, Marwala T (2017) Implications of the fourth industrial age for higher education. Thinker 73:10–15

Ziady H (2017) Work and the future. Financial Mail, October 12–October 18

Sell-Bot: An Intelligent Tool for Advertisement Synthesis on Social Media

Sonny Kabaso and Abejide Ade-Ibijola

Abstract Social Media Advertising is one of the fastest ways to reach a large group of targeted customers. In the present day, there are large volumes of advertisement campaigns designed for, and running on Social Media Platforms, yielding significant conversion rates for businesses. Advances in Artificial Intelligence (AI) have offered new techniques for tailoring advertisements towards targeted audiences. Some of these techniques include Machine Learning (ML) applications in the analysis of Big Data; acquired from customers spending, and/or buying patterns. Hence, software companies and researchers have created several tools that target specific customers in need of specific products and services. An aspect of AI that has not seen too many applications in Advertising is Natural Language Generation (NLG); the study of algorithms for generating natural or human languages. In this chapter, we present a NLG algorithm for the automatic generation of advertisements for Social Media platforms. This algorithm is implemented in a tool called `Sell-Bot` that is based on Context-free Grammars; a formal technique for describing or generating languages. This algorithm takes a list of available and currently discounted commercial products and services of an organisation, and intelligently composes a valid Social Media post (a tweet or Facebook post) that is often indistinguishable from a human expert marketer's tweet or post. An evaluation of Sell-Bot showed that 42% of the participants did not know whether the AI (`Sell-Bot`) or the human generated the adverts. `Sell-Bot` is expected to relieve humans of their efforts of posting advertisements on Social Media.

Keywords Synthesis of things · Social media · Advertising · Advert synthesis · Context-free grammar · Natural language generation · Procedural generation.

S. Kabaso · A. Ade-Ibijola (✉)
Formal Structures Algorithms and Industrial Applications Research Cluster, Department of Applied Information Systems, School of Consumer Intelligence and Information Systems, University of Johannesburg, Auckland Park Bunting Road Campus, Johannesburg, South Africa
e-mail: abejideai@uj.ac.za
URL: https://www.abejide.org

S. Kabaso
e-mail: smulenga.k@gmail.com

W. Doorsamy et al. (eds.), *The Disruptive Fourth Industrial Revolution*, Lecture Notes in Electrical Engineering 674,
https://doi.org/10.1007/978-3-030-48230-5_7

1 Introduction

Social Media can be defined as user-driven technology platforms that provide affordable content creation, communication and collaboration over the internet (Berthon et al. 2012; Kaplan and Haenlein 2012). The introduction of Web 2.0 changed aspects of advertising (Stelzner 2014). This is partly because Social Media was widely accepted, it enhanced user control and freedom over multi-way communication (Berthon et al. 2012; Boateng and Okoe 2015; Leung et al. 2015) and became a channel for marketing and advertising for the digital generation (Duffett 2015). Prior to this revolution, businesses could only advertise their products or services via billboards, television, radio, posters, and newspapers. These traditional methods of advertising only presented one-way communication at very high costs (Duffett 2015). As Social Media was introduced, it overcame some of these limitations (Mulhern 2009) and dramatically changed the modern advertising landscape (Dehghani and Tumer 2015; Otero and Gutiérrez 2015). Businesses had to revise their marketing strategies and invest in Social Media advertising (Chun et al. 2005), enabling them to create strong well-known brands and increasing coverage of the targeted population (Briggs and Hollis 1997). Nevertheless Social Media advertising did not replace the traditional methods but complemented them (Barreto 2013).

Studies have shown that the commercial effectiveness of Social Media advertising on platforms like Facebook can increase business exposures and brand awareness at minimal costs than advertising on non-virtual media (Harris and Rae 2009; Sajid 2016; Stelzner 2014; Vanden Bergh et al. 2011). Further benefits includes increased sales, improved customer satisfaction, improved search engine ranking and global market reach (Okazaki and Taylor 2013; Otero et al. 2014). It is in the best interest of businesses to value Social Media advertising on platforms like Twitter, Facebook, Instagram, LinkedIn and many others, to leverage on these benefits (Saxena and Khanna 2013). Many customers expect businesses to have a Social Media profile, to check the credibility of the brand and consult with many other users of the product through chats, comments and reviews before they make a purchase decision (Dehghani and Tumer 2015).

A study done by Lee and Kim (Doohwang Lee et al. 2011) determined that consumers are more likely to view and consider users comments and reactions, than the advert itself. For instance, Facebook allows for people with similar interests to easily communicate with each other in groups, creating the potential for people in the same groups to influence each other on purchasing decisions (Erkan et al. 2019). It is necessary that a business includes Social Media advertising into their marketing strategy and can not afford to ignore its benefits (Otero and Gutiérrez 2015). With Social Media advertising, a business can control the placement of adverts, having direct contact with potential customers. They can send out unique personalised promotions creating bonds and strong relationships with consumers.

Although Social Media came with several benefits in terms of advertising, it created new complexities. Reaching a large group of people did not mean that adverts would always reach the targeted audience. This brought about a need to identify specific consumers on these social platforms. Consumers increasingly express what they want, their needs, traits and values across many channels on Social Media. There is an explosive growth of consumers-generated data that is continuously growing in terms of volume, velocity, variety and veracity (Erevelles et al. 2016; Kietzmann et al. 2018). In turn, businesses turned to Data Science to obtain valuable customer insight from this data in order to make key marketing and media decisions without infringing on the privacy standards of the social platform and it users (Gazagne and Gösswein 2016; Kietzmann et al. 2018). Until recently, only a few publications have given attention to the problems faced in advertising (Leake and Gary 2008; Wierenga 2010).

One approach that has been focused on is real-time advertising (also known as programmatic advertising). It creates a more intelligent way for advertisers to target and communicate with consumers based on their demographics, interest, behaviour, timezone, geographical location and weather. Programmatic advertising focuses on the use of AI and data to boost marketing efficiency in real-time (Busch 2016). Publishers promote advertising space to marketers that want to reach the potential and right audience at the right time, taking advantage of the Big Data repositories of consumer data (AlSabeeh and Moghrabi 2017; Moghrabi and Al-Mohammed 2016).

In order to exploit the full potential of Social Media, marketers needed to focus on a data driven and social form of advertising. In this case the consumers become more important, making the marketing process simplistic (Seitz and Zorn 2016). Instead of creating broad advertising campaigns to many people, this form of advertising focuses on target marketing programmes. With Machine Learning (ML) a potential customer is identified based on their attributes, interests and browsing habits in order to customise a tailored advertisement and increase the chance of the consumers purchase from being exposed to the advertisement (Adams 2004; Chickering and Heckerman 2000; Langheinrich et al. 1999; Sun et al. 2016). Current cost-effective technologies in database storage and processing of real-world and online user data enable marketers to develop data-driven advertising campaigns, creating individuality in advertisement messages with ML and predictive approaches (Seitz and Zorn 2016).

AI has the potential of contributing to strategic marketing tactics focusing on saving money and driving growth. With consumer generated Big Data, marketers can use the power of computation, learning, heuristic functions and pattern recognition to make strategic decisions about prices, promotions and distribution channels of advertisements (Wierenga 2010). AI enabled Social Media advertising has changed the marketing rules and market dynamics, calling on a new marketing orientation and ecosystem.

With the vast amounts of products in and large customer database. A simple form of advertising on Social Media would be sending out special offers on prices to possible customers, hopefully persuading them to purchase the product (Chickering and Heckerman 2000). Even though this method is cost effective, its cumbersome when there is a need to advertise a large number of unique and personalised promotions of products that are on sale. It is time consuming to manually produce thousands of unique advertisements for each of the products in a short amount of time. We

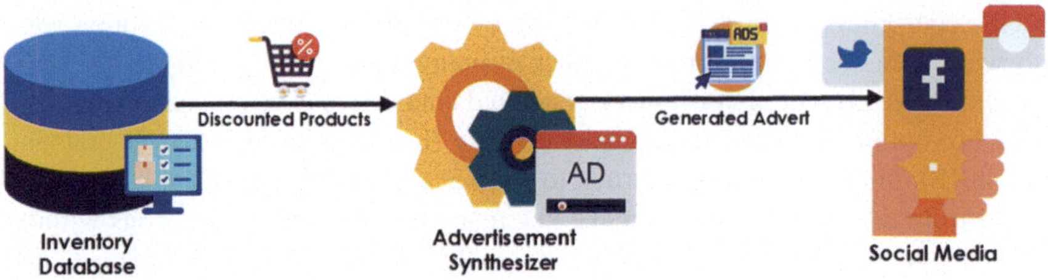

Fig. 1 Processes of advertisement synthesis using `Sell-Bot`

need to redefine and adjust the advertising approaches in a technologically enhanced marketing ecosystem that is cost effective and reduces the amount of time and effort.

One approach is to implement an automatic advertisement generator. This chapter presents an Intelligent Tool for the Synthesis of advertisements on Social Media. We do this by synthesising adverts in NL[1] using Context-Free Grammar (CFG). This involves generating advertisements about discounted products from a database. This processes is described in Fig. 1.

With the development of `Sell-Bot`, the following contributions were made. We have:

1. explored, and compared the implementation and impact of known AI techniques that have been previously applied in advertising, such as ML and Big Data Analytics,
2. investigated models of quantifying the impact of advertisements, and see how these models fair with Social Media advertisement campaigns,
3. designed an NLG algorithm for the automatic generation of advertisements for Social Media platforms. This is based on CFG; a formal technique for describing or generating languages,
4. built a software tool, called `Sell-Bot`, that implements the techniques described in (3) above, and
5. presented an evaluation of `Sell-Bot`, in a TT.[2] Users had to guess if adverts were written by a human marketing expert or generated by `Sell-Bot`. `Sell-Bot` was able to fool humans almost half the time.

The rest of this chapter is organised as follows. The second section of this chapter outlines the background, research questions, motivation and related works. The third section describes CFG for generating textual advertisements. The fourth section shows the implementation of `Sell-Bot` and some real examples of the generated adverts. The fifth section presents an evaluation of `Sell-Bot` from Social Media users and marketers. The last section concludes this chapter and shows future works.

[1]Natural Language.

[2]Turing Test: Used to determine whether a machine can think like a human-being (Luger 2005).

2 Background and Related Work

In this section, we introduce the problem, discussing the related work and motivation behind this work.

2.1 Research Questions

In this chapter, we have addressed a number of interesting research questions as follows.

1. What are the current implementations of Artificial Intelligence in Marketing?
2. Can we generate Social Media Advertisements about special promotions and discounts based on real time data about shelved products?
3. How can we design CFG rules to answer the above questions?

2.2 Motivation

Sending out advertisements for a large number of products is a time-consuming task by writing and is expensive. It is important that advertisers dedicate an efficient amount of time to ensuring that advertised content is frequently updated. Businesses have to devote large amounts of resources to Social Media advertising campaigns to avoid being out of touch (Harris and Rae 2009; Stelzner 2014; Vanden Bergh et al. 2011). To the best of our knowledge, there has not been any attempt to formalise the process of synthesising advertisements for Social Media using CFG. Instead of spending resources on manually writing Social Media advertisements, it is much more efficient to implement an intelligent tool that would be able to write advertisements like a human, but with minimal effort. This will allow businesses to focus resources on the core processes of their business strategies.

2.3 Artificial Intelligence Techniques in Advertising

There are a number of AI building blocks that can enable business to understand customer generated Big Data, allowing them to create effective and flexible data driven advertisement strategies. In this section, we present a few.

2.3.1 Image Recognition

This enables brand advertisers to gain information from the pictures and videos that users post on Social Media. Even if not directly motioned by the user, the pictures and videos they post can reveal many details about them (Kietzmann et al. 2018). For instance, in a posted selfie, image recognition can reveal information about identified brands that the user is interested in within the image. With this information advertisers can develop tailored adverts targeted to that specific users.

2.3.2 Big Data and Machine Learning

Consumers continuously generate vasts amounts of Big Data by surfing the internet (McAfee et al. 2012). From this data, marketing professionals are able to retrieve valuable customer insights and create customer profiles based on their attributes and their interactions, to make better marketing decisions (Erevelles et al. 2016; Hedin et al. 2011), monitor strategic decisions and identify the attitudes and opinions of the customers towards their brand, products and services (Amado et al. 2018; Fan et al. 2015; Tan et al. 2013). ML algorithms learn from the patterns in this customer data, and stores them in a knowledge base. Advertising software can produce useful insight about the users from the knowledge base and propose product options best suited for a potential customer (Adams 2017; Sundsøy et al. 2014). Processing more data about the customer produces more insightful information to advertisers. ML can also determine how likely a certain customer will purchase a product by predicting their lifetime value and conversation likelihood (Adams 2004; Chickering and Heckerman 2000; Langheinrich et al. 1999; Sun et al. 2016). This is used in Propensity Modelling to make real-word predictions of customer market segments and how to interact with a specific customer (Bailey et al. 2009).

2.3.3 Natural Language Understanding (NLU)

This allows for the analysis of the distinctions in human language. With Sentiment Analysis, advertisers can find meaning from large amounts of user generated data such as status updates, Facebook posts, Tweets and product reviews on Social Media (Cambria et al. 2013; Nasukawa and Yi 2003).

2.3.4 Natural Language Generation (NLG)

Using AI to generate human-like content, allows advertisers to generate original content such as personalised advertisements, emails, rich media and unique articles about products (Ramos-Soto et al. 2016). Similar techniques are used in image generation to synthesis lifelike rich media content like images and animations and also in the generation of speech to create stimulating voice overs (Kietzmann et al. 2018;

Reddy 1976). This technique can be used to supplement other AI enabled advertising techniques such as:

1. Advertisement Targeting and Re-Targeting: by running ML algorithms through consumer generated Big Data, a brand can identify and generate content that is relevant to each customer (Feng and Wei 2010; Manolescu and Raymond 2010; Sundsøy et al. 2014). The brand can also identify what content is more likely to return a existing customers based of the their historical shopping data (Lambrecht and Tucker 2013, 2011).
2. Smart Recommendations/Content Curation: this enables brands to generate and deliver content that is only relevant to the customer based on their previous interactions and shopping cart activities (Jiang et al. 2019).

There are a number of commercial and open-source market offered NLG ready-to-use tools that apply the techniques we explained above. We list a number of these tools below.

1. Wordsmith: Developed by Automated Insights, it is an NLG tool that uses an advanced template-based approach to generate natural human-like text from data (Wright 2015).
2. AX Semantics: This is an automated NLG tool. It generates unique meaningful written content from data in real-time (Madsack et al. 2018). It offer eCommerce, BI[3] and financial reporting, allowing businesses the generation of product descriptions, news articles, reports and documentation in over 100 languages.
3. Arria NLG Platform: This allows the generation of rich and varied narratives from data in several languages and formats without any human intervention. A good example is the weather forecast generation by Goldberg et al (Goldberg et al. 1994).
4. Yseop: Provides smart customer experience with an NLG platform that powers enterprise customer support, sales and reporting (Van and Emiel 2014).
5. Quill: This is an advanced NLG tool provided by Narrative Science (Wright 2015). It converts data into insightful narratives to create analytic reports.
6. Simplenlg: This is an open source NLG API written in JAVA originally developed by Ehud Reiter. It is an easy-to-us library and provides clear documentation for businesses or individuals that what to create NLG applications (Gatt and Reiter 2009).
7. NaturalOWL: is an open source NLG toolkit for generating OWL class descriptions. User do not need to have programming expertise to use NaturalOWL (Galanis and Androutsopoulos 2007).

The software tool presented in the chapter mainly focuses on formal methods of NLG. Although AI has been implemented in some areas of marketing, there are limited approaches that have employed NLG in advertising. Many automated advertising tools still use the notion of sending out pre-created digital advertisements on the

[3]Business Intelligence.

internet. There are a few organisations that have managed to build effective intelligent advertising platforms.

2.3.5 Facebook Advertising

Facebook plays a dominate role in Social Media advertising. The platform allows for the broadcasting adverts to large audience whilst driving business KPI's[4] of marketers (Dawson and Lamb 2016; Andrew et al. 2011). When marketers advertise on Facebook, they can reach consumers in a short amount of time with minimal costs. Also giving consumers the ability to openly interact with the advert by reacting to it with a *like*, *happy*, *sad*, *surprised* or *angry* emoticon, add a comment on it and also spread the advert further by sharing it. With this Big Data, Facebook has succeeded in developing an integrated programmatic advertising ecosystem by leveraging on it knowledge of users social relationships and interests (Seitz and Zorn 2016; Yaakop et al. 2013).

2.3.6 Instagram Advertising

Instagram is an image and video sharing Social Media platform with over 800 million users as of September 2017 (Statistica 2018). It enables brand to incorporate visual media into their advertising campaigns and links these visual adverts directly to brand's product page or website. It is also a popular platform for marketing influencers with a large number of followers. Brand can choose influencers aligned with their target audience to promote their products and services by posting, sharing statuses and hosting live videos (Evans et al. 2017; MediaKix 2019).

2.3.7 Twitter Targeted Advertising

Twitter is a micro-blogging platform that offers three customisable advertising programs that allow advertisers to connect with a global audience. These advert programmes include: Promoted Tweets, Promoted Accounts and Promoted Trends to help attract followers of the advertisers profile (Funk 2013). With user generated data and other information (such as profile, mobile device identifier, location or other authorised installed apps) collected by affiliates and partners, Twitter can personalise adverts for specific consumers (Twitter 2019).

[4]Key Performance Indicator.

2.3.8 LinkedIn

LinkedIn is a professional social network that offers a self-serve display advertising system that enables brands to target professionals based on rich profiles and behavioural data (Agarwal 2013). With ML and optimisation components such as response prediction, brands can deliver customised targeted adverts and predict the click-through rates (CTR) of when the advert is presented to the user.

2.3.9 Google Display Network

The Google platform consists of a number of popular partner web and mobile applications that include Gmail, YouTube, Blogger, Google Finance and Google Search. The platform allows for real-time target advertising to the right audiences at the right place. Marketers can find new customers that are likely to buy products or services and re-engage with existing customers using automation. With ML and predictive analysis, Google Ads learns which consumers are suitable to advertise to (Google 2000; Subhadra et al. 2018).

2.3.10 Adobe Marketing Cloud

Adobe delivers digital creative and personalised marketing in real time. Marketing professionals are able to create customer profiles based on their attribute and deliver effective targeted customer experience using a data analysis dashboard (Adobe 2018). Adobe also provides Social Media optimisation and campaign management.

2.3.11 AdRoll

AdRoll focuses online target and re-target advertising. AdRoll enables brand to understand their customers and deliver smart recommendations on a variety of Social Media channels (AdRoll 2007).

2.3.12 Mobvista

Mobvista is a content publishing advertising network. By leveraging on Big Data and ML, Mobvista delivers real time data analysis, and customised mobile and game advertisements to targeted users (Mobvista 2013).

2.4 *Related Work*

Natural language communication in computer systems is considered as a fundamental principle of an effective AI (Russell and Norvig 2016). There have been a number of publications that have made a significant impact on the applications of NLG. Here we present a few.

2.4.1 Synthesis of Things

An application presented by Obaido et al. (Obaido et al. 2019) proposed the generation of SQL[5] query narratives using CFGs, based on predefined templates. This application was proved to be beneficial for novices in introductory programming and non-technical experts in learning the SQL language. Ade-Ibijola (Ade-Ibijola 2017), also presents a tool that generates hypothetical user profiles by employing probabilistic CFG using Social Media as a test case. These profiles are beneficial in anonymising sensitive data (Stevens 2012). Another application by Kumagai et al. (2016) employed Monte Carlo Tree Search (MCTS) for probabilistic natural language generation. This approach also employed context-grammar rules for search operations and evaluation of assumed text generations which focused only on the syntactic structure. After several experiments, it was confirmed that this approach was able to generate sentences with multiple words and different phrasing. Bauer and Koller (2010), implemented TAG[6] based grammars in the generation of sentences by combining sentence planing and surface realization similar to SPUD[7] systems. Ryan et al. (2016), also presented an in-game text generator using expressive free text markup and CFGs.

2.4.2 Generation of Problems and Solution

Other publications implement NLG within an educational setting in Introductory Programming. Ade-Ibijola (2019), proposed generation of practice python exercises and solutions for novice programmers using CFGs. He also proposed the generation of regular expression exercises and solutions for novice programmers (Ade-Ibijola 2018). These applications were further expanded by Kabaso and Ade-Ibijola (2020) on Social Media in the automatic generation of academic feedback and exercises on Twitter for programming students. Students received feedback and practice exercise and solution that were generated by Ade-Ibijola (2018, 2019).

[5]Structured Query Language.

[6]Tree-Adjoining Grammar.

[7]Sentence Planning Using Descriptions.

2.5 *Definition of Terms*

In this section we define some unfamiliar terms used in this chapter.

Definition 1 (*Advertising* (Alaimo and Kallinikos 2018)) The action of calling for attention or making someone aware of a particular commercial product or service.

Definition 2 (*Programmatic Advertising* (Busch 2016; Statler 2016)) The automated execution of advertising campaigns in real-time intelligently.

Definition 3 (*Context-free Grammar (CFG)* (Aho et al. 1986)) A CFG is a 4-tuple $G = (N, \Sigma, R, S)$, where:

1. N is a set of non-terminal variables.
2. Σ is an alphabet of terminal symbols where Σ is disjoint from N.
3. R is a set of rules, with the form $N \longrightarrow (N \cup \Sigma)$.
4. $S \in N$, it is the starting non-terminal variable.

3 Grammar Design for Advertisement Synthesis

In this section, we present the CFG rules for the generation of tweets. The terminals are represented as "*string literals*" and the non-terminals are delineated with <angle_brackets>. We first had to identify the differences, similarities, limitations and regulations of the social platforms on which we wish to broadcast these advertisements. Even though some platforms such as Instagram and YouTube, are limited to pictures and videos and others allow text and rich media, many social platforms have similar ideas on how people create and share content. Our test case focuses on two mainstream platforms: Facebook and Twitter. Both a Facebook post and a Tweet have a similar structure, allowing for content to be shared similarly. The difference is that Twitter only allows 280 characters (Twitter 2019), while Facebook gives the freedom of about 63206 characters including all spaces. In this paper, we define four main attributes of focus:

1. **text**: the short message that a user wants to deliver, which targets the context of a specified audience (i.e. <msg>),
2. **mention**: a Facebook username or Twitter handle preceded with the "at" (@) symbol. This attribute is added to directly inform a specific user, it is called mentioning (i.e. <mention>),
3. **url**: a web link that redirects the customer to the source of the advertisement (i.e. <url>), and
4. **hashtag**: is any wording appended with the hashtag (#) symbol, which is a metadata tag that indicates the topic or associated community/group; this makes it easy for users to find the message (i.e. <hashtag>). With mentioning and hashtagging, we can directly target specific groups or users.

Focusing on these similarities, with $\mathcal{L}(G)$, we start with the building blocks in Sect. 3.1.

3.1 Building Block

To begin, we start by defining similar CFG rules as Kabaso and Ade-Ibijola (2020). In Rule 1, we define an `<entity>` via the regular expression $[A-Za-z0-9\_]^+$, this would be any word presenting a Social Media *@username* or *#hashtag* in Rules 2 and 3. Rule 4 defines any other optional text outside the defined rules. Rule 6 defines `<url>` a web-link that redirects a user to an external website from the social platform. We also specify in Rule 7 any Unicode emoji visual representation of emotions, objects or symbols related to the context of a synthesised advert.

$$\texttt{<entity>} \longrightarrow [A-Za-z0-9\_]^+ \tag{1}$$

$$\texttt{<mention>} \longrightarrow @\texttt{<entity>} \tag{2}$$

$$\texttt{<hashtag>} \longrightarrow \#\texttt{<entity>} \tag{3}$$

$$\texttt{<opt\_txt>} \longrightarrow \texttt{<entity>} \mid \varepsilon \tag{4}$$

$$\texttt{<discount>} \longrightarrow [1-99]\,\% \tag{5}$$

$$\texttt{<url>} \longrightarrow \textit{usual web address string of maximally } 280 \textit{ characters} \tag{6}$$

$$\texttt{<emoji>} \longrightarrow \textit{unicode visual emoji} \tag{7}$$

3.2 Advert Synthesis

Rule 9 defines the content of the advert message, consisting of the period in which the promotion is limited, the name and price of the product, its category, discount, percentage and the end date of the spacial offer from the database. This rule can generate all possible permutations as defined in Σ including `<opt_txt>` optional text. When tested, this generated not less than 5040 possible unique combinations. In Rule 8 we define one or more hashtags on the advertisement.

$$\texttt{<tag>} \longrightarrow \texttt{<hashtag>}^+ \mid \varepsilon \tag{8}$$

$$\begin{aligned}\texttt{<msg>} \longrightarrow\ & \texttt{<opt\_txt><discount><opt\_txt><category><opt\_txt>} \\ & \texttt{<product\_name> <opt\_txt><period><opt\_txt>} \\ & \texttt{(<special>} \mid \varepsilon\texttt{)<opt\_txt><price><opt\_txt>} \\ & \texttt{<emoji><end\_date><tag>}\end{aligned} \tag{9}$$

$$\texttt{<category>} \longrightarrow \textit{category from database} \tag{10}$$

$$\texttt{<product\_name>} \longrightarrow \textit{product\_name from database} \tag{11}$$

$$\texttt{<special>} \longrightarrow \textit{special from database} \tag{12}$$

$$\texttt{<price>} \longrightarrow \textit{price from database} \tag{13}$$

$$\texttt{<end\_date>} \longrightarrow \textit{end\_date from database} \tag{14}$$

The period is the time in which the promotion is viable. This could be either the weather season, month, or public holiday in Rule 15.

$$\texttt{<period>} \longrightarrow \textit{season} \mid \textit{month} \mid \textit{public\_holiday} \mid \varepsilon \tag{15}$$

To derive an advert, first we consider the rules of the social platform on which it will be advertised. In our case, considering Facebook and Twitter. Both a tweet and a post have the same attributes. Even though this is true, a tweet can only contain 280 characters or less while Facebook has no such limitation in Rule 16.

$$\begin{aligned}\texttt{<advert>} \longrightarrow\ & \texttt{<msg>} \mid \texttt{<msg><url>} \mid \\ \longrightarrow\ & (\texttt{<mention><msg>} \mid \texttt{<msg><mention>})\texttt{<url>} \mid \\ \longrightarrow\ & \texttt{<url>}(\texttt{<mention><msg>} \mid \texttt{<msg><mention>}) \mid \\ \longrightarrow\ & \texttt{<msg>}(\texttt{<mention><url>} \mid \texttt{<url><mention>}) \mid \\ \longrightarrow\ & \texttt{<mention>}(\texttt{<msg><url>} \mid \texttt{<url><msg>}) \\ & \textit{whereby the entire advert must} \\ & \textit{not exceed } 280 \textit{ characters if it is a tweet}\end{aligned} \tag{16}$$

Example 1 (Derivation of an Advert) In this example, we derive an example of a product using the defined grammar rules.

$$\texttt{<advert>} \Longrightarrow \texttt{<msg><url>} \tag{17}$$

$$\begin{aligned}&\Longrightarrow \textit{Spend } \texttt{<discount><opt\_txt><category><opt\_txt> <product\_name>} \\ &\texttt{<opt\_txt><period><opt\_txt>(<special>} \mid \varepsilon\texttt{)<opt\_txt><price>} \\ &\texttt{<opt\_txt><emoji><end\_date><tag><url>}\end{aligned} \tag{18}$$

$$\begin{aligned}&\Longrightarrow \textit{Spend } 15\% \texttt{ <opt\_txt><category><opt\_txt> <product\_name>} \\ &\texttt{<opt\_txt><period><opt\_txt>(<special>} \mid \varepsilon\texttt{)<opt\_txt><price>} \\ &\texttt{<opt\_txt><emoji><end\_date><tag><url>}\end{aligned} \tag{19}$$

$$\begin{aligned}&\Longrightarrow \textit{Spend } 15\% \textit{ off on all } \texttt{<category><opt\_txt> <product\_name>} \\ &\texttt{<opt\_txt><period><opt\_txt>(<special>} \mid \varepsilon\texttt{)<opt\_txt><price>} \\ &\texttt{<opt\_txt><emoji><end\_date><tag><url>}\end{aligned} \tag{20}$$

$$\begin{aligned}&\Longrightarrow \textit{Spend } 15\% \textit{ off on all Grains/Cereal } \texttt{<opt\_txt> <product\_name>} \\ &\texttt{<opt\_txt><period><opt\_txt>(<special>} \mid \varepsilon\texttt{)<opt\_txt><price>} \\ &\texttt{<opt\_txt><emoji><end\_date><tag><url>}\end{aligned} \tag{21}$$

$$\begin{aligned}&\Longrightarrow \textit{Spend } 15\% \textit{ off on all Grains/Cereal products like } \texttt{<product\_name>} \\ &\texttt{<opt\_txt><period><opt\_txt>(<special>} \mid \varepsilon\texttt{)<opt\_txt><price>} \\ &\texttt{<opt\_txt><emoji><end\_date><tag><url>}\end{aligned} \tag{22}$$

$$\begin{aligned}&\Longrightarrow \textit{Spend } 15\% \textit{ off on all Grains/Cereal products like cornflakes} \\ &\textit{this}\texttt{<period><opt\_txt>(<special>} \mid \varepsilon\texttt{)<opt\_txt><price>}\end{aligned}$$

<opt_txt><emoji><end_date><tag><url> (23)

$\Longrightarrow$ *Spend* 15% *off on all Grains/Cereal products like cornflakes*
this (*season* | *month* | *public*$_h$*oliday* | ε)<opt_txt>
(<special> | ε)<opt_txt><price>
<opt_txt><emoji><end_date><tag><url> (24)

$\Longrightarrow$ *Spend* 15% *off on all Grains/Cereal products like cornflakes*
this winter on (<special> | ε)<opt_txt><price>
<opt_txt><emoji><end_date><tag><url> (25)

$\Longrightarrow$ *Spend* 15% *off on all Grains/Cereal products like cornflakes*
this winter on Big Deals <opt_txt><price>
<opt_txt><emoji><end_date><tag><url> (26)

$\Longrightarrow$ *Spend* 15% *off on all Grains/Cereal products like cornflakes*
this winter on Big Deals from<price>
<opt_txt><emoji><end_date><tag><url> (27)

$\Longrightarrow$ *Spend* 15% *off on all Grains/Cereal products like cornflakes*
*this winter on Big Deals from R*25
<opt_txt><emoji><end_date><tag><url> (28)

$\Longrightarrow$ *Spend* 15% *off on all Grains/Cereal products like cornflakes*
*this winter on Big Deals from R*25
till the<emoji><end_date><tag><url> (29)

$\Longrightarrow$ *Spend* 15% *off on all Grains/Cereal products like cornflakes*
*this winter on Big Deals from R*25
till the 21*st of June*<tag><url> (30)

$\Longrightarrow$ *Spend* 15% *off on all Grains/Cereal products like cornflakes*
*this winter on Big Deals from R*25
till the 21*st of June #SpecialWinter* <url> (31)

$\Longrightarrow$ *Spend* 15% *off on all Grains/Cereal products like cornflakes*
*this winter on Big Deals from R*25
till the 21*st of June #SpecialWinter https : //urlzs.com/5zSKV* (32)

In Example 1, we derive a simple advert. You can notice that the winter season for <period> was been chosen and <emoji> was left out as optional.

	CategoryName	ProductName	QuantityPerUnit	UnitPrice	Discount	Special	EndDate
1	Beverages	Apples	10 per Bag	19.00	5	Special Giveaways	2019-09-24
2	Condiments	Original Frankfurter grüne Soße	12 boxes	13.00	25	Big Deals	2019-10-25
3	Beverages	Lakkalikööri	500 ml	18.00	10	Special Giveaways	2019-09-24
4	Beverages	Rhönbräu Klosterbier	24 - 0.5 l bottles	7.75	10	Special Giveaways	2019-09-24
5	Produce	Longlife Tofu	5 kg pkg.	10.00	30	Specail Season	2019-01-04
6	Seafood	Röd Kaviar	24 - 150 g jars	15.00	30	Give Back Weekends	2019-06-01
7	Dairy Products	Mozzarella di Giovanni	24 - 200 g pkgs.	34.80	35	Week Specials	2019-01-04
8	Dairy Products	Flotemysost	10 - 500 g pkgs.	21.50	35	Week Specials	2019-01-04
9	Beverages	Outback Lager	24 - 355 ml bottles	15.00	10	Special Giveaways	2019-09-24
10	Dairy Products	Gudbrandsdalsost	10 kg pkg.	36.00	35	Week Specials	2019-01-04
11	Confections	Scottish Longbreads	10 boxes x 8 pieces	12.50	15	Payless Special	2019-01-04
12	Beverages	Laughing Lumberjack Lager	24 - 12 oz bottles	14.00	10	Special Giveaways	2019-09-24
13	Condiments	Louisiana Hot Spiced Okra	24 - 8 oz jars	17.00	25	Big Deals	2019-10-25
14	Condiments	Louisiana Fiery Hot Pepper S...	32 - 8 oz bottles	21.05	25	Big Deals	2019-10-25
15	Grains/Cereals	Wimmers gute Semmelknödel	20 bags x 4 pieces	33.25	30	Big Deals	2019-08-11
16	Condiments	Vegie-spread	15 - 625 g jars	43.90	25	Big Deals	2019-10-25

Fig. 2 Snapshot of products database (Microsoft 2019)

4 Implementation and Results

4.1 How It Works

This chapter has discussed the implementation of a tool called `Sell-bot`. `Sell-bot` is a Social Media advertisement web application on the .Net platform. By implementing similar techniques as Kabaso and Ade-Ibijola (2020), it synthesises advertisements on discounted products from an inventory database.

Our test case uses Microsoft's sample Northwind databases (Microsoft 2019), focusing mainly on the products and category entities. We then added the necessary additional columns for the special name,[8] product discount percentage and the end date of the promotion. In order to advertise, it is necessary that we have a specified discount and end date for each product. With this, we can easily return only the required product rows using SQL queries shown in Fig. 2.

4.2 Advert Synthesis

Figure 3 represents the process of how an advert is generated and posted. To ensure that the generated text advertisements are dynamic and narrative, we implemented the Google Cloud Calendar API within the engine (Google 2019). The calendar API determines the current month and public holidays depending on the geographic location where the advert is generated. We then synthesise advertisements about products that would relate to that specific period. For instance, synthesising adverts about cold beverages in the summer season. With #hashtags, we can also target

[8]The special name is the theme of the promotion, this could be related to an event or weather season.

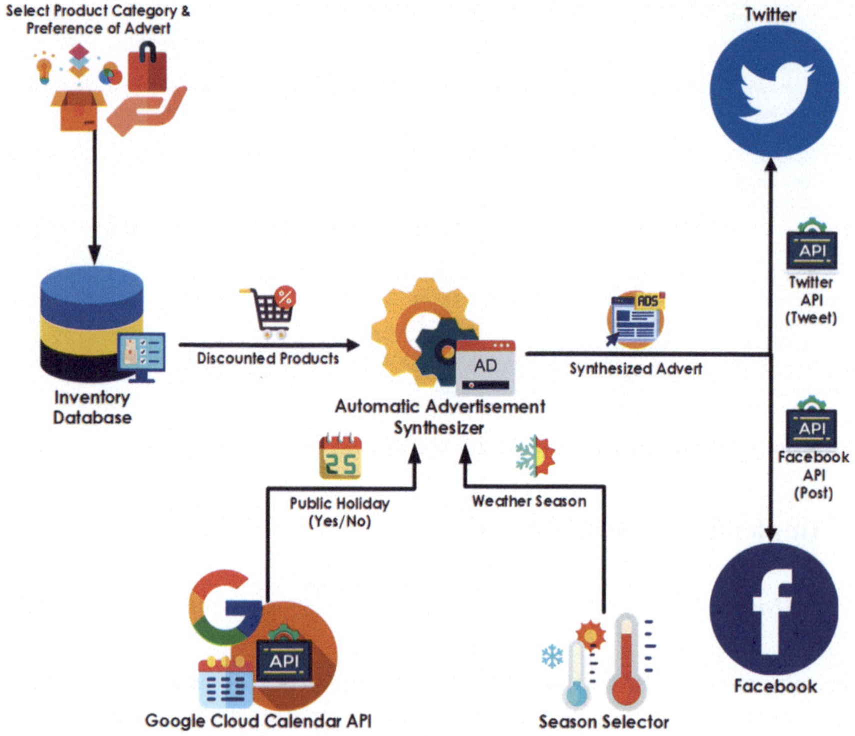

Fig. 3 Processes of advertisement generation

specific groups and communities. The generated adverts are then posted to popular Social Media platforms such as Facebook and Twitter through their relevant APIs.

4.3 *Using Sell-Bot*

To post an advert we need to be logged into the `Sell-Bot` web application. This gives us the option to add a new product or send out synthesised advertisements automatically or manually. Figures 4 and 5 shows screen-shots of the advertisement and add new product options. Automatic adverts are sent out depending on the season and time period by selecting products that are suitable for the weather. To manually broadcast adverts, we have to select the category of the products, the date on which the first advert will be sent out, the time interval between each advert and the number of adverts. Figure 6a shows a generated advert.

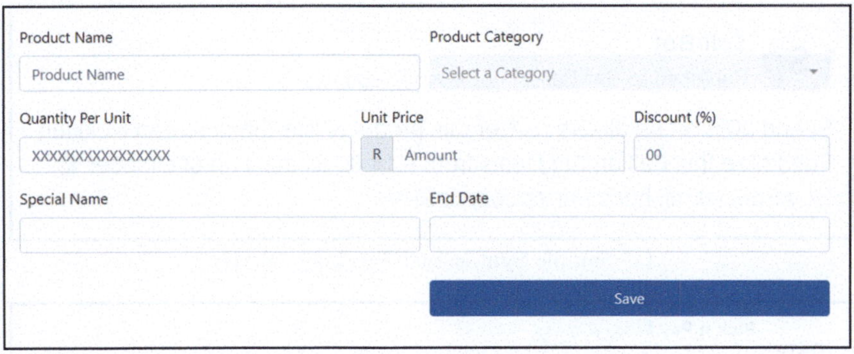

Fig. 4 Adding a new product to the database

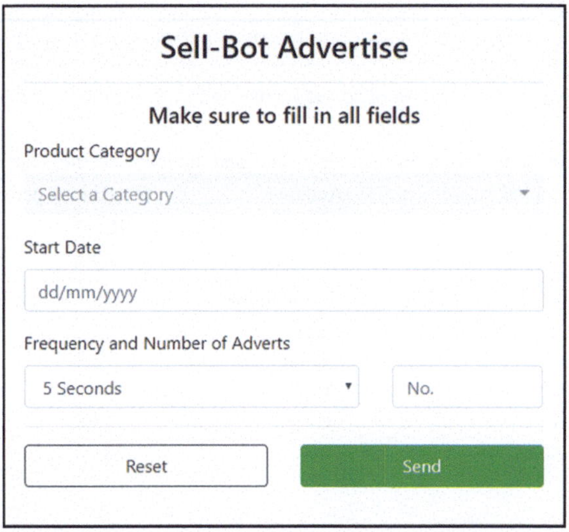

Fig. 5 Sending advertisements using `Sell-Bot`

5 Evaluation of Sell-Bot

5.1 Social Media Users' Perception

In this section, we present the results of an online survey that evaluates `Sell-Bot`. We have surveyed the perception of Social Media users on automatically generated advertisements. The survey was conducted in Johannesburg, South Africa. A total of 319 individuals participated in the survey. The participants had a wide range of qualifications, ranging from diplomas to doctorates, with 68.8% having diplomas and 2.8% having doctorates. 75.2% of the 319 participants stated that they had digital marking experience. The majority (98.7%) of all participants stated that they were frequently exposed to Social Media advertisements. When the participants were

(a) Sample Synthesised Facebook Advert

(b) Advertisement by Pick n Pay Stores on Twitter 16 May 2019 [88]

Fig. 6 Comparison of advert synthesised by `Sell-Bot` and digital marketing expert

Table 1 Turing test of `Sell-Bot` and digital marketing expert

Ad No	Author	Human perception (%)	
		Right	Wrong/Neutral
1	Sell-Bot	63.3	36.7
2	Digital marketing expert	22.3	77.7
3	Digital marketing expert	72.4	27.6
4	Sell-bot	73.4	26.6
Sum		231.4	168.6

asked if they knew any automated advertising software, 72.4% claimed *Yes*, while 27.6% stated *No*.

Throughout the survey, participants were shown some adverts synthesised by `Sell-Bot` and others created by a Digital Marketing Expert from Pick n Pay Stores to identify which advert was created by `Sell-Bot` or a human without them having any prior knowledge about who, what or where the advert was taken from (see Fig. 6). The arrangement of the adverts was randomised to ensure the participants did not follow a specific sequence of answering the survey.

In one of the cases, 77.7% of the users were unable to tell the difference between a `Sell-Bot` advertisement and a Digital Marketing Expert. Table 1 shows a Turing Test of the results in (see Fig. 7). Table 1 also presents the average of when

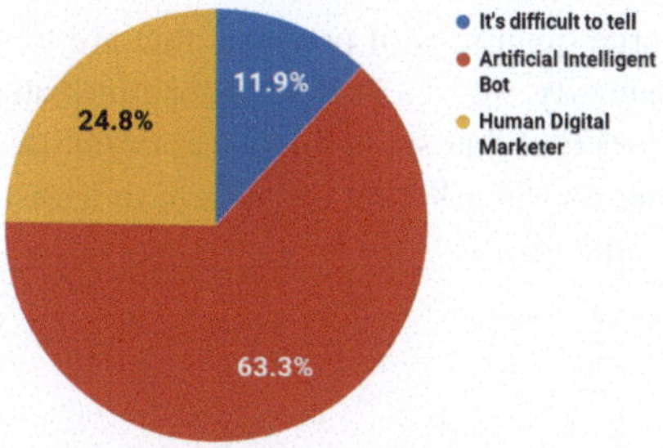

(a) Advert Generatedby `Sell-Bot`

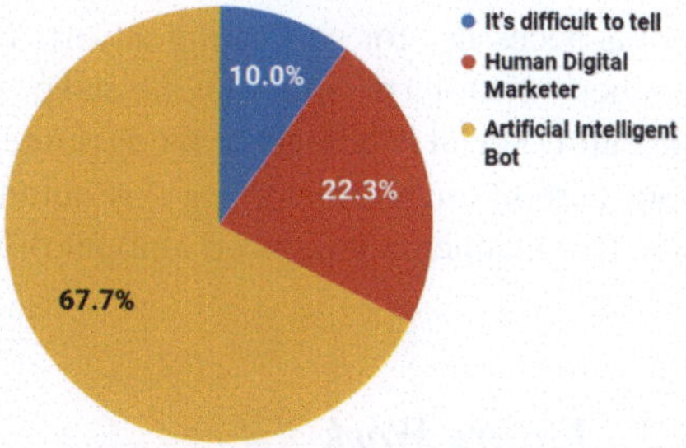

(b) Advert Created by Digital Marketing Expert

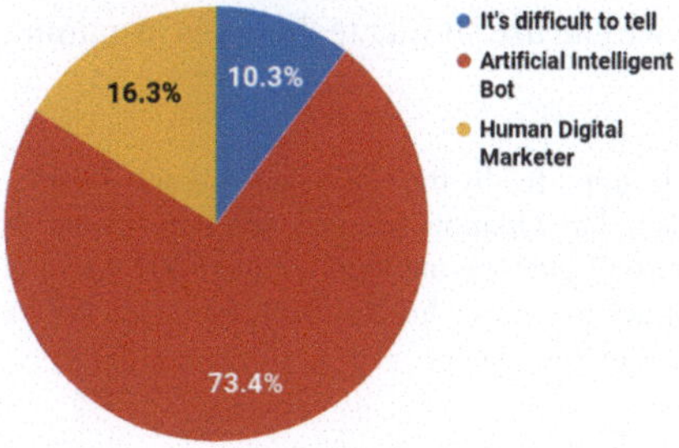

(c) Advert Generated by `Sell-Bot`

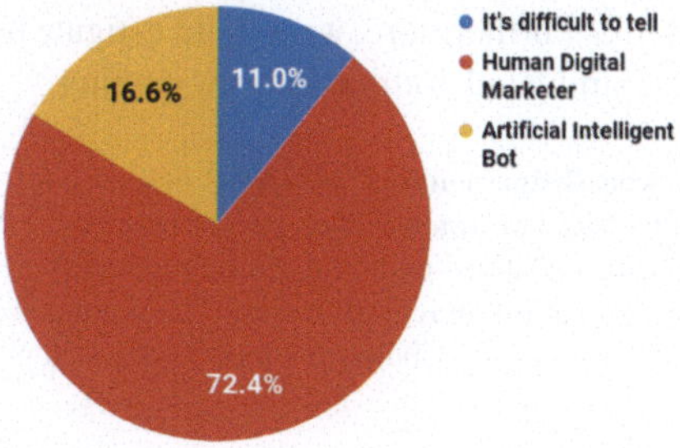

(d) Advert Created by Digital Marketing Expert

Fig. 7 Survey: `Sell-Bot` or human

the participants chose the correct option between `Sell-Bot` or Digital Marketing Expert. An average of $\frac{\sum x}{4} = \frac{168}{4} = 42.2\%$ participates were wrong about who created the advert while $\frac{\sum y}{4} = \frac{231.4}{4} = 57.8\%$ chose the right option.

6 Conclusion and Future Work

6.1 Conclusion

In this chapter, we have discussed Social Media as an advertising platform. Leveraging on its network capabilities, personal extensibility and information demonstrability, business can effectively, and flexibly reach their customers. We identified and compared some of the current techniques of programmatic advertising in real-time on platforms like Google, Facebook, LinkedIn and Twitter. We have also presented a Social Media advertising tool called `Sell-Bot` based of CFG's and Natural Language Generation illustrating a method for automated advertising of products on Social Media. In our evaluation, 95% of all the participants, stated that `Sell-Bot` would be a useful tool for advertising in the Fourth Industrial Revolution. Such a tool is necessary for generating adverts for large quantities of products that are too cumbersome for a person to create and post manually. `Sell-Bot` does not eliminate the entire Social Media advertising strategy, it is tool that supplements an organisations current marketing strategy by automating their marketing ecosystem in a cost effective manner at a reduced amount of time and effort.

6.2 Future Work

In the future, we will explore the systems of rich media advertisements of Social Media. Furthermore, we will investigate how we can use similar techniques in gaming and simulated commentary in gaming.

Acknowledgements This work is based on research supported by the National Research Foundation (NRF) of South Africa (Grant Number: 119041). Any Opinion, findings and conclusions or recommendations expressed in this material are those of authors and therefor the NRF does not accept liability in regard thereto. The second author acknowledges his Research Assistant, Nikita Patel, for her contribution in drawing the rich pictures in this chapter.

References

Adams T (2017) Ai-powered social bots. CoRR. http://arxiv.org/abs/1706.05143

Adams R (2004) Intelligent advertising. AI Soc 18(1):68–81

Ade-Ibijola A (2017) Synthesis of social media profiles using a probabilistic context-free grammar. In: 2017 pattern recognition association of South Africa and robotics and mechatronics (PRASA-RobMech), pp 104–109

Ade-Ibijola A (2018) Synthesis of regular expression problems and solutions. Int J Comput Appl 1–17

Ade-Ibijola A (2019) Syntactic generation of practice novice programs in python. In: Kabanda S, Suleman H, Gruner S (eds) ICT Education. Springer International Publishing, Cham, pp 158–172

Adobe (2018) Adobe marketing cloud. http://www.adobe.com/solutions/digital-marketing.html. Accessed: 2019-07-01

AdRoll (2007) One growth platform, multiple solutions. https://www.adroll.com/platform. Accessed: 2019-07-02

Agarwal D (2013) Computational advertising: the linkedin way. In: Proceedings of the 22nd ACM international conference on conference on information & knowledge management, CIKM '13, New York, NY, USA. ACM, pp 1585–1586

Aho AV, Sethi R, Ullman JD (1986) Compilers, principles, techniques. Addison Wesley 7(8):9

Alaimo C, Kallinikos J (2018) Objects, metrics and practices: an inquiry into the programmatic advertising ecosystem. In: Schultze U, Aanestad M, Mähring M, Østerlund C, Riemer K (eds) Living with monsters? social implications of algorithmic phenomena, hybrid agency, and the performativity of technology. Springer International Publishing, Springer, Cham, pp 110–123

AlSabeeh DA, Moghrabi IA (2017) Programmatic advertisement and real time bidding utilization. In: Kar AK, Ilavarasan PV, Gupta MP, Dwivedi YK, Mäntymäki M, Janssen M, Simintiras A, Al-Sharhan S (eds) Conference on e-Business, e-Services and e-Society. Springer International Publishing, Cham, pp 289–297

Amado A, Cortez P, Rita P, Moro S (2018) Research trends on big data in marketing: a text mining and topic modeling based literature analysis. Eur Res Manag Bus Econ 24(1):1–7

Bailey C, Baines PR, Wilson H, Clark M (2009) Segmentation and customer insight in contemporary services marketing practice: why grouping customers is no longer enough. J Mark Manag 25(3-4):227–252

Barreto AM (2013) Do users look at banner ads on facebook? J Res Interact Mark 7(2):119–139

Bauer D, Koller A (2010) Sentence generation as planning with probabilistic LTAG. In: Proceedings of the 10th international workshop on tree adjoining grammar and related frameworks (TAG+10), pp. 127–134

Berthon PR, Pitt LF, Plangger K, Shapiro D (2012) Marketing meets web 2.0, social media, and creative consumers: Implications for international marketing strategy. Bus Horiz 55(3):261–271

Boateng H, Okoe AF (2015) Consumers attitude towards social media advertising and their behavioural response: the moderating role of corporate reputation. J Res Interact Mark 9(4):299–312

Briggs R, Hollis N (1997) Advertising on the web: is there response before click-through? J Advert Res 37(2):33–46

Busch O (2016) The programmatic advertising principle. In: Programmatic advertising. Springer, Berlin, pp 3–15

Cambria E, Schuller B, Xia Y, Havasi C (2013) New avenues in opinion mining and sentiment analysis. IEEE Intell Syst 28(2):15–21

Chickering DA, Heckerman D (2000) A decision theoretic approach to targeted advertising. In: Proceedings of the sixteenth conference on uncertainty in artificial intelligence. Morgan Kaufmann Publishers Inc., pp 82–88

Chun R, Da Silva R, Davies G, Roper S (2005) Corporate reputation and competitiveness. Routledge

Dawson P, Lamb M (2016) Enhanced success with programmatic social advertising. Springer International Publishing, Cham, pp 103–110

Dehghani M, Tumer M (2015) A research on effectiveness of facebook advertising on enhancing purchase intention of consumers. Comput Hum Behav 49:597–600

Duffett RG (2015) Facebook advertisings influence on intention-to-purchase and purchase amongst millennials. Internet Res 25(4):498–526

Erevelles S, Fukawa N, Swayne L (2016) Big data consumer analytics and the transformation of marketing. J Bus Res 69(2):897–904

Erkan I, Gokerik M, Acikgoz F (2019) The impacts of facebook ads on brand image, brand awareness, and brand equity. In: Handbook of research on entrepreneurship and marketing for global reach in the digital economy. IGI Global, pp 442–462

Evans NJ, Phua J, Lim J, Jun H (2017) Disclosing instagram influencer advertising: the effects of disclosure language on advertising recognition, attitudes, and behavioral intent. J Interact Advert 17(2):138–149

Fan S, Lau RY, Zhao JL (2015) Demystifying big data analytics for business intelligence through the lens of marketing mix. Big Data Res 2(1):28–32

Feng J, Wei J (2010) System and method for deep targeting advertisement based on social behaviors, 14 Oct 2010. US Patent App. 12/420,638

Funk T (2013) Advertising on twitter and other networks. Springer, Berkeley, CA, pp 103–124

Galanis D, Androutsopoulos I (2007) Generating multilingual descriptions from linguistically annotated owl ontologies: the NaturalOWL system. In: Proceedings of the eleventh European workshop on natural language generation. Saarbrücken, Germany. Association for Computational Linguistics, DFKI GmbH, pp 143–146

Gatt A, Reiter E (2009) Simplenlg: a realisation engine for practical applications. In: Proceedings of the 12th European workshop on natural language generation (ENLG 2009), pp 90–93

Gazagne G, Gösswein A (2016) Redefining retargeting. Springer International Publishing, Cham, pp 233–242

Goldberg E, Driedger N, Kittredge RI (1994) Using natural-language processing to produce weather forecasts. IEEE Expert 9(2):45–53

Google (2000) About the google display network. https://support.google.com/google-ads/answer/2404190?hl=en. Accessed: 2019-04=20
Google (2019) Google calendar API. https://developers.google.com/calendar/. Accessed: 2019-04-20
Harris L, Rae A (2009) Social networks: the future of marketing for small business. J Bus Strat 30(5):24–31
Hedin H, Irmeli H, Vaarnas M (2011) The handbook of market intelligence: understand, compete and grow in global markets. Wiley
Jiang L, Cheng Y, Yang L, Li J, Yan H, Wang X (2019) A trust-based collaborative filtering algorithm for e-commerce recommendation system. J Ambient Intell HumIzed Comput 10(8):3023–3034
Kabaso S, Ade-Ibijola A (2020) Synthesis of social media messages and tweets as feedback medium in introductory programming. In: Tait B, Kroeze J, Gruner S (eds) ICT Education, vol 1136. Springer International Publishing, Cham, pp 3–16
Kaplan AM, Haenlein M (2012) Social media: back to the roots and back to the future. J Syst Inf Technol 14(2):101–104
Kietzmann J, Paschen J, Treen E (2018) Artificial intelligence in advertising: how marketers can leverage artificial intelligence along the consumer journey. J Advert Res 58(3):263–267
Kumagai K, Kobayashi I, Mochihashi D, Asoh H, Nakamura T, Nagai T (2016) Human-like natural language generation using Monte Carlo Tree Search. In: Proceedings of the INLG 2016 workshop on computational creativity in natural language generation. Association for Computational Linguistics, Edinburgh, UK, pp 11–18
Lambrecht A, Tucker C (2013) When does retargeting work? information specificity in online advertising. J Mark Res 50(5):561–576
Lambrecht A, Tucker C (2011) When does retargeting work?: timing information specificity. Citesee
Langheinrich M, Nakamura A, Abe N, Kamba T, Koseki Y (1999) Unintrusive customization techniques for web advertising. Comput Netw 31(11–16):1259–1272
Leake DB, Gary J (2008) AI magazine poster: the AI landscape. AI Mag 29(2):3–3
Lee D, Kim HS, Kim JK (2011) The impact of online brand community type on consumer's community engagement behaviors: Consumer-created versus marketer-created online brand community in online social-networking web sites. Cyberpsychology, Behav, Soc Netw 14(1-2):59–63
Leung XY, Bai B, Stahura KA (2015) The marketing effectiveness of social media in the hotel industry: a comparison of facebook and twitter. J Hosp Tour Res 39(2):147–169
Luger GF (2005) Artificial intelligence: structures and strategies for complex problem solving. Pearson education
Madsack A, Cavallo A, Heininger J, Weißgraeber R (2018) AX Semantics' submission to the conll-sigmorphon 2018 shared task. In: Proceedings of the CoNLL SIGMORPHON 2018 shared task: Universal Morphological Reinflection, pp 43–47
Manolescu D, Laghaeian R (2010) Social media driven advertisement targeting, 4 Feb 2010. US Patent App. 12/184,254
McAfee A, Brynjolfsson E, Davenport TH, Patil DJ, Barton D (2012) Big data: the management revolution. Harv Bus Rev 90(10):60–68
MediaKix (2016) How do instagram influencers make money? http://mediakix.com/2016/03/instagram-influencersmaking-money/#gs.null. Accessed: 2019-11-30
Microsoft (2019) Get the sample databases for ado.net code samples. https://docs.microsoft.com/en-us/dotnet/framework/data/adonet/sql/linq/downloading-sample-databases. Accessed: 2019-03-15
Mobvista (2013) https://www.mobvista.com/en/. Accessed: 2019-07-02
Moghrabi IAR, Al-Mohammed AR (2016) Social media or social business networks? In: Dwivedi YK, Mäntymäki M, Ravishankar MN, Janssen M, Clement M, Slade EL, Rana NP, Al-Sharhan S, Simintiras AC (eds) Social media: the good, the bad, and the ugly. Springer International Publishing, Cham, pp 42–52
Mulhern F (2009) Integrated marketing communications: from media channels to digital connectivity. J Mark Commun 15(2–3):85–101

Nasukawa T, Yi J (2003) Sentiment analysis: capturing favorability using natural language processing. In: Proceedings of the 2nd international conference on Knowledge capture. ACM, pp. 70–77

Obaido G, Ade-Ibijola A, Vadapalli H (2019) Generating narrations of nested SQL queries using context-free grammars. In: 2019 conference on information communications technology and society (ICTAS). IEEE, pp 1–6

Okazaki S, Taylor CR (2013) Social media and international advertising: theoretical challenges and future directions. Int Mark Rev 30(1):56–71

Otero EL, Gallego PAM, Pratt RME (2014) Click-and-mortar SMEs: attracting customers to your website. Bus Horiz 57(6):729–736

Otero EL, Gutiérrez RC (2015) Using social media advertising to increase the awareness, promotion and diffusion of public and private entities. In: Distributed computing and artificial intelligence, 12th international conference. Cham, Springer, pp 377–384

Pick n Pay Stores (2019) Advert. https://twitter.com/PicknPay/status/1129272839410978816. Accessed: 2019-05-15

Ramos-Soto A, Bugarín A, Barro S (2016) On the role of linguistic descriptions of data in the building of natural language generation systems. Fuzzy Sets Syst 285:31–51

Reddy DR (1976) Speech recognition by machine: a review. Proc IEEE 64(4):501–531

Russell SJ, Norvig P. (2016) Artificial intelligence: a modern approach. Pearson Education Limited

Ryan J, Seither E, Mateas M, Wardrip-Fruin N (2016) Expressionist: an authoring tool for in-game text generation. In: Nack F, Gordon AS (eds) International conference on interactive digital storytelling. Springer International Publishing, Cham, pp 221–233

Sajid SI (2016) Social media and its role in marketing

Saxena A, Khanna U (2013) Advertising on social network sites: a structural equation modelling approach. Vision 17(1):17–25

Seitz J, Zorn S (2016) Perspectives of programmatic advertising. Springer International Publishing, Cham, pp 37–51

Smock AD, Ellison NB, Lampe C, Wohn DY (2011) Facebook as a toolkit: a uses and gratification approach to unbundling feature use. Comput Hum Behav 27(6):2322–2329

Statistica (2018) Number of active monthly instagram users from january 2013 to september 2018. https://www.statista.com/statistics/253577/number-of-monthly-active-instagram-users/. Accessed: 30 Nov 2019

Statler S (2016) A programmatic advertising primer. Apress, Berkeley, CA, pp 255–264

Stelzner MA (2014) How marketers are using social media to grow their businesses. Soc Media Mark Ind Rep

Stevens GM (2012) Data security breach notification laws

Subhadra V, Urvashi K, Pranav F, Tarjni V (2018) Google adwords: a window into the google display network. In: Information and communication technology for sustainable development. Singapore, Springer, pp 345–355

Sun Z, Dawande M, Janakiraman G, Mookerjee VS (2016) The making of a good impression: information hiding in ad exchanges. MIS Q 40(3):717–739

Sundsøy P, Bjelland J, Iqbal AM, de Montjoye YA, et al. (2014) Big data-driven marketing: how machine learning outperforms marketers' gut-feeling. In: Kennedy WG, Agarwal N, Yang SJ (eds) Social computing, behavioral-cultural modeling and prediction. Springer International Publishing, Cham, pp. 367–374

Tan W, Blake MB, Saleh I, Dustdar S (2013) Social-network-sourced big data analytics. IEEE Internet Comput (5):62–69

Twitter (2019) Documentation. https://developer.twitter.com/en/docs. Accessed: 2019-03-05

Twitter (2019) How twitter ads work. https://business.twitter.com/en/help/troubleshooting/how-twitter-ads-work.html. Accessed: 2019-03-05

Van der Kaa H, Krahmer E (2014) Journalist versus news consumer: the perceived credibility of machine written news. In: Proceedings of the computation+ journalism conference, Columbia university, New York, vol 24. New York, p 25

Vanden Bergh BG, Lee M, Quilliam ET, Hove T (2011) The multidimensional nature and brand impact of user-generated ad parodies in social media. Int J Advert 30(1):103–131

Wierenga B (2010) Marketing and artificial intelligence: great opportunities, reluctant partners. In: Marketing intelligent systems using soft computing. Springer, Berlin, pp 1–8

Wright A (2015) Algorithmic authors. Commun ACM 58(11):12–14

Yaakop A, Anuar MM, Omar K (2013) Like it or not: issue of credibility in facebook advertising. Asian Soc Sci 9(3):154

Advancements and Role of Emotion Recognition in the 4th Industrial Revolution

Tanmoy Roy, Tshilidzi Marwala, and Snehashish Chakraverty

Abstract Emotions are ubiquitous to human life. Previous studies suggest that ninety percent of the life experience of humankind is affected by at least one emotion, and that emotional state guides their thoughts and behavior. The neurological study suggests that the design of human brain ensures that the *emotional brain* almost always influences the *rational brain*. Which means every rational decision is affected by the underlying emotional state. For decades, the influence of emotional states on human life has been overlooked by the AI researchers until recently. For the last twenty years, the study on different aspects of emotions and its detection for the more significant cause is gaining considerable attention and importance. With the advent of the Fourth Industrial Revolution (4IR), advancements in Artificial Intelligence (AI) is already touching new heights and in this context successful detection of emotional states by AI systems will be another big leap towards more realistic social AI agents. This chapter will discuss in detail the importance and recent technological advancements towards emotion recognition (ER) in the backdrop of the Fourth Industrial Revolution. ER uses different types of human physical and behavioral expressions, such as facial features, speech features, and their variants. Deep Learning (DL) has emerged as one of the prominent pillars of 4IR. ER task extensively uses DL techniques for classifying different emotions. So, this chapter includes a discussion of relevant technical aspects of DL. Finally, it includes a brief description of a newly proposed Speech Emotion Recognition (SER) model as a case study of the ER systems.

T. Roy (✉) · T. Marwala
University of Johannesburg Electrical and Electronic Engineering, Johannesburg, South Africa
e-mail: tanmoy@tanmoy.in

T. Marwala
e-mail: tmarwala@gmail.com

S. Chakraverty
Department of Mathematics, National Institute of Technology, Rourkela, India
e-mail: sne_chak@yahoo.com

W. Doorsamy et al. (eds.), *The Disruptive Fourth Industrial Revolution*,
Lecture Notes in Electrical Engineering 674,
https://doi.org/10.1007/978-3-030-48230-5_8

Keywords Emotion recognition · 4th Industrial revolution · Deep learning · Feature extraction

1 Introduction

Present Artificial Intelligence (AI) systems' unique characteristic is their ability to execute spoken conversations with humans efficiently. Examples like Siri, Cortana, and Alexa are the technical marvels which not only can hear us comfortably, but they can reply to us with equal comfort as well. However, are these spoken conversations with AI agents natural enough? To be specific, are these AI agents intelligent enough to read our moods or feelings? The answer is still 'NO'. AI agents still cannot read emotions expressed by humans. The first significant work on *emotion detection* was reported way back in 1978 by Williamson (1978) but the challenge to detect emotions from the speech in realtime is still on.

Industrial Revolution refers to the abrupt socio-economic change that takes place in human lives across the national boundaries due to the successful implementation of significant innovations and technical advancements into forms usable by humans. It is prominent from the earlier three industrial revolutions that innovations and inventions are the driving forces behind those revolutions. The present industrial revolution called the *4th industrial revolution* (4IR) is also based on the innovations in the field of *artificial intelligence* (AI), robotics, the internet of things (IoT), nanotechnology, biotechnology, and quantum computing to name a few. A profound shift is being observed in all industries with these innovations (Schwab 2016). The technological advancements taking place in the present revolution is unprecedented, mainly because of two reasons: (1) technological advancements now have a much wider reach, the digital platform has penetrated almost every household across the world to deliver the advancements taking place in different sectors, and (2) technology is changing very rapidly for good. Initiation of the 4IR already having a remarkable impact on the social and political practices. For example, the rise of social media has changed some critical aspects of society, like the way of communication, political awareness, and information availability are few of them. Moreover, the confluence of data with massive computational storage and cognitive power is transforming the industry and society at every level, creating opportunities that were once unimaginable from health and education to agriculture, manufacturing, and services (Schwab et al. 2018). One crucial development in 4IR is the advancements of *machine learning* techniques, specifically the *deep learning* (DL) models. DL models are providing state-of-the-art performance to some of the toughest technical challenges humankind is facing like *automatic speech recognition, computer vision, self-driving cars*.

Emotions are ubiquitous to human life, and a study shows that 90% time of our everyday life, we feel at least one emotion (Trampe et al. 2015). Once we experience any emotion, it guides our thoughts and behavior (Clore and Huntsinger 2007), to communicate those thoughts, and we express them in different forms, such as facial expression, speech utterances, and physical movements. Researchers are trying to detect the emotional state from speech and facial expressions. Another input the researchers are using is the brain signals of the subject during the emotional states. A detailed discussion is dedicated to describing the features of emotion classification. So, emotion recognition is a *machine learning* (ML) problem where emotional states are classified based on the features describing those emotional states. The DL techniques are being applied to solve the emotion detection problem.

Successful emotion detection will have an impact on both the industry as well as society. Emotional state detection can revolutionize the following fields: (1) the health care and medical facilities (France et al. 2000; Picard 1997; Schelinski and von Kriegstein 2019; Shu et al. 2018); (2) the human machine interfacing systems, social security and the AI systems (Dellaert et al. 1996; Schuller et al. 2004; Shu et al. 2018); (3) and the entertainment industry (Charles et al. 2009; Hansen and Cairns 1995; Nakatsu et al. 1999; Schuller et al. 2004). The human-machine interaction will attain a new high with the advent of sensitive, sociable robots (Cowie et al. 2001). These advanced AI systems can play the role of vital agents of social change by building a more natural relationship with human beings. An artificial caregiver can become a very loyal companion to persons in need.

2 Emotion: The Hidden Force Which Drives Us

Recent researches suggest that only ten percent of the life experience of humankind is devoid of any emotional experience, and when an emotion is experienced, it guides people's thoughts and behavior. Darwin (1948) was one of the first to make the case that the study of emotional expressions is beneficial and also discussed the effects of emotions. However, only in the 1990s, this study is recognized as an essential aspect of any study related to humankind. Subsequent studies revealed that emotional states have diverse effects on our everyday life (Trampe et al. 2015). From the very first day of our lives, different emotions play a significant role in our survival and further progress, which researchers such as Freedman et al. (1967), Frank (1988), Damasio (1994) explored in detail.

2.1 How Emotion Influences Our Every Move?

The human brain controls every move we make from blinking of eyes to applying breaks while driving a car. Every external or internal stimulation (also called interrupt) has to be managed by the brain. The interrupt could be an audio-visual input, a change in the surrounding environment, an occurrence of a disease, or any other event. The modern human brain came to the present stage through different phases of evolution and is having a layered structure. The *reptilian brain* (Fig. 1), the oldest layer (Massey 2002), carry out the instinctive and autonomic functions such as breathing, heartbeat and sucking reflex. The *amygdala*, the critical component of the next layer called the emotional brain or *limbic system*, handles the emotional stimuli. The outermost layer is *neocortex*, also known as the *rational* brain, consisting of a

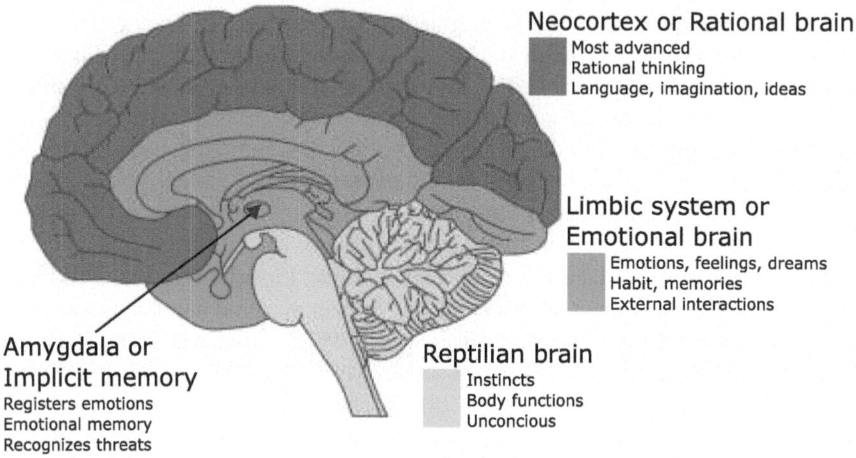

Fig. 1 Three primary layers of the human brain

layer of gray matter. The *neocortex* is responsible for conscious processing of sensory stimuli. Figure 1 shows three primary layers of the human brain and the location of the amygdala along with the mention of the major functionalities of them in brief.

Now the question is why human emotional states influence every move they make? The internal structure of the human brain has the most vital clue.

- The *limbic system* or *emotional* brain and the *neocortex* or the *rational* brain work in collaboration to respond to any incoming stimulation. Both the brains are interconnected, and the connection between the amygdala and cortex contains a much larger number of neurons than in reverse direction, from the cortex to the amygdala (Carter 1998; LeDoux 1996; Panksepp 1998). As a result, there is a subconcious transfer of emotional state influence from the emotional brain to the rational brain. The emotional brain receives the stimuli before the rational brain, and due to a much stronger neural connection from emotional to rational brain, the collective response gets significantly affected by the underlying emotional state (Fig. 2).
- Moreover, the timing of neural signal's arrival in the emotional and rational brains is different; the emotional brain receives it first and gets activated relatively early.

These two structural arrangements of the human brain explain the influence of emotional states in our every move.

3 Emotion Recognition System

Human emotions are reflected through various modes like the choice of words in spoken utterances, voice tone, facial expression, physical gesture, skin temperature, respiration rate, heartbeat, muscle tension (Picard et al. 2001). Combination of multi-

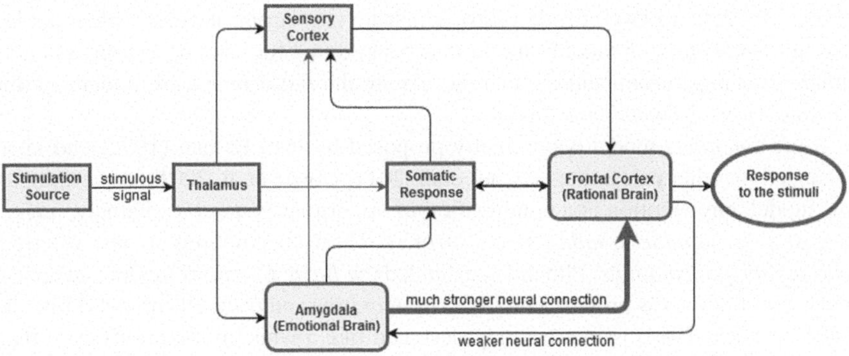

Fig. 2 Flow diagram to show the processing of a stimulus in the human brain

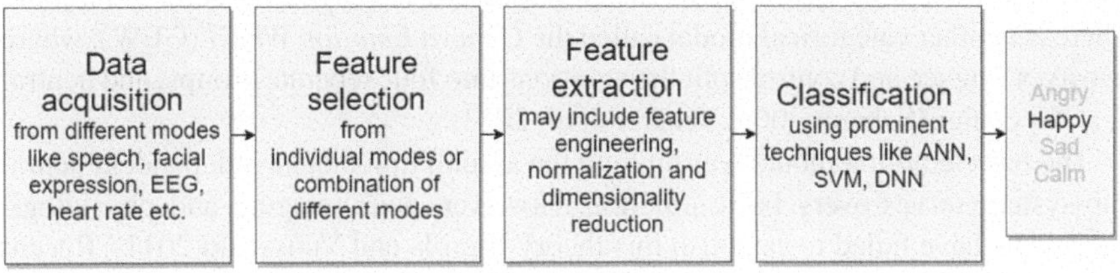

Fig. 3 Block diagram of emotion recognition system

ple modes can improve the emotion recognition accuracy rate. However, it should be mentioned here that human beings cannot accurately recognize emotions, and quite often fail fail to identify their own emotional states Picard et al. (2001). So, expecting a machine to identify an emotional state correctly is not practical, thus researchers have to assume that there is a reasonably sufficient window of error within which emotion recognition can be achieved at an acceptable success rate (Fig. 3).

The following subsections will explain in detail different aspects of emotion recognition systems.

3.1 Psychological Models of Emotion

Psychology of emotions can be viewed as a combined experience of consciousness (psychology), bodily sensation (physiology), and behavior (action-speech). Emotions are relatively brief episodes of synchronized responses that produce noticeable changes in the functioning of an organism. Such changes are brought about by triggering significant events (Scherer 2000).

There are around 300 emotions identified by researchers (Connor and Arnold 1973; Schubiger 1958), but most researchers agree upon at least regarding some emotions, including anger, sadness, joy, fear, shame, pride, surprise, disgust, and guilt

(Ekman 1992; Izard 1991; Frijda 1986), which are very strong and easily identifiable. Psychological emotion models are used as background for labeling emotional data, and the two most prominent of such models are the *Basic Emotions* model and the *Valence-Arousal-Dominance* model.

Basic Emotions model was initially proposed by Paul Ekman (1992) and later enhanced by other researchers (Ayadi et al. 2011; Cowie et al. 2001). According to this model, any emotion is a combination of six primary emotions, namely, *anger, disgust, fear, happiness, sadness, and surprise*. These six emotions are also referred to as *archetypal emotions*. Plutchik's extended the *Basic Emotions* model and added two more emotions as primary emotions: *anticipation* and *trust*. His model (Plutchik and Kellerman 1980) is interestingly designed like a wheel of emotions such that similar emotions are located side by side and very different (bipolar) emotions on the opposite side. This model is called Plutchik's wheel, which is like a color wheel where the intensities of the different emotions are displayed by color saturation. There is another categorical model called the *Geneva Emotion Wheel* (GEW), where two axes valence and control split the emotions into four separate groups, and neutral is at the center (Scherer 2005; Scherer et al. 2013).

The basic emotion model was built on the assumption that an independent neural sub-system serves every basic emotion. However, neuroimaging and physiological studies have failed to establish this theory (Eerola and Vuoskoski 2011). Recent research shows that discrete dimensional models of emotion are gaining more importance. A two-dimensional circumplex model proposes that all affective states arise from two independent neurophysiological systems: one related to valence (a pleasure-displeasure continuum) and the other to arousal (activation-deactivation). That is, varying degrees of both valence and arousal represents different emotions (Posner et al. 2005; Russell 1980). In another approach, different underlying dimensions of effect are chosen: energetic arousal and tense arousal (Eerola and Vuoskoski 2011). Introduction of another dimension "stance", which defines attention-rejection, into the 2-D model (Fig. 4, Kim and André 2008) resulted in a 3-D representation of the emotions (Kim and André 2008; Scholsberg 1954). One interesting point raised by Schimmack and Grob (2000) is that in the 3-dimensional model, the axes need not be orthogonal to each other (Fig. 4) in actual affect data.

This psychological aspect of emotion is an essential step towards dataset creation or selection for SER research. Synthetic data creation process required to decide on the model to follow for labeling purposes. Two-dimensional emotional models are extensively used in most of the synthetic SER datasets, and mostly the underlying emotions are re-created (enacted). The synthetic speech datasets used for this work also use the 2-dimensional model where the basic emotions are enacted.

3.2 Features of Emotions

Human beings express their emotions through various mechanisms, such as *facial expression*, *physical movement*, and *tone of voice*. However, other physiological

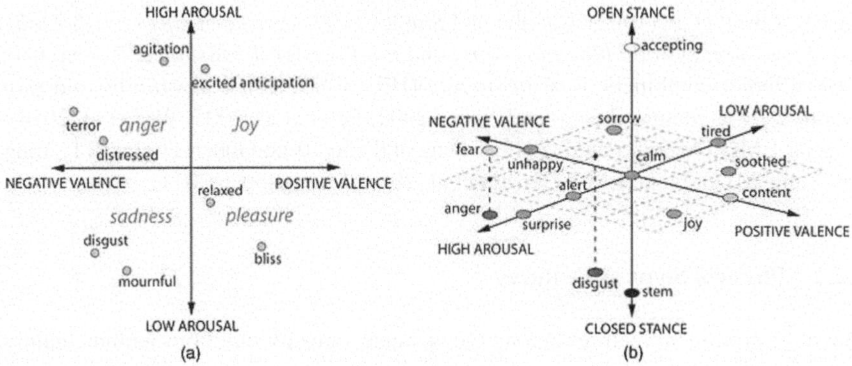

Fig. 4 The section **a** shows the valence-arousal 2-dimensional model. And the 2-dimensional is extended to 3-dimensional section (**b**)

changes that may occur due to an emotional state change, for example, changes in blood pressure, heartbeat rate, degree of sweating, shaking, and changes in skin coloration Iliev et al. (2010). The psychological signal, such as brain signal, is another source from where human emotions can be captured. Thus, two broad categories for collection of emotions exist: (1) physical signs like speech, facial expression and gestures, and (2) psychological signals like electro-encephalogram (EEG), heartbeat rate, and blood pressure.

3.2.1 Psychological Signals as Features

Researchers are using different psychological signals, which include electro-encephalogram (EEG), electro-cardiogram (ECG), electro-myogram (EMG), galvanic skin response (GSR), and respiration (RSP) (Shu et al. 2018). Researchers observed that it is relatively difficult to detect emotional changes using a single physiological signal precisely (Shu et al. 2018).

Researchers using EEG signals for emotion recognition considered have applied different approaches. For example, (1) Wu et al. (2017) used two channels of frontal EEG signals. (2) Lan et al. (2016) used combination of *fractal dimension*, *five statistics*, and *four band power*. (3) Qiao et al. (2017) used deep learning model to extract high level feature. (4) Candra et al. (2015) applied *discrete wavelet transform* for feature extraction. (5) Zheng et al. (2017) used *power spectral density* of EEG signals. (6) Mehmood and Lee (2015) employed Hjorth Parameters.

ECG signal also carries vital information regarding emotions. Following researchers extracted various features ECG signals: (1) few linear and nonlinear features are extracted by Valenza et al. (2014). (2) Valenza et al. (2011) extracted both the time domain and frequency domain features. (3) Agrafioti et al. (2011) extracted Hilbert instantaneous frequency as feature. (4) Cheng et al. (2017) combined the linear, nonlinear, and time-domain features. (5) Some researchers Kreibig

(2010), Adsett et al. (1962), Bandler and Shipley (1994) tracked *heart rate, diastolic blood pressure, systolic blood pressure*, and *total periferal resistance*. Researchers used different combinations of *heart rate* (HR), GSR, *skin temperature*, and *skin conductivity* as feature (Lisetti and Nasoz 2004; Peter et al. 2005; Wen et al. 2014). Regular EMG (AlZoubi et al. 2012; Cheng and Liu 2008; Gouizi et al. 2011; Yang and Yang 2011), facial EMG (Jerritta et al. 2014; Murugappan 2011).

3.2.2 Physical Signs as Features

The most prominent features among the physical signs for emotions include mainly *speech signal* and *facial expression*. Speech features are mainly of three types : (1) the prosodic features, (2) the excitation source features, and (3) the spectral or vocal tract features. While speaking, humans can utilize different tools available in this system to vary the duration, pitch, and intensity of the spoken utterances, called prosody alteration, to express their different feelings in words. Researchers used different derivatives of pitch and energy as various prosody features (Ayadi et al. 2011; Bänziger and Scherer 2005; Cowie et al. 2001; Cowie and Cornelius 2003; Lee and Narayanan 2005; Murray and Arnott 1993; Rao and Yegnanarayana 2006). The features used to represent glottal activity, mainly the vibration of glottal folds, are known as the source or excitation source features (Airas and Alku 2004; Cowie et al. 2001; Davitz and Beldoch 1964; Makhoul 1975; Murray and Arnott 1993; Scherer 1986; Sundberg et al. 2011; Sun and Moore 2012). Spectral features are the characteristics of various sound components generated from different cavities of the vocal tract system. Spectral features include 1. ordinary linear predictor coefficients (LPC) (Rabiner and Schafer 1978), 2. one-sided autocorrelation linear predictor coefficients (OSALPC) (Hernando and Nadeu 1997), 3. shorttime coherence method (SMC) (Bouquin 1996), and 4. least-squares modified Yule-Walker equations (LSMYWE) (Bou-Ghazale and Hansen 2000).

Facial expression is used by researchers to detect emotions (Ali et al. 2015; Busso et al. 2004; Drume and Jalal 2012; Quintero et al. 2018; Vivek and Reddy 2015; Wang et al. 2018; Zhang et al. 2016). Defining the face model is one of the primary tasks of using the facial expression for emotion detection. The distance between two centers of both eyes is used as the generic distance measure for computing various characteristics of the face. For example, the width of the nose, the vertical distance between eyes and eyebrows, and the vertical distance between the eyes and center of the mouth (Wang et al. 2018) are used as facial attributes.

3.3 Classification Technique

Researchers have applied different classification techniques for emotion recognition. Hidden Markov Model (HMM) was the workhorse for speech-related applications, and it was eventually used for ER as well (Bitouk et al. 2010; Kwon et al. 2003;

Lin et al. 2014; Philippou-Hübner et al. 2010; Zhou et al. 2001). However, the classification accuracy achieved by them was not satisfactory enough. So, other powerful classification techniques such as Gaussian Mixture Model (GMM) (Atassi and Esposito 2008; Jeon et al. 2011; Luengo et al. 2005; Neiberg et al. 2006; Schuller et al. 2004; Zhou et al. 2009), Support Vector Machine (SVM) (Hassan and Damper 2010; Lee et al. 2004; Pierre-Yves 2003; Schuller et al. 2004; Sun and Moore 2012; Wu et al. 2011), k-Nearest Neighbors (kNN) (Dellaert et al. 1996; Lee et al. 2001; Pao et al. 2005; Wang and Guan 2004), Artificial Neural Network (ANN) (Fernandez and Picard 2003; Nakatsu et al. 1999; Schuller et al. 2004; Zhu and Luo 2007) and more recently Deep Learning (DL) architectures (Chen et al. 2018; Neumann and Vu 2017; Parthasarathy and Tashev 2018; Trigeorgis et al. 2016; Zhang et al. 2018) came into the picture. Classification performance has improved, but solutions suitable for practical business use are not yet achieved. The main reason is that classification performance across databases varies significantly, which means that a model performs poorly when trained using one dataset and when tested with another dataset, and the accuracy is significantly low (Table 1).

3.3.1 Deep Learning Classification for ER

Different variants of deep learning models exist now, but Convolutional Neural Networks (CNNs) (LeCun et al. 1989; Lecun et al. 1998) and Recurrent Neural Networks (RNNs) (Rumelhart et al. 1986) are the most successful ones. CNNs are neural networks that use convolution in place of general matrix multiplication in at least one of their layers. Whereas, when feedforward neural networks are extended to include feedback connections, they are called recurrent neural networks. RNNs are specialized in processing sequential data.

ER researchers have used CNNs (Chen et al. 2018; Neumann and Vu 2017; Parthasarathy and Tashev 2018; Trigeorgis et al. 2016; Zhang et al. 2018), RNNs (Mirsamadi et al. 2017; Trigeorgis et al. 2016; Wöllmer et al. 2008), or a combination of the two extensively. Shallow 1-layer or 2-layer CNN structures may not be able to learn effectively the features which are discriminative enough to distinguish the subjective emotions (Zhang et al. 2018). So, researchers largely recommend a deep structure. Some researchers (Huang and Narayanan 2016; Huang and Narayanan 2017; Mirsamadi et al. 2017; Neumann and Vu 2017) have studied the effectiveness of attention mechanisms in ER.

Researchers are applying end-to-end deep learning systems for ER problems (Han et al. 2017a, b; Schmitt et al. 2016; Trigeorgis et al. 2016; Tzirakis et al. 2018), and most of them use the arousal-valence model of emotions. Although using end-to-end deep learning, the average classification accuracy for arousal is 78.16%, which is sufficient, but for valence, it is pretty low at 43.06%. Recently, a maximum accuracy of 87.32% was achieved in other DNN techniques. This was achieved by deploying a fine-tuned Alex-Net on Emo-DB (Zhang et al. 2018). Han et al. (2014) used

Table 1 Prominent classification models for emotion recognition along with few literature references

No.	Classifiers	References
1.	Hidden Markov model	Bitouk et al. (2010), Fernandez and Picard (2003), Zhou et al. (2001), Nwe et al. (2003), Schuller et al. (2003), Kwon et al. (2003), Lee et al. (2004), Philippou-Hübner et al. (2010), Lin et al. (2014)
2.	Gaussian Mixture model	Jeon et al. (2011), Lugger and Yang (2007), Atassi and Esposito (2008), Schuller et al. (2004), Slaney and McRoberts (2003), Neiberg et al. (2006), Wang and Guan (2004), Zhou et al. (2009), Mubarak et al. (2005), Luengo et al. (2005), Breazeal and Aryananda (2002)
3.	k-Nearest Neighbor	Dellaert et al. (1996), Murugappan (2011), Pao et al. (2005), Wang and Guan (2004), Yu et al. (2001), Mehmood and Lee (2015), Lee et al. (2001)
4.	Support vector machine	Yang and Yang (2011), Candra et al. (2015), Vivek and Reddy (2015), Drume and Jalal (2012), Schuller et al. (2004), Sun and Moore (2012), Zhang et al. (2016), Ali et al. (2015), Wu et al. (2011), Lee et al. (2004), Cheng et al. (2017), Kwon et al. (2003), Hassan and Damper (2010), Rozgic et al. (2012), Yeh and Chi (2010), Mehmood and Lee (2015)
5.	Artificial neural network	Schuller et al. (2004), Cheng and Liu (2008), Wang et al. (2018), Yang and Yang (2011), Zhu and Luo (2007), Wang and Guan (2004), Petrushin (2000), Nakatsu et al. (1999), Nicholson et al. (1999), Tato et al. (2002), Fernandez and Picard (2003)
6.	Bayes classifier	Dellaert et al. (1996), Lugger and Yang (2007), Wang et al. (2008), Fernandez and Picard (2003)
7.	Linear discriminant analysis	Yildirim et al. (2004), Murugappan (2011), McGilloway et al. (2000), Ververidis et al. (2004), Lee and Narayanan (2005), Lee et al. (2001)
8.	Deep neural network	Parthasarathy and Tashev (2018), Trigeorgis et al. (2016), Chen et al. (2018), Zhang et al. (2018), Neumann and Vu (2017), Hassan et al. (2019), Mirsamadi et al. (2017), Huang and Narayanan (2017), Huang and Narayanan (2016), Tzirakis et al. (2018), Han et al. (2017a), Qiao et al. (2017), Schmitt et al. (2016)

Extreme Learning Machine (ELM) for classification, where a DNN takes as input the popular acoustic features within a speech segment and produces segment-level emotion state probability distributions, from which utterance-level features are constructed.

3.4 Challenges in ER Research

Emotion recognition mystery is not yet solved, and it has proved to be complicated. Here are the prominent difficulties faced by the researchers.

- The topic called *emotion* is inherently uncertain. As the real experience of emotion is very subjective, its expression varies largely from person to person. Moreover, there is little consensus over the definition of emotion. These are the major hurdle to proceed with the research (Schröder and Cowie 2006). For example, several studies (Amir et al. 2001; Grimm et al. 2007; Jeon et al. 2011; Kim et al. 2010; Lin et al. 2014; Lugger and Yang 2007; Ververidis and Kotropoulos 2005) reported that there is confusion between anger and happiness emotional in expression.
- ER is challenging because of the affective gap between the subjective emotions and low-level features (Zhang et al. 2018). Also, the feature analysis in the ER is less studied (Han et al. 2014), and researchers are still actively looking for the best feature set.
- Speaker and language dependency of classification results are a concern for building more generic speech emotion recognition systems (Koolagudi and Rao 2010). The same model gives very different classification results with different datasets. Studies reported the phenomenon of speaker dependency and tried to address that issue (Grimm et al. 2007; Kim et al. 2011; Lugger and Yang 2007; McGilloway et al. 2000).
- Standard databases are not available so that new models can be effectively benchmarked. Moreover, the absence of useful quality natural emotional databases is hindering real-life implementation.
- Cross-corpora recognition results are low (Schuller et al. 2010; Shami and Verhelst 2007; Sun and Moore 2012), which indicates that existing models are not generalizing enough for real-life implementation.
- Classification between high-arousal and low-arousal emotions is achieved more accurately, but for other cases, it is low, which needs to be addressed. Moreover, the accuracy of n-way classification with all the emotions in the database is still very low.
- An emotional state is considered as a physical condition, and any external intervention for emotion detection could be considered as a privacy breach. So, on ethical ground, detecting someone's emotion without permission could lead to uneven consequences.
- Finally, before implementing the emotion detection systems in people's daily life, researchers also need to be clear about the unintended consequences that could take place. For example, the vulnerability of different people at different emotional states could give rise to fraud and criminal intentions. So, precautions need to be adopted before implementing ER systems.

4 Role of Emotion Recognition in 4IR

4IR has already started impacting the social lives of human beings with the advent of social media platforms, artificial intelligence (AI), machine learning, and deep learning techniques. Voice assistants like Siri, Alexa, Google Home requires minimal human interventions. The ability to detect human emotional states by the AI systems can significantly boost the sociability of the voice assistants and other AI systems. The human-machine bond can attain new heights with an emotional touch, and the intelligent systems conceptualized in science-fiction can be a reality (Picard 1997). Emotion detection can change some crucial aspects of human life: (1) socio-political; (2) socio-economic; (3) health-care and medical; (4) technological; and (5) entertainment.

In the following subsections these aspects are discussed.

4.1 Socio-Political Impact

Social media platforms like Facebook, Twitter, and Instagram started connecting people socially in the later phase of the Third Industrial Revolution. In the 4IR, these platforms have advanced many folds. Now, these platforms started affecting the lives of people by manipulating their choices through sentiment analysis. Social media platforms are now impacting political (Aday et al. 2010; Ceron et al. 2014; Stieglitz and Dang-Xuan 2012) as well as non-political social movements (Hussein 2018; Khan et al. 2014), and the emotion or sentiment detection can work as a catalyst in those movements. Emotional state or mood-based content can be a reality soon when social media content is used for, say, anger management.

4.2 Socio-Economic Impact

Consumer feedback could be more realistic when there is prior information about the emotional state of the customer so that feedback time and practice can be reviewed. The effects of specific emotions on human behavior during financial decision making or economic judgment is an exciting topic of research. Research shows that human emotional states limit human rationality (Damasio 1994; Frijda 1986; Scherer et al. 1984), and that leads to the proposal of artificial agents in the human decision-making process (Marwala and Hurwitz 2015).

4.3 Impact on Health-Care

Research confirms that different emotional states have a significant impact on our health. Some emotions negatively affect the functioning of the heart and also increase the risk of heart disease (Gianaros et al. 2014); these emotions are called negative emotions such as anxiety, anger, and depression. On the other hand, positive emotions can help one ot recover from cardiovascular ailments (Fredrickson et al. 2000). McAllister et al. (2007) have shown that patients with genetic diseases can be facilitated by managing emotional conditions. New evidence show that positive emotions may help limit cancer growth (Ben-Shaanan et al. 2018), while negative emotions could contribute to cancer incidence (Antoni et al. 2006; Chida et al. 2008). It is also been demonstrated, through research, that emotions could influence the onset, course, and remission of the disease (Parsons and Marcer 2005). Emotional state tracking through speech can help doctors or relatives to manage patient conditions remotely (France et al. 2000), and patient assistant systems are being developed to help patients according to their emotional needs. Researchers are trying to model the emotions of autistic people in order to improve understanding of their behaviour and help improve communications with them (Picard 1997; Schelinski and von Kriegstein 2019).

4.4 Impact on Technology

Smart call center agents could decide on the need to transfer a customer call to a human agent, depending on the mood of the customer (Lee and Narayanan 2005; Nakatsu et al. 1999). Within the sphere of learning, the advent of virtual learning platforms could additionally be improved by adequately tracking the emotional state of the learner in order to regulate or customise learning content. Virtual assistants such as Siri, Alexa, or Google home will be more sensitive to our moods and can become more sensible human companion. SER could immensely help drivers (Schuller et al. 2004) and pilots (Hansen and Cairns 1995) to manage critical circumstances by reporting in real-time their emotional states to respective authorities.

4.5 Impact on Entertainment

The introduction of Virtual reality (VR) in the entertainment world like Gaming and Movies is changing the perception and definition of entertainment. People are now longing for experiences not usually available in real life, for example, visiting places they have physically never been in a virtual tour. Entertainment is becoming more and more personal with personalized offerings from the entertainment world. Introduction of emotion detection capabilities in the entertainment sector will provide more realistic experiences which will revolutionize the sector and the concepts like interactive movie (Nakatsu et al. 1999) and story-telling (Charles et al. 2009) will be more accessible.

5 Speech Emotion Recognition: A Case Study of ER

Emotion classification from speech utterances is a complex classification problem, and researchers have used different Machine Learning (ML) techniques to solve it. In this section, a new SER model is being discussed where the feature set used is extracted by applying a discrete wavelet transform.

5.1 Feature Extraction

The proposed SER model uses a feature set designed specifically for speech emotion classification (Tanmoy et al. 2018). The feature set is described in brief here. A discrete speech signal S can be defined as follows:

$$S = \{s_n\}_{n \in \mathbb{N}}, \text{ where } \{s_n\} = \{s_1, s_2, \ldots, s_n\}, \{s_n\} \in \mathbb{R} \tag{1}$$

where N is the length of S. Signal S is decomposed into trend and fluctuation sub-signals using Discrete wavelet transform (DWT) (Walker 2008) and can be represented as a mapping $S \longmapsto (t_1 | f_1)$ where t_1 and f_1 are the level-1 trend and fluctuation sub-signals respectively. Then, at level-2, DWT generates another set of trend and fluctuation signals from t_1 such that $t_1 \longmapsto (t_2 | f_2)$. So, level-2 transformation is defined as $S \longmapsto (t_2 | f_2 | f_1)$ where t_2 and f_2 are level-2 trend and fluctuation signals respectively of length $N/4$. So, in similar way at level-4 the DWT (Walker 2008) is defined as:

$$S \longmapsto ((t_4 | f_4 | f_3 | f_2 | f_1)) \tag{2}$$

where, f_3 is the level-3 fluctuation sub-signal of length $N/8$ and t_4, f_4 are level-4 trend and fluctuation sub-signals respectively of length $N/16$. DWTs considered for this model are $db6$, $db8$, $db10$, $db12$, and $db14$ which gives total of 25 sub-signals from each utterance.

Next, the information content of the sub-signals is computed by measuring the *information entropy* (IE). IE of a signal is high in case of random activity, whereas low IE is observed in case of ordered activity (Quiroga et al. 2001). Information entropy of a n length signal Q can be represented as (see Kullback 1959):

$$E(Q) = \sum_{i=1}^{n} p(q_i) log_{10} p(q_i), \tag{3}$$

where $Q = \{q_1, q_2, \ldots, q_n\}$ is a set of random phenomena, and $p(q_i)$ is the probability of a random phenomenon q_i. Next, the similarity is computed between the entropy values of the two states of the same speaker; one is the neutral state, and another is the respective emotional state. Here euclidean distance is computed as similarity measure. The covariance and correlation coefficient of the entropy sequences are also computed as part fo feature extraction process which adds $25 + 25 = 50$ more features. The gender of the speaker is a key feature of the feature set. And finally, the length ratio between the emotional and neutral utterance is also added to feature list. So, the extracted feature set has 77 features from each speech utterance.

5.2 *Classification Models*

Here the SER problem is considered as a binary classification problem by classifying the emotions in pairs. Four well-known classification models have been trained to identify the underlying emotions, and the results are compared to find the best performing model. The four classification models are (1) *gaussian naive bayes* (GNB); (2) *k-nearest neighbours* (KNN); (3) *support vector classifier* (SVC) with *radial basis function* (RBF) kernel; and (4) *feed-forward neural network* (FNN).

Bayes theorem based Naive Bayes (NB) classifier is one of the easy to build models which usually works with categorical data. The likelihood of the classes is computed from the posterior probability with the help of the Bayes theorem equation. However, for numerical (non-categorical) data, the Gaussian distribution can be assumed to compute the likelihood using the posterior probability, and this approach is called the Gaussian Naive Bayes. The problem at hand is the right candidate for GNB.

The K-Nearest Neighbor is a similarity measure based classification model. Most computations are carried out while classifying unseen data, and that is why it called the lazy learning method. KNN is a simple yet effective classification technique. Classification is performed based on the majority vote of the neighbors, and the number of neighbors to consider for voting is determined by the predefined value of the parameter k. Choosing the appropriate value for the parameter k is critical towards getting the desired classification performance.

When two classes are linearly separable in the feature space, linear SVC can be applied to determine the hyperplane that separates two classes. However, since linearly separable classes are rare in the practical scenario, classes separable by nonlinear regions can be handled by SVC with nonlinear kernel functions. Radial Basis Function is one such kernel function that is commonly used with SVC (Hsu et al. 2003). This work applies SVC with the RBF kernel for emotion classification. SVC with RBF shows a tendency to overfit, which is managed by carefully chosen values for hyperparameters (Cortes and Vapnik 1995).

Another classification technique used in this system is a Feed-forward Neural Network (FNN) (Chakraverty and Mall 2017). FNNs are quintessential deep neural networks, and the deep neural network framework has been used for this work. Input data are transformed into the hidden layers to search for a mapping function, and finally, the predicted results are produced. Gradients of the loss function provide essential input to the optimizer to update the weights further to find the best possible weights. The present model contains a single hidden layer and there are 100 processing nodes in that hidden layer. The hidden layer uses the (rectifier linear unit) (ReLu)

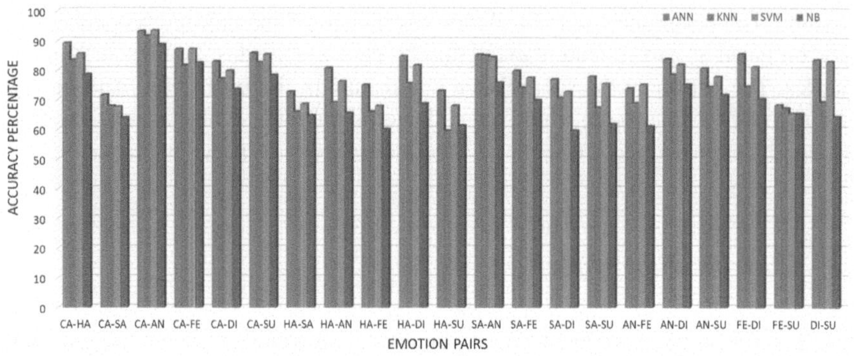

Fig. 5 Comparison of classification accuracy of the four models. Abbreviations: *AN* angry, *CA* calm, *DI* disgust, *HA* happy, *SA* sad, *SU* surprise

as an activation function, whereas the output layer uses a *sigmoid* function as an activation function. The loss function used in this case is a *binary cross-entropy* loss. This type of model uses the *error backpropagation* to update the weights, and the complete optimization process of updating weights throughout iterations is carried out by specialized algorithms called optimizers. Here, the *adam* optimizer is used, which is based on the *stochastic gradient descent*. The FNN model requires some hyper-parameters to be chosen carefully to get a satisfactory performance. For the present model, the epoch is 100, and the batch size is 40.

5.3 Results and Discussions

In this section, the classification performance of the four models is compared. The models demonstrate encouraging results compared to contemporary research results (Fayek et al. 2017; Han et al. 2014; Lee and Tashev 2015; Shen et al. 2011). Average accuracy achieved is more than 80%, and in some cases, it exceeded 90%, whereas (Han et al. 2014; Lee and Tashev 2015; Yang and Lugger 2010) attained 63.89 and 57.91% maximum accuracy. Figure 5 depicts the results, and it can be observed that the FNN model outperforms the other three models.

Sometimes, the human ear can get deceived in separating emotions from speech (Engberg et al. 1997; Slaney and McRoberts 2003), and the results also uphold that theory. It is prominent from the results that, emotions like happiness and fear are hard to classify since classification accuracy is low compared to other emotion pairs.

Very high feature space dimension is a real challenge while applying the DL architectures for SER, and the number of features sometimes runs into thousands (Han et al. 2014; Wöllmer et al. 2008). The proposed model can address this high dimensionality issue by reducing the number of features to 77. Finally, the models are trained on low-end computation resources in comparison to DL architectures, which promotes a lower carbon footprint to complex problems.

6 Conclusion

This chapter elaborates on the importance and role of emotion detection in the ongoing *fourth industrial revolution*; Sect. 4 discusses this in detail. The feature sets and classification techniques are vital components of the ER systems, and the latest advancements are discussed in Sects. 3 and 4. SER is a special case of emotion recognition, where the speech signal is used for emotion detection. A new feature set is used with a neural network based classification model to classify the emotions concealed in speech signals.

Although the ER is an unsolved problem, the advent of many new deep learning based classification models and advanced feature extraction techniques are boosting the classification performance. The new SEGRT feature set is going to be used with many deep learning multi-class classification models. Moreover, the SEGRT feature set has a scope of further enhancement to improve classification performance. So, we can conclude that *emotion recognition* has a significant role to play in the *fourth industrial revolution* to impact specifically in fields like AI, health care, and entertainment.

References

Aday S, Farrell H, Lynch M, Sides J, Kelly J, Zuckerman E (2010) Blogs and bullets: new media in contentious politics. United States Inst Peace 65: 1–31

Adsett CA, Schottstaedt WW, Wolf SG (1962) Changes in coronary blood flow and other hemodynamic indicators induced by stressful interviews. Psychosomatic Med 24(4): 331–336

Agrafioti F, Hatzinakos D, Anderson AK (2011) ECG pattern analysis for emotion detection. IEEE Trans Affect Comput 3(1): 102–115

Airas M, Alku P (2004) Emotions in short vowel segments: effects of the glottal flow as reflected by the normalized amplitude quotient. In: André E, Dybkjær L, Minker W, Heisterkamp P (eds) Affective dialogue systems. Springer Berlin Heidelberg, pp 13–24

Ali H, Hariharan M, Yaacob S, Adom AH (2015) Facial emotion recognition based on higher-order spectra using support vector machines. J Med Imaging Health Inf 5(6): 1272–1277

AlZoubi O, D'Mello SK, Calvo RA (2012) Detecting naturalistic expressions of nonbasic affect using physiological signals. IEEE Trans Affect Comput 3(3): 298–310

Amir N, Kerret O, Karlinski D (2001) Classifying emotions in speech: a comparison of methods. Interspeech

Antoni MH, Lutgendorf SK, Cole SW, Dhabhar FS, Sephton SE, McDonald PG, Stefanek M, Sood AK (2006) The influence of bio-behavioural factors on tumour biology: pathways and mechanisms. Nat Rev Cancer 6: 240–248. https://doi.org/10.1038/nrc1820

Atassi H, Esposito A (2008) A speaker independent approach to the classification of emotional vocal expressions. In: 2008 20th IEEE international conference on tools with artificial intelligence, vol 2, pp 147–152. https://doi.org/10.1109/ICTAI.2008.158

Ayadi M, Kamel MS, Karray F (2011) Survey on speech emotion recognition: features, classification schemes, and databases. Pattern Recogn 44(3): 572–587

Bandler R, Shipley MT (1994) Columnar organization in the midbrain periaqueductal gray: modules for emotional expression? Trends Neurosci 17(9): 379–389

Bänziger T, Scherer KR (2005) The role of intonation in emotional expressions. Speech Commun 46(3): 252–267. https://doi.org/10.1016/j.specom.2005.02.016

Ben-Shaanan TL, Schiller M, Azulay-Debby H, Korin B, Boshnak N, Koren T, Krot M, Shakya J, Rahat MA, Hakim F, Rolls A (2018) Modulation of anti-tumor immunity by the brain's reward system. Nat Commun 9(1). https://doi.org/10.1038/s41467-018-05283-5

Bitouk D, Verma R, Nenkova A (2010) Class-level spectral features for emotion recognition. Speech Commun 52:613–625

Bou-Ghazale SE, Hansen JHL (2000) A comparative study of traditional and newly proposed features for recognition of speech under stress. IEEE Trans Speech Audio Process 8(4): 429–442. https://doi.org/10.1109/89.848224

Breazeal C, Aryananda L (2002) Recognition of affective communicative intent in robot-directed speech. Auton Robots 12(1): 83–104. https://doi.org/10.1023/A:1013215010749

Busso C, Deng Z, Yildirim S, Bulut M, Lee SM, Kazemzadeh A, Lee S, Neumann U, Narayanan S (2004) Analysis of emotion recognition using facial expressions, speech and multimodal information. In: Proceedings of the 6th international conference on multimodal interfaces. ACM, pp 205–211

Candra H, Yuwono M, Chai R, Handojoseno R, Elamvazuthi I, Nguyen HT, Su S (2015) Investigation of window size in classification of EEG-emotion signal with wavelet entropy and support vector machine. In: 2015 37th annual international conference of the IEEE engineering in medicine and biology society (EMBC). IEEE, pp 7250–7253

Carter Rita (1998) Mapping the mind. University of California Press, Berkeley, CA

Ceron A, Curini L, Iacus SM, Porro G (2014) Every tweet counts? how sentiment analysis of social media can improve our knowledge of citizens' political preferences with an application to Italy and France. New Media Soc 16(2): 340–358. https://doi.org/10.1177/1461444813480466

Chakraverty S, Mall S (2017) Artificial neural networks for engineers and scientists: solving ordinary differential equations. CRC Press, Taylor & Francis Group

Charles F, Pizzi D, Cavazza M, Vogt T, Andre E (2009) Emoemma: emotional speech input for interactive storytelling. In: 8th international conference on autonomous agents and multiagent systems (AAMAS 2009), vol 2, pp 1381–1382

Cheng B, Liu G (2008) Emotion recognition from surface EMG signal using wavelet transform and neural network. In: Proceedings of the 2nd international conference on bioinformatics and biomedical engineering (ICBBE), pp 1363–1366

Cheng Z, Shu L, Xie J, Philip Chen CL (2017) A novel ECG-based real-time detection method of negative emotions in wearable applications. In: 2017 international conference on security, pattern analysis, and cybernetics (SPAC). IEEE, pp 296–301

Chen M, He X, Yang J, Zhang H (2018) 3-D convolutional recurrent neural networks with attention model for speech emotion recognition. IEEE Signal Process Lett 25(10): 1440–1444. https://doi.org/10.1109/LSP.2018.2860246

Chida Y, Hamer M, Wardle J, Steptoe A (2008) Do stress-related psychosocial factors contribute to cancer incidence and survival? Nat Clin Pract Oncol 5: 466–475. https://doi.org/10.1038/ncponc1134

Clore GL, Huntsinger JR (2007) How emotions inform judgment and regulate thought. Trends Cognitive Sci 11(9):393–9

Connor J, Arnold G (1973) Intonation of colloquial English. Longman UK

Cortes C, Vapnik V (1995) Support-vector networks. Mach Learn 20: 273–297. https://doi.org/10.1007/BF00994018

Cowie R, Douglas-Cowie E, Tsapatsoulis N, Votsis G, Kollias S, Fellenz W, Taylor JG (2001) Emotion recognition in human-computer interaction. IEEE Signal Process Mag 18(1):32–80. https://doi.org/10.1109/79.911197

Cowie R, Cornelius RR (2003) Describing the emotional states that are expressed in speech. Speech Commun 40(1): 5–32. https://doi.org/10.1016/S0167-6393(02)00071-7. http://www.sciencedirect.com/science/article/pii/S0167639302000717

Damasio A (1994) Descartes' error. Grosset/Putnam, New York

Darwin C (1948) The expression of emotion in man and animals. Watts

Davitz JR, Beldoch M (1964) The communication of emotional meaning. McGraw-Hill series in psychology. Greenwood Press. https://books.google.co.za/books?id=1ggRAQAAIAAJ
Dellaert F, Polzin T, Waibel A (1996) Recognizing emotion in speech. In: Proceedings of the fourth international conference on spoken language, vol 3, pp 1970–1973. https://doi.org/10.1109/ICSLP.1996.608022
Drume D, Jalal AS (2012) A multi-level classification approach for facial emotion recognition. In: 2012 IEEE international conference on computational intelligence and computing research, pp 1–5. https://doi.org/10.1109/ICCIC.2012.6510279
Eerola T, Vuoskoski JK (2011) A comparison of the discrete and dimensional models of emotion in music. Psychol Music 39(1): 18–49. https://doi.org/10.1177/0305735610362821
Ekman P (1992) An argument for basic emotions. Cogn Emot 6:169–200
Engberg IS, Hansen AV, Andersen O, Dalsgaard P (1997) Design, recording and verification of a Danish emotional speech database. In: Proceedings of the 5th European conference on speech communication and technology
Fayek HM, Lech M, Cavedonb L (2017) Evaluating deep learning architectures for speech emotion recognition. Neural Netw 92: 60–68
Fernandez R, Picard RW (2003) Modeling drivers' speech under stress. Speech Commun 40:145–159
France DJ, Shiavi RG, Silverman S, Silverman M, Wilkes M (2000) Acoustical properties of speech as indicators of depression and suicidal risk. IEEE Trans Biomed Eng 47(7): 829–837. https://doi.org/10.1109/10.846676
Frank RH (1988) Passions within reason: the strategic role of emotions. Norton
Fredrickson BL, Mancuso RA, Branigan C, Tugade MM (2000) The undoing effect of positive emotions. Motiv Emot 24(4): 237–258
Freedman DG, Loring CB, Martin RM (1967) Emotional behavior and personality development. Infancy Early Childhood, 429–502
Gianaros PJ, Marsland AL, Kuan DCH, Schirda BL, Jennings JR, Sheu LK, Hariri AR, Gross JJ, Manuck SB (2014) An inflammatory pathway links atherosclerotic cardiovascular disease risk to neural activity evoked by the cognitive regulation of emotion. Biol Psychiatry 75: 738–745. https://doi.org/10.1016/j.biopsych.2013.10.012
Gouizi K, Reguig FB, Maaoui C (2011) Analysis physiological signals for emotion recognition. In: International workshop on systems, signal processing and their applications, WOSSPA. IEEE, pp 147–150
Grimm M, Kroschel K, Mower E, Narayanan S (2007) Primitives-based evaluation and estimation of emotions in speech. Speech Commun 49(10): 787–800. https://doi.org/10.1016/j.specom.2007.01.010. http://www.sciencedirect.com/science/article/pii/S0167639307000040
Han K, Dong Y, Tashev I (2014) Speech emotion recognition using deep neural network and extreme learning machine. In: Proceedings of the interspeech
Hansen JHLL, Cairns DA (1995) Icarus: Source generator based real-time recognition of speech in noisy stressful and Lombard effect environments. Speech Commun 16(4): 391–422. https://doi.org/10.1016/0167-6393(95)00007-B
Han J, Zhang Z, Ringeval F, Schuller B (2017a) Reconstruction-error-based learning for continuous emotion recognition in speech. In: 2017 IEEE international conference on acoustics, speech and signal processing (ICASSP), pp 2367–2371. https://doi.org/10.1109/ICASSP.2017.7952580
Han J, Zhang Z, Ringeval F, Schuller B (2017b) Prediction-based learning for continuous emotion recognition in speech. In: 2017 IEEE international conference on acoustics, speech and signal processing (ICASSP), pp 5005–5009. https://doi.org/10.1109/ICASSP.2017.7953109
Hassan MM, Alam MGR, Uddin MZ, Huda S, Almogren A, Fortino G (2019) Human emotion recognition using deep belief network architecture. Inf Fusion 51: 10–18. https://doi.org/10.1016/j.inffus.2018.10.009. http://www.sciencedirect.com/science/article/pii/S1566253518301301
Hassan A, Damper RI (2010) Multi-class and hierarchical SVMS for emotion recognition. Interspeech

Hernando J, Nadeu C (1997) Linear prediction of the one-sided autocorrelation sequence for noisy speech recognition. IEEE Trans Speech Audio Process 5(1): 80–84. https://doi.org/10.1109/89.554273

Hsu CW, Chang C-C, Lin C-J (2003) A practical guide to support vector classification

Huang C-W, Narayanan S (2016) Attention assisted discovery of sub-utterance structure in speech emotion recognition. Interspeech

Huang C, Narayanan SS (2017) Deep convolutional recurrent neural network with attention mechanism for robust speech emotion recognition. In: 2017 IEEE international conference on multimedia and expo (ICME), pp 583–588. https://doi.org/10.1109/ICME.2017.8019296

Hussein DMEM (2018) A survey on sentiment analysis challenges. J King Saud Univ Eng Sci 30(4): 330–338. https://doi.org/10.1016/j.jksues.2016.04.002. http://www.sciencedirect.com/science/article/pii/S1018363916300071

Iliev AI, Scordilis MS, Papa JP, Falcã AX (2010) Spoken emotion recognition through optimum-path forest classification using glottal features. Comput Speech Lang 24(3): 445–460. https://doi.org/10.1016/j.csl.2009.02.005. http://www.sciencedirect.com/science/article/pii/S0885230809000084

Izard CE (1991) The psychology of emotions. Springer US

Jeon JH, Xia R, Liu Y (2011) Sentence level emotion recognition based on decisions from subsentence segments. In: 2011 IEEE international conference on acoustics, speech and signal processing (ICASSP), pp 4940–4943. https://doi.org/10.1109/ICASSP.2011.5947464

Jerritta S, Murugappan M, Wan K, Yaacob S (2014) Emotion recognition from facial EMG signals using higher order statistics and principal component analysis. J Chin Inst Eng 37(3): 385–394

Khan K, Baharudin B, Khan A, Ullah A (2014) Mining opinion components from unstructured reviews: a review. J King Saud Univ Comput Inf Sci 26(3): 258–275

Kim J, André E (2008) Emotion recognition based on physiological changes in music listening. IEEE Trans Pattern Anal Mach Intell 30: 2067–2083. https://doi.org/10.1109/TPAMI.2008.26

Kim J, Lee S, Narayanan SS (2010) An exploratory study of manifolds of emotional speech. In: 2010 IEEE international conference on acoustics, speech and signal processing, pp 5142–5145. https://doi.org/10.1109/ICASSP.2010.5495032

Kim J, Park J, Oh Y (2011) On-line speaker adaptation based emotion recognition using incremental emotional information. In: 2011 IEEE international conference on acoustics, speech and signal processing (ICASSP), pp 4948–4951. https://doi.org/10.1109/ICASSP.2011.5947466

Koolagudi SG, Rao KS (2010) Real life emotion classification using VOP and pitch based spectral features. In: 2010 annual IEEE India conference (INDICON), pp 1–4. https://doi.org/10.1109/INDCON.2010.5712728

Kreibig SD (2010) Autonomic nervous system activity in emotion: a review. Biol Psychol 84(3): 394–421

Kullback S (1959) Digital signal processing. Wiley

Kwon O-W, Chan K, Hao J, Lee T-W (2003) Emotion recognition by speech signals. In: Eighth European conference on speech communication and technology

Lan Z, Sourina O, Wang L, Liu Y (2016) Real-time EEG-based emotion monitoring using stable features. Vis Comput 32(3): 347–358

Le Bouquin R (1996) Enhancement of noisy speech signals: Application to mobile radio communications. Speech Commun 18(1): 3 – 19. https://doi.org/10.1016/0167-6393(95)00021-6. http://www.sciencedirect.com/science/article/pii/0167639395000216

LeCun Y, Boser B, Denker JS, Henderson D, Howard RE, Hubbard W, Jackel LD (1989) Backpropagation applied to handwritten zip code recognition. Neural Comput 1(4): 541–551. https://doi.org/10.1162/neco.1989.1.4.541

Lecun Y, Bottou L, Bengio Y, Haffner P (1998) Gradient-based learning applied to document recognition. Proc IEEE 86(11): 2278–2324. https://doi.org/10.1109/5.726791

LeDoux J (1996) The emotional brain: the mysterious underpinnings of emotional life. Simon and Schuster, New York

Lee CM, Narayanan S (2005) Toward detecting emotions in spoken dialogs. IEEE Trans Speech Audio Process 13: 293–303

Lee CM, Narayanan S, Pieraccini R (2001) Recognition of negative emotions from the speech signal. In: IEEE workshop on automatic speech recognition and understanding. ASRU '01, pp 240–243. https://doi.org/10.1109/ASRU.2001.1034632

Lee CM, Yildirim S, Bulut M, Kazemzadeh A, Busso C, Deng Z, Lee S, Narayanan S (2004) Emotion recognition based on phoneme classes. In: Proceedings of the ICSLP, pp 889–892

Lee J, Tashev I (2015) High-level feature representation using recurrent neural network for speech emotion recognition. In: Proceedings of the interspeech

Lin J, Wei W, Wu C, Wang H (2014) Emotion recognition of conversational affective speech using temporal course modeling-based error weighted cross-correlation model. In: Signal and information processing association annual summit and conference (APSIPA), 2014 asia-pacific, pp 1–7. https://doi.org/10.1109/APSIPA.2014.7041621

Lisetti CL, Nasoz F (2004) Using noninvasive wearable computers to recognize human emotions from physiological signals. EURASIP J Appl Signal Process 1672–1687

Luengo I, Navas E, Hernáez I, Sánchez J (2005) Automatic emotion recognition using prosodic parameters. Interspeech

Lugger M, Yang B (2007) The relevance of voice quality features in speaker independent emotion recognition. In: 2007 IEEE international conference on acoustics, speech and signal processing-ICASSP '07, vol 4, pp IV–17–IV–20. https://doi.org/10.1109/ICASSP.2007.367152

Makhoul J (1975) Linear prediction: a tutorial review. Proc IEEE 63(4): 561–580. https://doi.org/10.1109/PROC.1975.9792

Marwala T, Hurwitz E (2015) Artificial intelligence and asymmetric information theory. CoRR. arXiv:1510.02867

Massey DS (2002) A brief history of human society: the origin and role of emotion in social life: 2001 presidential address. Am Sociol Rev 67(1): 1–29. http://www.jstor.org/stable/3088931

McAllister M, Davies L, Payne K, Nicholls S, Donnai D, MacLeod R (2007) The emotional effects of genetic diseases: Implications for clinical genetics. Am J Med Genet A 143A(22): 2651–2661. https://doi.org/10.1002/ajmg.a.32013

McGilloway S, Cowie R, Cowie ED, Stan G, Westerdijk M, Stroeve S (2000) Approaching automatic recognition of emotion from voice: a rough benchmark. In: Proceedings of the ISCA ITRW on speech and emotion

Mehmood RM, Lee HJ (2015) Emotion classification of EEG brain signal using SVM and KNN. In: 2015 IEEE international conference on multimedia and expo workshops (ICMEW). IEEE, pp 1–5

Mirsamadi S, Barsoum E, Zhang C (2017) Automatic speech emotion recognition using recurrent neural networks with local attention. In: 2017 IEEE international conference on acoustics, speech and signal processing (ICASSP), pp 2227–2231. https://doi.org/10.1109/ICASSP.2017.7952552

Mubarak OM, Ambikairajah E, Epps J (2005) Analysis of an MFCC-based audio indexing system for efficient coding of multimedia sources. In: Proceedings of the eighth international symposium on signal processing and its applications, vol 2, pp 619–622. https://doi.org/10.1109/ISSPA.2005.1581014

Murray I, Arnott J (1993) Toward the simulation of emotion in synthetic speech: a review of the literature on human vocal emotion. J Acoust Soc Am 93: 1097–108. https://doi.org/10.1121/1.405558

Murugappan M (2011) Electromyogram signal based human emotion classification using KNN and IDA. In: 2011 IEEE international conference on system engineering and technology. IEEE, pp 106–110

Nakatsu R, Solomides A, Tosa N (1999) Emotion recognition and its application to computer agents with spontaneous interactive capabilities. In: Proceedings IEEE international conference on multimedia computing and systems, vol 2, pp 804–808. https://doi.org/10.1109/MMCS.1999.778589

Neiberg D, Elenius K, Laskowski K (2006) Emotion recognition in spontaneous speech using GMMS. In: Proceedings of the interspeech

Neumann M, Vu NT (2017) Attentive convolutional neural network based speech emotion recognition: a study on the impact of input features, signal length, and acted speech. CoRR. arXiv:1706.00612

Nicholson J, Takahashi K, Nakatsu R (1999) Emotion recognition in speech using neural networks. In: ICONIP'99. ANZIIS'99 ANNES'99 ACNN'99. 6th international conference on neural information processing. Proceedings (Cat. No.99EX378)

Nico HF (1986) The emotions. Cambridge University Press

Nwe TL, Foo SW, De Silva LC (2003) Speech emotion recognition using hidden markov models. Speech Commun 41: 603–623

Panksepp J (1998) Affective neuroscience: the foundations of human and animal emotions. Oxford University Press, New York

Pao T-L, Chen Y-T, Yeh J-H, Liao W-Y (2005) Combining acoustic features for improved emotion recognition in mandarin speech. In: Tao J, Tan T, Picard RW (eds) Affective computing and intelligent interaction. Springer Berlin Heidelberg, pp 279–285. ISBN 978-3-540-32273-3

Parsons J, Marcer N (2005) Osteopathy: models for diagnosis. Treatment and Practice, Elsevier Health Sciences

Parthasarathy S, Tashev I (2018) Convolutional neural network techniques for speech emotion recognition. In: 2018 16th international workshop on acoustic signal enhancement (IWAENC), pp 121–125. https://doi.org/10.1109/IWAENC.2018.8521333

Peter C, Ebert E, Beikirch H (2005) A wearable multi-sensor system for mobile acquisition of emotion-related physiological data. In: International conference on affective computing and intelligent interaction. Springer, pp 691–698

Petrushin V (2000) Emotion in speech: recognition and application to call centers. Proc Artif Neural Netw Eng 01

Philippou-Hübner D, Vlasenko B, Grosser T, Wendemuth A (2010) Determining optimal features for emotion recognition from speech by applying an evolutionary algorithm. Interspeech, pp 2358–2361

Picard RW, Vyzas E, Healey J (2001) Toward machine emotional intelligence: analysis of affective physiological state. IEEE Trans Pattern Anal Mach Intell 23(10): 1175–1191. https://doi.org/10.1109/34.954607

Pierre-Yves O (2003) The production and recognition of emotions in speech: features and algorithms. Int J Human-Comput Stud 59(1): 157–183. https://doi.org/10.1016/S1071-5819(02)00141-6. http://www.sciencedirect.com/science/article/pii/S1071581902001416

Plutchik R, Kellerman H (1980) Theories of emotion. Academic Press

Posner J, Russell JA, Peterson BS (2005) The circumplex model of affect: an integrative approach to affective neuroscience, cognitive development, and psychopathology. Dev Psychopathol 17(3):715–734

Qiao R, Qing C, Zhang T, Xing X, Xu X (2017) A novel deep-learning based framework for multi-subject emotion recognition. In: 2017 4th international conference on information, cybernetics and computational social systems (ICCSS). IEEE, pp 181–185

Quintero LAM, Muñoz-Delgado J, Sánchez-Ferrer JC, Fresán A, Brüne M, de Montis IA (2018) Facial emotion recognition and empathy in employees at a juvenile detention center. Int J Offender Therapy Comp Criminol 62(8): 2430–2446

Quiroga RQ, Rosso OA, Basar E, Schurman M (2001) Wavelet entropy in event-related potentials: a new method shows ordering of EEG oscillations. Biol Cybern 84: 291–299

Rabiner LR, Schafer RW (1978) Digital Processing of Speech Signals. Prentice-Hall, Prentice-Hall signal processing series. ISBN 9780132136037. https://books.google.co.za/books?id=YVtTAAAAMAAJ

Rao KS, Yegnanarayana B (2006) Prosody modification using instants of significant excitation. IEEE Trans Audio Speech Lang Process 14(3): 972–980. https://doi.org/10.1109/TSA.2005.858051

Rosalind WP (1997) Affective computing. MIT Press, Cambridge, MA, USA
Rozgic V, Ananthakrishnan S, Saleem S, Kumar R, Aravind NV, Rohit P (2012) Emotion recognition using acoustic and lexical features. Interspeech
Rumelhart D, Hinton G, Williams R (1986) Learning representations by back-propagating errors. Nature 323:533–536
Russell JA (1980) A circumplex model of affect. J Pers Soc Psychol 39(6):1161–1178
Schelinski S, von Kriegstein K (2019) The relation between vocal pitch and vocal emotion recognition abilities in people with autism spectrum disorder and typical development. J Autism Develop Disord 49(1): 68–82. https://doi.org/10.1007/s10803-018-3681-z
Scherer K (1986) Vocal affect expression: a review and a model for future research. Psychol Bullet 99: 143–65. https://doi.org/10.1037//0033-2909.99.2.143
Scherer KR (2000) Psychological models of emotion. Neuropsychol Emotion 137(3): 137–162
Scherer KR (2005) What are emotions? and how can they be measured? Soc Sci Inf 44(4): 695–729. https://doi.org/10.1177/0539018405058216
Scherer KR et al (1984) On the nature and function of emotion: a component process approach. Approaches Emot 2293(317): 31
Scherer KR, Shuman V, Fontaine JRJ, Cristina S (2013) The grid meets the wheel: assessing emotional feeling via self-report. In: Components of emotional meaning: a sourcebook. Series in affective science. Oxford University Press, pp 281–298. https://doi.org/10.1093/acprof:oso/9780199592746.003.0019
Schimmack U, Grob A (2000) Dimensional models of core affect: a quantitative comparison by means of structural equation modeling. Eur J Pers 14(4):325–345
Schmitt M, Ringeval F, Schuller BW (2016) At the border of acoustics and linguistics: bag-of-audio-words for the recognition of emotions in speech. Interspeech
Scholsberg H (1954) Three dimensions of emotion. Psychol Rev. 61:81–88
Schröder M, Cowie R (2006) Issues in emotion-oriented computing towards a shared understanding. In: Workshop on emotion and computing. Humaine
Schubiger M (1958) English intonation: its form and function. Niemeyer, Germany
Schuller B, Rigoll G, Lang M (2003) Hidden markov model-based speech emotion recognition. In: 2003 international conference on multimedia and expo. ICME '03. Proceedings (Cat. No.03TH8698)
Schuller B, Rigoll G, Lang M (2004) Speech emotion recognition combining acoustic features and linguistic information in a hybrid support vector machine-belief network architecture. In: Proceedings of the IEEE ICASSP, vol 1, pp 577–580. https://doi.org/10.1109/ICASSP.2004.1326051
Schuller B, Vlasenko B, Eyben F, Wollmer M, Stuhlsatz A, Wendemuth A, Rigoll G (2010) Cross-corpus acoustic emotion recognition: Variances and strategies. IEEE Trans Affect Comput 1(2): 119–131. https://doi.org/10.1109/T-AFFC.2010.8
Schwab K (2016) The fourth industrial revolution. World Economic Forum
Schwab K, Nadella S, Davis N (2018) Shaping future of the fourth industrial revolution. World Economic Forum
Shami M, Verhelst W (2007) An evaluation of the robustness of existing supervised machine learning approaches to the classification of emotions in speech. Speech Commun 49(3): 201–212. https://doi.org/10.1016/j.specom.2007.01.006. http://www.sciencedirect.com/science/article/pii/S016763930700009X
Shen P, Changjun Z, Chen X (2011) Automatic speech emotion recognition using support vector machine. In: Proceedings of the international conference on electronic mechanical engineering and information technology, vol 2, pp 621–625. https://doi.org/10.1109/EMEIT.2011.6023178
Shu L, Xie J, Yang M, Li Z, Li Z, Liao D, Xu X, Yang X (2018) A review of emotion recognition using physiological signals. Sensors 18(7). https://doi.org/10.3390/s18072074
Slaney M, McRoberts G (2003) Babyears: a recognition system for affective vocalizations. Speech Commun 39: 367–384

Stieglitz S, Dang-Xuan L (2012) Political communication and influence through microblogging–an empirical analysis of sentiment in twitter messages and retweet behavior. In: 2012 45th Hawaii international conference on system sciences, pp 3500–3509. https://doi.org/10.1109/HICSS.2012.476

Sundberg J, Patel S, Bjorkner E, Scherer KR (2011) Interdependencies among voice source parameters in emotional speech. IEEE Trans Affect Comput 2(3): 162–174. https://doi.org/10.1109/T-AFFC.2011.14

Sun R, Moore E (2012) A preliminary study on cross-databases emotion recognition using the glottal features in speech. Interspeech

Tanmoy R, Snehashish C, Tshilidzi M, Paul S (2018) Introducing new feature set based on wavelets for speech emotion classification. In: 2018 IEEE applied signal processing conference (ASPCON), pp 124–128. https://doi.org/10.1109/ASPCON.2018.8748666

Tato R, Santos R, Kompe R, Pardo JM (2002) Emotional space improves emotion recognition. Interspeech

Trampe D, Quoidbach J, Taquet M (2015) Emotions in everyday life. PLOS One 10(12): 1–15. https://doi.org/10.1371/journal.pone.0145450

Trigeorgis G, Ringeval F, Brueckner R, Marchi E, Nicolaou MA, Schuller B, Zafeiriou S (2016) Adieu features? end-to-end speech emotion recognition using a deep convolutional recurrent network. In: 2016 IEEE international conference on acoustics, speech and signal processing (ICASSP), pp 5200–5204. https://doi.org/10.1109/ICASSP.2016.7472669

Tzirakis P, Zhang J, Schuller BW (2018) End-to-end speech emotion recognition using deep neural networks. In: 2018 IEEE international conference on acoustics, speech and signal processing (ICASSP), pp 5089–5093. https://doi.org/10.1109/ICASSP.2018.8462677

Valenza G, Citi L, Lanatá A, Scilingo EP, Barbieri R (2014) Revealing real-time emotional responses: a personalized assessment based on heartbeat dynamics. Sci Rep 4: 4998

Valenza G, Lanata A, Scilingo EP (2011) The role of nonlinear dynamics in affective valence and arousal recognition. IEEE Trans Affect Comput 3(2): 237–249

Ververidis D, Kotropoulos C (2005) Emotional speech classification using gaussian mixture models. In: 2005 IEEE international symposium on circuits and systems, vol. 3, pp 2871–2874. https://doi.org/10.1109/ISCAS.2005.1465226

Ververidis D, Kotropoulos C, Pitas I (2004) Automatic emotional speech classification. In: 2004 IEEE international conference on acoustics, speech, and signal processing, vol 1, pp I–593. https://doi.org/10.1109/ICASSP.2004.1326055

Vivek TV, Reddy GRM (2015) A hybrid bioinspired algorithm for facial emotion recognition using CSO-GA-PSO-SVM. In: 2015 fifth international conference on communication systems and network technologies, pp 472–477. https://doi.org/10.1109/CSNT.2015.124

Walker JS (2008) A primer on WAVELETS and their scientific applications. Taylor and Francis Group, LLC

Wang S-H, Phillips P, Dong Z-C, Zhang Y-D(2018) Intelligent facial emotion recognition based on stationary wavelet entropy and jaya algorithm. Neurocomputing 272: 668–676

Wang Y, Du S, Zhan Y (2008) Adaptive and optimal classification of speech emotion recognition. In: 2008 fourth international conference on natural computation, vol 5, pp 407–411. https://doi.org/10.1109/ICNC.2008.713

Wang Y, Guan L (2004) An investigation of speech-based human emotion recognition. In: IEEE 6th workshop on multimedia signal processing, pp 15–18

Wen W, Liu G, Cheng N, Wei J, Shangguan P, Huang W (2014) Emotion recognition based on multi-variant correlation of physiological signals. IEEE Trans Affect Comput 5(2): 126–140

Williamson J (1978) Speech analyzer for analyzing pitch or frequency perturbations in individual speech pattern to determine the emotional state of the person

Wöllmer M, Eyben F, Reiter S, Schuller BW, Cox C, Douglas-Cowie E, Cowie R (2008) Abandoning emotion classes—towards continuous emotion recognition with modelling of long-range dependencies. Interspeech

Wu S, Falk TH, Chan W-Y (2011) Automatic speech emotion recognition using modulation spectral features. Speech Commun 53(5): 768–785. https://doi.org/10.1016/j.specom.2010.08.013. http://www.sciencedirect.com/science/article/pii/S0167639310001470

Wu S, Xu X, Shu L, Hu B (2017) Estimation of valence of emotion using two frontal EEG channels. In: 2017 IEEE international conference on bioinformatics and biomedicine (BIBM). IEEE, pp 1127–1130

Yang B, Lugger M (2010) Emotion recognition from speech signals using new harmony features. Signal Process 90:1415–1423

Yang S, Yang G (2011) Emotion recognition of EMG based on improved IM BP neural network and SVM. JSW 6(8): 1529–1536

Yeh L-Y, Chi T-S (2010) Spectro-temporal modulations for robust speech emotion recognition. Interspeech

Yildirim S, Bulut M, Chul ML, Abe K, Zhigang D, Sungbok L, Shrikanth N, Carlos B (2004) An acoustic study of emotions expressed in speech. Interspeech

Yu F, Chang E, Xu Y-Q, Shum H (2001) Emotion detection from speech to enrich multimedia content. In: IEEE Pacific rim conference on multimedia

Zhang Y-D, Yang Z-J, Hui-Min L, Zhou X-X, Phillips P, Liu Q-M, Wang S-H (2016) Facial emotion recognition based on biorthogonal wavelet entropy, fuzzy support vector machine, and stratified cross validation. IEEE Access 4: 8375–8385

Zhang S, Zhang S, Huang T, Gao W (2018) Speech emotion recognition using deep convolutional neural network and discriminant temporal pyramid matching. IEEE Trans Multimedia 20(6): 1576–1590. https://doi.org/10.1109/TMM.2017.2766843

Zheng W-L, Zhu J-Y, Bao-Liang L (2017) Identifying stable patterns over time for emotion recognition from EEG. IEEE Trans Affect Comput

Zhou G, Hansen JHL, Kaiser JF (2001) Nonlinear feature based classification of speech under stress. IEEE Trans Speech Audio Process 9:201–216

Zhou Y, Sun Y, Zhang J, Yan Y (2009) Speech emotion recognition using both spectral and prosodic features. In: 2009 international conference on information engineering and computer science, pp 1–4. https://doi.org/10.1109/ICIECS.2009.5362730

Zhu A, Luo Q (2007) Study on speech emotion recognition system in e-learning. In: Jacko JA (ed) Human-computer interaction. HCI intelligent multimodal interaction environments. Springer Berlin Heidelberg, pp 544–552. ISBN 978-3-540-73110-8

Noisy, Intermediate-Scale Quantum Computing and Industrial Revolution 4.0

Makhamisa Senekane, Motobatsi Maseli, and Molibeli Benedict Taele

Abstract The fourth Industrial Revolution (4IR) integrates digital, physical and biological systems to create Cyber-Physical Systems (CPS) that blur the line between the digital, physical and biological systems. One of the drivers of the 4IR is quantum computing; which harnesses quantum mechanical concepts such as entanglement, superposition and tunneling to perform computation. A full-scale quantum computer has not yet been realized. However, Noisy Intermediate-Scale Quantum (NISQ) computers are already in use. In this chapter, we explore NISQ as a disruptive technology of the fourth Industrial Revolution. Furthermore, we discuss the NISQ disruptors together with the applications of the NISQ frameworks. Finally, a novel privacy-preserving quantum machine learning scheme is introduced in this chapter. In essence, the various applications discussed in this chapter show various sectors that are disrupted by NISQ computing. Therefore, the overall objective of this chapter is to provide an exposition to NISQ computing, and demonstrate a link between NISQ computing and the fourth Industrial Revolution.

Keywords Quantum computing · Fourth industrial revolution · Qubit · Noisy intermediate-scale quantum (NISQ) · Quantum machine learning · Differential privacy

1 Introduction

Following the scientific advances in the field of thermodynamics, the first Industrial Revolution (1IR) involved mechanization of products by harnessing the steam power (Maynard 2015; Schwab 2017). On the other hand, scientific advances in the field of

M. Senekane (✉) · M. B. Taele
Department of Physics and Electronics, National University of Lesotho, Roma, Lesotho
e-mail: mc.senekane@nul.ls

M. Maseli
Department of Mathematics and Computer Science, National University of Lesotho, Roma, Lesotho

W. Doorsamy et al. (eds.), *The Disruptive Fourth Industrial Revolution*, Lecture Notes in Electrical Engineering 674,
https://doi.org/10.1007/978-3-030-48230-5_9

electromagnetism resulted in the second Industrial Revolution (2IR); which involves the harnessing of electricity in mass productions (Maynard 2015). Furthermore, the success of the first quantum revolution, which resulted in the invention of a transistor, provided a platform for the third Industrial Revolution (3IR). The third Industrial Revolution involved the use of electronics and information technology to automate production. The current industrial revolution is the fourth Industrial Revolution (4IR). Figure 1 shows a schematic diagram of the four industrial revolutions.

The fourth Industrial Revolution (4IR), also referred to as Industry 4.0, is characterized by paradigm shift from the Information and Communication society of 3IR to the super-intelligent society (Maynard 2015; Schwab 2017; Schwab and Davis 2018; Gleason 2018; Um 2019). It fuses together various technologies in such a way as to blur the lines between and across the digital, physical and biological spheres, resulting in Cyber-Physical Systems (CPS) (Schwab 2017). The digital drivers of 4IR include the Internet of Things (IoT), Artificial Intelligence (AI), blockchain technology and quantum computing. On the other hand, physical drivers of the fourth Industrial Revolution include additive manufacturing, three-dimensional (3D) printing, robotics and new materials such as graphene. Finally, biological drivers of 4IR include nanotechnology, biotechnology and synthetic biology techniques such as gene editing, gene therapy and artificial gene synthesis.

Quantum computing heralds an era of the second quantum revolution (Dowling and Milburn 2003; Aspect 2007). Quantum computing harnesses quantum mechanical concepts such as superposition, entanglement and tunneling in order to process information (Lo et al. 1998; Bennett and DiVincenzo 2000; Nielsen and Chuang 2010; Preskill 1998; Mermin 2007; Yanofsky and Mannucci 2008; Aaronson 2013; Wilde 2017). A quantum computer provides a disruptive and paradigm shift in how computing is done. A conventional computer uses a binary digit (bit) as a unit of information. A bit can be in a state of either '0' or '1'. On the other hand, a quantum

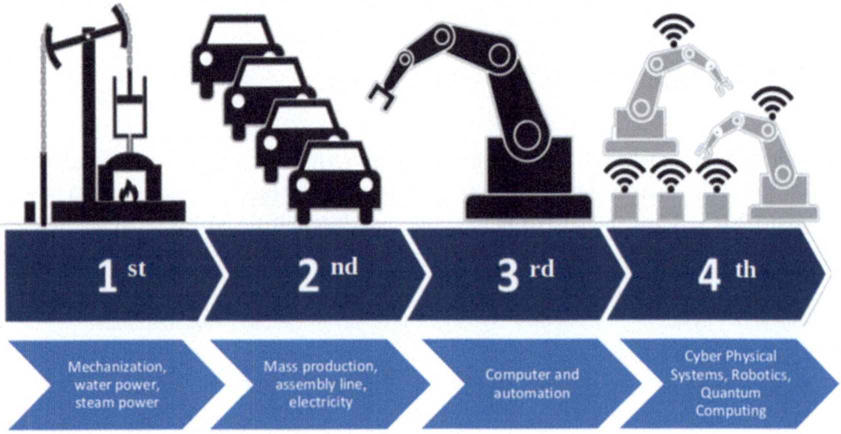

Fig. 1 A schematic diagram of the four industrial revolutions. The Figure is adopted from: https://www.allaboutlearn.com/industry-4-0/

computer uses a quantum binary digit (qubit) as a unit of information. Unlike a bit, a qubit can be in a superposition of states '0' and '1' (Nielsen and Chuang 2010; Wilde 2017). By harnessing the principles of quantum mechanics, quantum computers can solve some of the problems that are intractable to the conventional computers (Nielsen and Chuang 2010; Mermin 2007; Aaronson 2013).

The full-scale quantum computers do not exist yet. However, Noisy Intermediate-Scale Quantum (NISQ) computers are already in existence (Preskill 2018; Boixo et al. 2018; Schuld and Petruccione 2018). NISQ computers are characterized by a low number of qubits (less than 100 qubits), and the noisy nature of the computation (Preskill 2018). In this chapter, we provide an exposition to quantum computing, with emphasis on noisy Intermediate-scale quantum computing. The key contributions of this Chapter are an exposition to quantum computing and NISQ frameworks, an exposition to NISQ players/disruptors, an exposition to NISQ computing applications, and a demonstration of a novel differentially private machine learning scheme on a NISQ platform.

The remainder of this chapter is structured as follows. The next section provides background information on quantum information technology, quantum computing models, qubit technologies, quantum machine learning and differential privacy. This is followed by Sect. 3, which provides and discusses NISQ computing disruptors. Section 4 provides the applications that are computed on NISQ platforms. Finally, Sect. 5 provides a summary of this Chapter.

2 Background Information

2.1 Quantum Information Theory

For conventional computing devices, the basic unit of information is a bit, which can be in a state of either '0' or '1'. However, in the quantum realm, the basic unit of information is a qubit (Nielsen and Chuang 2010; Wilde 2017; Desurvire 2009; Hayashi 2017). Unlike a conventional bit, a qubit can exist in a linear combination (superposition) of two states $|0\rangle$, and $|1\rangle$, and these two states are known as computational basis states. Additionally, a state of a qubit is a vector in a two-dimensional complex vector space (Nielsen and Chuang 2010). Mathematically, a qubit $|\psi\rangle$, is given as

$$|\psi\rangle = \alpha|0\rangle + \beta|1\rangle, \tag{1}$$

where α and β, which are complex numbers known as probability amplitudes, satisfy the condition

$$|\alpha|^2 + |\beta|^2 = 1, \tag{2}$$

and $|.\rangle$ is a Dirac notation of a quantum state (Nielsen and Chuang 2010; Wilde 2017). In matrix notation (Heisenberg notation), the computational basis states are given as $|0\rangle = \begin{pmatrix} 1 & 0 \end{pmatrix}^T$ and $|1\rangle = \begin{pmatrix} 0 & 1 \end{pmatrix}^T$. From Eq. 2, it should be noted that upon measurement, the quantum state can collapse to '0' with probability $|\alpha|^2$ and to '1' with probability $|\beta|^2$, hence why these probabilities add up 1.

Geometrically, a qubit can be represented using a Bloch sphere, as shown in Fig. 2. From the figure, it can be seen that the probability amplitudes α and β are given as (Nielsen and Chuang 2010; Wilde 2017; Schuld and Petruccione 2018)

$$\alpha = \cos\frac{\theta}{2} \tag{3}$$

and

$$\beta = e^{i\varphi}\sin\frac{\theta}{2}, \tag{4}$$

where $0 \leq \theta \leq \pi$ and $0 \leq \varphi \leq 2\pi$.

In essence, there are two operations that can be performed on the qubit. The first operation is the unitary transformation, which is also known as unitary gate operation (quantum gates will be covered later in this chapter). Another operation that can be performed on a qubit is a quantum measurement (Nielsen and Chuang 2010; Mermin 2007). Unlike the unitary gate operation, qubit measurement is not unitary, hence not reversible. Once the quantum state (a qubit) is measured, it collapses to the classical state of either '0' or '1'.

Qubits can be combined together to form complex composite systems. For instance, two quantum states $|0\rangle$ and $|1\rangle$ can be combined to form a new composite state $|\psi_{01}\rangle$, which is just a tensor product of the two states:

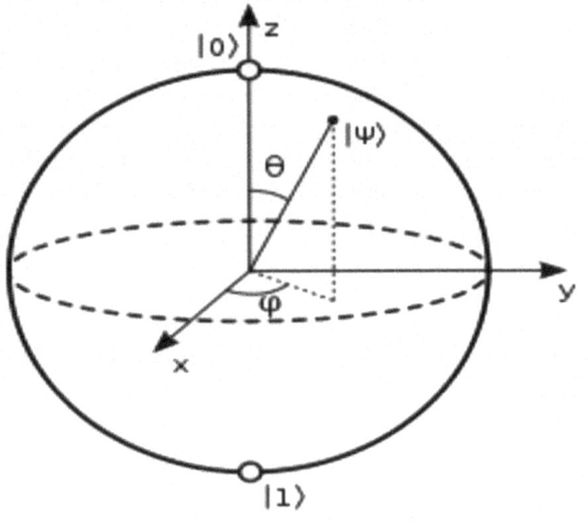

Fig. 2 A geometric representation of a qubit using Bloch sphere

$$|\psi_{01}\rangle = |01\rangle = |0\rangle \otimes |1\rangle. \tag{5}$$

The quantum state $|\psi_{01}\rangle$ can be separated to the constituent states $|0\rangle$ and $|1\rangle$. However, not all quantum states are separable. The quantum states that are not separable are said to be entangled (Nielsen and Chuang 2010; Yanofsky and Mannucci 2008; Wilde 2017). Together with superposition and tunneling (which will be discussed later in this chapter), entanglement is one of the resources of quantum computation.

2.2 *Models of Quantum Computing*

Quantum computing can be realized using any of the following models of computation: circuit model, measurement-based (cluster state) model, topological quantum computation model, dissipative model (Verstraete et al. 2009; Kliesch et al. 2011; Sinayskiy and Petruccione 2012), adiabatic quantum computation model and continuous variable quantum computation model. Quantum computing models are computationally equivalent, up to a polynomial factor (Nielsen and Chuang 2010; Aharonov et al. 2008).

2.2.1 Circuit-Based Model of Quantum Computing

The quantum circuit model is the most commonly used model of quantum computing (Nielsen and Chuang 2010). It is the model which is conceptually closest to the conventional manner of computation, since it too uses the quantum gates; which are the unitary operators, to perform computation. Quantum circuits consist of a series of input quantum states, a network of unitary transformations (quantum gates) which evolve input quantum states in order to realize a computation, and output quantum states, which can then be measured (Nielsen and Chuang 2010; Preskill 1998; Wilde 2017; Senekane et al. 2013).

With the quantum circuit model, quantum algorithms can be designed by using a sequence of quantum gates. Quantum gates can either act on a single qubit, on two qubits, or on three qubits. Examples of one-qubit gates include the Pauli gates (Identity, X, Y, and Z gates) and the Hadamard gates. On the other hand, an example of a two-qubit gate is a controlled NOT (CNOT) gate, while three-qubit gates include Toffoli and Fredkin gates (Nielsen and Chuang 2010). A Hadamard gate is given as

$$H = \frac{1}{\sqrt{2}}\begin{pmatrix} 1 & 1 \\ 1 & -1 \end{pmatrix}, \tag{6}$$

Identity gate (I) is given as

$$I = \begin{pmatrix} 1 & 0 \\ 0 & 1 \end{pmatrix}, \tag{7}$$

X gate as

$$X = \begin{pmatrix} 0 & 1 \\ 1 & 0 \end{pmatrix}, \tag{8}$$

and CNOT as

$$CNOT = \begin{pmatrix} 1 & 0 & 0 & 0 \\ 0 & 1 & 0 & 0 \\ 0 & 0 & 0 & 1 \\ 0 & 0 & 1 & 0 \end{pmatrix}. \tag{9}$$

Further details about quantum gates can be found in (Nielsen and Chuang 2010; Preskill 1998; Wilde 2017). However, it's worth briefly discussing how these gates act on the qubits. For instance, an X gate flips the quantum state. Thus, $|0\rangle$ is flipped to $|1\rangle$ and vice versa. On the other hand, a Hadamard gate introduces quantum superposition such that

$$H|0\rangle = \frac{1}{\sqrt{2}}(|0\rangle + |1\rangle) \tag{10}$$

and

$$H|1\rangle = \frac{1}{\sqrt{2}}(|0\rangle + |1\rangle). \tag{11}$$

Finally, the controlled NOT (CNOT) gate, which is a two-qubit quantum gate, flips the second qubit if the first qubit is in state '1'. Thus,

$$\begin{aligned} &CNOT|00\rangle \mapsto |00\rangle, CNOT|01\rangle \mapsto |01\rangle, \\ &CNOT|10\rangle \mapsto |11\rangle, CNOT|11\rangle \mapsto |10\rangle. \end{aligned} \tag{12}$$

2.2.2 Cluster State Quantum Computing Model

The previous model of computation (circuit model) is based on the sequence of unitary gate operations. On the other hand, a cluster state quantum computing model, which is also called measurement-based quantum computing model, operates differently. It mainly uses entanglement as a resource for quantum computation. Initially, highly entangled multi-particle quantum states (cluster states) are prepared, followed by the single-qubit measurements, in order to perform quantum computation (Raussendorf and Briegel 2001; Raussendorf 2009; Raussendorf et al. 2007; Walther et al. 2005).

Unlike in the case of a circuit model, where the quantum computation is reversible (due to the use of unitary gates), quantum computation in a cluster state model is irreversible (one-way). This owes to the prominent role that measurement plays in this model. It should be recalled that quantum measurement collapses a quantum state to the classical one, hence why it (measurement) is an irreversible operation. Just like the circuit model, measurement-based model is universal for quantum computation (Raussendorf 2009).

2.2.3 Topological Quantum Computation

Topological quantum computing model uses topological phases of matter to realize hardware-based fault-tolerant quantum computation (Preskill 1998; Collins 2006; Nayak et al. 2008; Pachos 2012; Stanescu 2016; Lahtinen and Pachos 2017). It performs computation by making use of topological properties of physical systems. The key benefit of topological quantum computing is that due to topological properties of the physical systems, topological physical systems remain invariant under local perturbations, resulting in robustness against noise (Collins 2006; Pachos 2012; Stanescu 2016).

Topological quantum computing model stores and manipulates information through the use of two-dimensional exotic particles called non-Abelian anyons (Preskill 1998; Pachos 2012; Stanescu 2016; Lahtinen and Pachos 2017). Majorana fermions are the simplest non-Abelian anyons (Alicea 2012; Beenakker 2013). As will be discussed later, at least one NISQ disruptor uses a Majorana fermion as a qubit technology for quantum computing. The computation in this model of computing is performed by first creating pairs of anyons, and then braiding together these anyons to perform computation (Preskill 1998; Collins 2006; Pachos 2012; Stanescu 2016). Like the previous models, topological quantum computing model is universal for quantum computation (Freedman et al. 2002).

2.2.4 Adiabatic Model of Quantum Computing

This model of quantum computing makes use of quantum tunneling, which is a quantum phenomenon in which a quantum system can tunnel through the barrier with a non-zero probability (Schuld and Petruccione 2018; Razavy 2013). Also, it is based on adiabatic theorem, which states that if the quantum state at ground state is evolved slowly enough, it will remain at the ground state throughout the entirety of its evolution (Farhi et al. 2000). Additionally, adiabatic quantum computation is universal (Farhi et al. 2000; Biamonte and Love 2008).

In order to perform computation using adiabatic quantum computing, the ground state of the Hamiltonian (Hamiltonian is the energy of the system), H_p, is found.

Then a simple Hamiltonian (H_i) is prepared initialized (Farhi et al. 2000). This simple Hamiltonian is then slowly (adiabatically) evolved until it converges to H_p. For a final Hamiltonian H_f, adiabatic quantum computation can be represented as

$$H_f = (1 - s)\left[H_i + H_p\right], \tag{13}$$

for evolution time factor s, where $0 \leq s \leq 1$.

2.2.5 Continuous Variable Quantum Computing

Unlike the models of quantum computing discussed thus far, this model of computation uses features of quantum systems (observables) which are continuous in nature (Schuld and Petruccione 2018; Weedbrook et al. 2012; Braunstein and Pati 2012; Adesso et al. 2014; Serafini 2017). Additionally, as opposed to discrete-variable quantum systems (which encode in a discrete quantum system such as a two-level qubit), continuous-variable quantum systems encode information in a quantum state with a continuous basis (Schuld and Petruccione 2018). Thus, discrete-variable quantum systems reside in the discrete-dimensional complex Hilbert space, while their continuous-variable counterparts reside in the infinite-dimensional complex Hilbert space. However, like the models discussed earlier, continuous-variable quantum computing model is also universal for quantum computation (Braunstein and Pati 2012).

One area where quantum computing is disruptive is in the deployment of technology used to perform quantum computing. Unlike a conventional computer; which uses semiconductor technology for computing, quantum computing uses a variety of technologies to perform quantum computation. These technologies are discussed in the next subsection.

2.3 *Qubit Technologies*

Technically, any two-level quantum system is a potential candidate for the qubit (Lo et al. 1998). However, in practice, some quantum systems are easier to prepare and manipulate than others. Therefore, in this chapter, the reporting on the qubit technologies will be limited to the cases where such technologies have already been investigated in the laboratories. The aim is not to provide the extensive literature review on quantum technologies, but rather, to provide basic exposition to such technologies.

Quantum systems that can be used to realize qubits include nuclear magnetic resonance systems (Nielsen and Chuang 2010; Gruska 1999), trapped neutral atoms and ions (Nielsen and Chuang 2010; Briegel et al. 2000; Saffman 2016), superconducting circuits (Chaikin et al. 1995; Sidebottom 2012), photons/light (Knill et al. 2001), quantum dots (Loss and DiVincenzo 1998; DiVincenzo 2000; Eriksson et al.

2004; Trauzettel et al. 2007), nitrogen-vacancies in diamonds (Prawer and Greentree 2008; Weber et al. 2010; Childress and Hanson 2013), and non-Abelian anyons (Nayak et al. 2008; Alicea 2012; Beenakker 2013).

One application where quantum computing is having a disruptive impact is in the area of artificial intelligence, particularly in machine learning. The next subsection provides a brief introduction to machine learning and goes further to discuss quantum machine learning.

2.4 Machine Learning and Quantum Machine Learning

Machine learning is a branch of artificial intelligence (one of the drivers of 4IR) which enables computers to learn from data (past experience) without being explicitly programmed (Schuld and Petruccione 2018; Russell and Norvig 2016). Machine learning can be broadly classified into three categories, namely; supervised learning, unsupervised learning and reinforcement learning. In supervised learning, the input data (training data) is provided together with the output data (labels). The task of supervised learning models is to create a mapping that maps the features to the labels (Russell and Norvig 2016). Supervised learning is applied in classification and regression problems (Schuld and Petruccione 2018).

In unsupervised learning, only the features are provided, without the labels. The task of an unsupervised machine learning model is to generate come up with the distribution (structure) that generates the underlying data (the features). Unsupervised learning is applied in clustering and dimensionality reduction problems. Finally, reinforcement learning is a category of machine learning that determines the action to be taken within a specific environment in order to

maximize the reward (Russell and Norvig 2016; Sutton and Barto 2018). Applications of reinforcement learning include robotics and optimization problems.

Quantum machine learning fuses together the conventional machine learning techniques with the concepts from quantum information theory. The objective is to provide a 'quantum advantage' by using quantum information theory concepts (such as entanglement, superposition and tunneling) in order to improve the performance of machine learning models (Schuld and Petruccione 2018; Wittek 2014). However, it should be noted that the pursuit of 'quantum advantage' can prove to be elusive in some cases (Tang 2018; Gilyén et al. 2018).

In general, machine learning can be summarized as shown in Fig. 3, depending on the kind of data being used, and the computing platform. The 'CQ' (classical data, classical computer) approach is the conventional approach to quantum computing; where classical data is processed on the conventional classical computer. Additionally, the 'QC' (quantum data, classical computer) approach uses quantum data on the classical computer. Furthermore, 'CQ' (classical data, quantum computer) approach processes classical data on the quantum platform. Finally, the 'QQ' (quantum data, quantum computer) approach focuses on the processing of the quantum data on the quantum computer. Both 'CQ' and 'QQ' can be considered as quantum machine

Fig. 3 Different approaches to machine learning

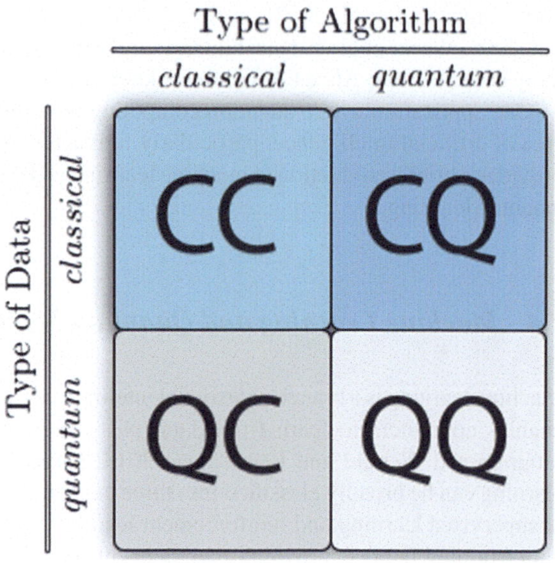

learning (Schuld and Petruccione 2018). Quantum machine learning is introduced in this subsection because it will be revisited later in this chapter, when discussing the differentially private quantum machine learning (which is a novel and potentially disruptive application of quantum computing).

2.5 *Differential Privacy*

Later in this chapter, a quantum differential privacy will be discussed. It is therefore imperative to provide a brief introduction to the conventional differential privacy (DP) in this subsection. Differential privacy was first formulated as a mathematically rigorous definition of security/privacy by Dwork et al. (2006; Dwork 2006; Senekane 2019). It provides a guarantee that the risk of an individual (or a group of individuals) participating in a statistical database will not increase as a result of participating in such a statistical database. Additionally, DP is useful for quantifying and bounding privacy loss (Dwork 2011).

Differential privacy can be obtained using two approaches. The first approach generates ε-DP. This DP is known as a multiplicative DP, and is the strongest definition of security. However, the guarantee of a ε-DP can be relaxed, in order to have (ε, δ)-DP (Dwork et al. 2014). The latter DP is said to be additive DP, and is also known to be asymptotically more accurate than the former (Dwork et al. 2014).

Definition 1 A randomized function K gives an ε-differential privacy if for all datasets D_1 and D_2 differing in at most one element, and all $S \subseteq Range(K)$,

$$\mathbb{P}[\mathrm{K}(D_1) \in \mathrm{S}] \leq exp(\epsilon) \times \mathbb{P}[\mathrm{K}(D_2) \in \mathrm{S}], \tag{14}$$

where $\epsilon \ll 1$.

Definition 2 A randomized function K gives an, (ε, δ)-differential privacy if for all datasets D_1 and D_2 differing in at most one element, and all $S\ Range(K)$,

$$\mathbb{P}[\mathrm{K}(D_1) \in \mathrm{S}] \leq exp(\epsilon) \times \mathbb{P}[\mathrm{K}(D_2) \in \mathrm{S}] + \delta, \tag{15}$$

where $\epsilon \ll 1$ and δ is cryptographically small.

This section provided the basic information on quantum information processing and differential privacy. This information will be used in later sections of this chapter. As already discussed earlier in this section, quantum computing has a disruptive impact in the deployment of technologies used to perform computing. However, technology impact is not the only disruption that quantum computing has. The other disruptive impacts are both economic and social. The next section discusses economic and also social impacts of quantum computing. This is done by discussing a variety of quantum computing's disruptive players.

3 NISQ Computing Disruptive Players

Noisy intermediate-scale quantum computing disruptors portray a fusion of prominent technology giants such as Google, Microsoft and International Business Machines (IBM) with start-ups such as Rigetti Computing and Xanadu. In this section, we provide an overview of NISQ players. However, since NISQ computing is a rapidly growing market, the list of companies discussed in this section is by no means exhaustive. However, an observation that can be made is that to date, the NISQ market is dominated by companies from North America and Europe.

The NISQ players discussed in this section address two disruptive impacts of quantum computing, namely economic and social impacts. The economic impact is mainly due to the creation of quantum computing-related jobs. On the other hand, the social impact focuses on how inclusive each company is, in terms of sharing its quantum computing resources with the community.

3.1 IBM

IBM is arguably the most advanced NISQ player, both in terms of quantum hardware and quantum software. It uses the circuit model of quantum computing and uses superconducting circuits as qubit technology. Additionally, IBM uses a python-based software tool called Quantum Information Science Kit (QISKit) (Cross 2018; Silva

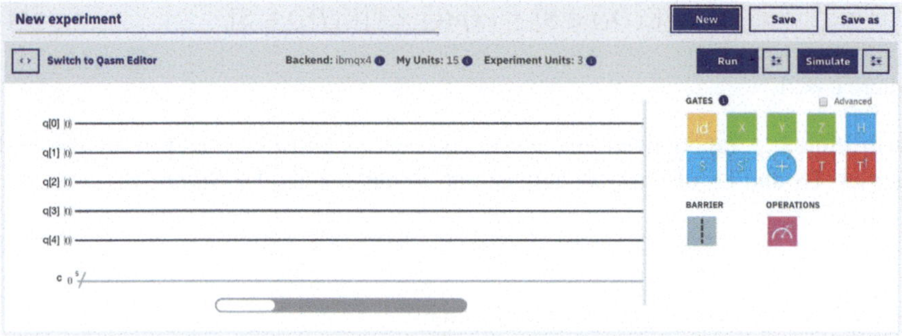

Fig. 4 A layout of IBM's 5-qubit quantum experience platform

2018; LaRose 2019). Furthermore, IBM free provides cloud-based access to its NISQ computer through the Quantum Experience platform (Silva 2018).

IBM Quantum Experience platform's layout is shown in Fig. 4. The platform enables circuit synthesis (quantum experiment design) using either Open quantum Assembly language (OpenQASM) or the composer (Silva 2018; LaRose 2019). The composer uses a drag-and-drop feature, whereby in order to build quantum circuits, gates can be dragged from the right into the composer on the left. Additionally, the platform provides an option to either run quantum experiments on the actual NISQ processor or simulate them using IBM's supercomputing facilities. Currently, IBM offers access to a NISQ processor with up to 5 qubits, for researchers who subscribe for free.

If one decides to design quantum experiments on the local machine and later send them to the IBM NISQ infrastructure, they can use the QISKit tool. QISKit is a python-based Software Development Kit (SDK) for quantum programming (Silva 2018). QISKit comprises of four components/elements, namely QISKit Terra (for composing quantum algorithms at circuit level), QISKit Aqua (for building complex algorithms), QISKit Ignis (for addressing noise and errors) and QISKit Aer (for acceleration of the development of quantum computers through the improved simulations, emulations and debugging) (Wille et al. 2019; Barabasi et al. 2019).

3.2 Google

Like IBM, Google uses the circuit model of quantum computing and uses superconducting circuits as qubit technology. The software platforms used by Google are Cirq and OpenFermion-Cirq (LaRose 2019; McClean et al. 2017). Cirq is a software library for writing, manipulating, and optimizing quantum circuits and then running them against quantum computers and simulators (LaRose 2019). The key design philosophy of Cirq is that instead of abstracting away the hardware details of quantum computing circuits, such details should be exposed.

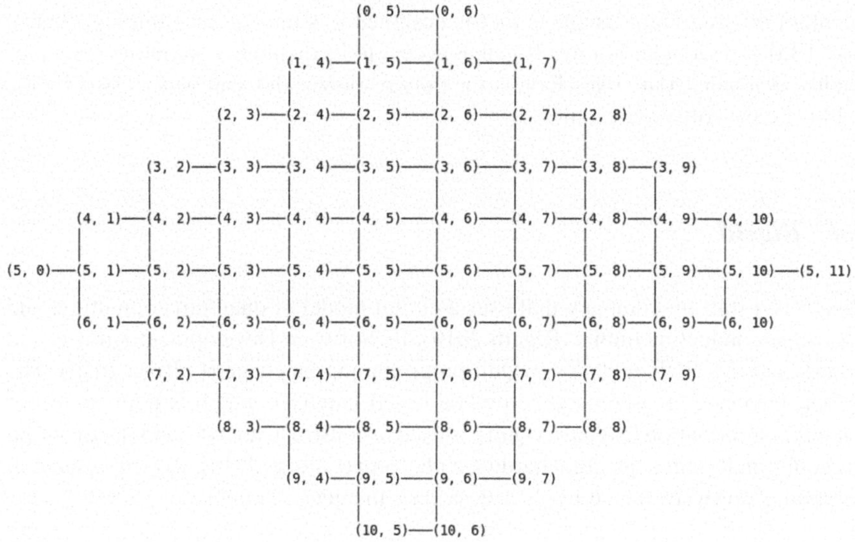

Fig. 5 Bristlecone NISQ processor's qubit arrangement

Additionally, Cirq is intended to be an interface for researchers to access Google's 22-qubit Foxtail and 72-qubit Bristlecone NISQ processors. Figure 5 shows an architecture (qubit arrangement) for a 72-qubit Bristlecone NISQ processor.

OpenFermion-Cirq is a software library for developing algorithms for chemistry and material science problems. It is based on the open-source, python-based software library called OpenFermion; which is used to simulate fermionic systems and quantum chemistry problems on a quantum computer (McClean et al. 2017). OpenFermion-Cirq is a Google plug-in that extends the functionality of OpenFermion by providing routines and tools for using Cirq to compile and compose circuits for quantum simulation algorithms.

Like IBM, Google makes its quantum computing resources freely accessible to the community. However, the access is only to the software resources (Cirq and OpenFermion-cirq), not the hardware.

3.3 Microsoft

Unlike IBM and Google, Microsoft uses a topological model of quantum computing and uses a Majorana fermion (non-Abelian anyon) as a qubit technology (LaRose 2019; Castelvecchi 2017). On the software side, Microsoft uses a software development kit called Quantum Development Kit (QDK), which features Microsoft's Q# programming language (LaRose 2019; Häner et al. 2018). The Q# language supports the simulation of up to 30 qubits on a local machine. Q# is a domain-specific C#-like quantum programming language that is designed to simulate quantum algorithms. Like IBM and Google, Microsoft makes access to its quantum computing resources readily available. Also, like Google, the only resources that can be accessed by the public are the software resources.

3.4 Rigetti

Rigetti is a start-up company that uses a circuit model of quantum computing, and uses superconducting qubits (Rigetti 2019). Its Software Development Kit is called Forest, and this SDK includes pyQuil programming language (LaRose 2019). The pyQuil language, an open-source python-based language which is built on top of Quantum information language (Quil), is used to construct, analyze and run quantum programs on Rigetti's quantum hardware platform (LaRose 2019). Rigetti's quantum computing software can be freely accessed by the general public.

3.5 D-Wave

Unlike most NISQ computing companies, D-Wave does not use a circuit model of quantum computing. Rather, it uses adiabatic quantum computation (D-Wave 2019). When it comes to qubit technology used, D-Wave uses superconducting circuits. For the development of quantum algorithms, D-Wave uses Ocean SDK. Furthermore, D-Wave has a cloud-based quantum platform called Leap. However, access to Leap cloud service is limited to researchers from around 33 countries, which are mostly European countries (D-Wave 2019).

3.6 Xanadu

Xanadu is a start-up company from Canada, which uses continuous-variable model of quantum computing. It has developed two python-based quantum software tools, namely PennyLane (Bergholm et al. 2018) and Strawberry Fields (Killoran et al. 2019). PennyLane is intended for applications in optimization and machine learning problems. On the other hand, Strawberry Fields is a python-based software kit intended for the design and simulation of quantum algorithms using a photonic continuous-variable quantum computer. Xanadu offers the public free access to its software resources.

3.7 *Other NISQ Players*

NISQ computing market is the rapidly growing market. As a result, it is virtually impossible to discuss all the NISQ computing players. The list of disruptive players discussed above is but a handful of the current players. Other NISQ computing players include ProjectQ (LaRose 2019), ProteinQure (ProteinQure 2019), 1QBit (QBit 2019), IonQ (IonQ uses trapped ions as qubits) (IonQ 2019) and Zapata Computing (Zapata Computing is a start-up that sprung from the research laboratory at Harvard University) (Computing 2019).

4 NISQ Applications

NISQ computing is creating a rapidly growing market niche. As such, Noisy Intermediate-Scale Quantum computing applications cut across a wide range of sectors. These sectors include simulation and verification of quantum experiments, optimization, artificial intelligence, quantum machine learning, chemistry and materials science (Schuld and Petruccione 2018; LaRose 2019; Coles et al. 2018).

This section covers some (and by no means not all) of the applications. Furthermore, a new application, namely, the differentially- private quantum machine learning on NISQ computing, is discussed.

4.1 *NISQ Computing Application in Quantum Algorithms*

Quantum computing has the potential to be one of the most disruptive technologies of the fourth Industrial Revolution. This owes to the fact that quantum computing uses quantum mechanics, which enables massive parallelism and hence increased computational capabilities, compared to conventional computing paradigm. One of the most important applications of quantum computing is in the design of algorithms that outperform their conventional counterparts.

A quantum algorithm for testing for entanglement of quantum systems is known as Bell-test algorithm (Nielsen and Chuang 2010; Wilde 2017). The quantum circuit for implementing the Bell-test algorithm on IBM's 5-qubit Quantum Experience platform is shown in Fig. 6. Furthermore, Fig. 7 shows the results obtained from this experiment after 1024 iterations.

Another prominent quantum algorithm is the Grover's search algorithm (Nielsen and Chuang 2010). In essence, this algorithm can be likened to searching for a needle in a hay-stack. Compared to the best-known conventional algorithm, Grover's algorithm provides a quadratic speed-up for unstructured search problems. Figure 8 shows the details of the implementation of Grover's algorithm using Google's Cirq.

Fig. 6 A quantum circuit for Bell-test algorithm using IBM quantum experience

Fig. 7 Results obtained for Bell-test experiment after 1024 iterations

```
Secret bit sequence: [0, 1]
Circuit:
(0, 0): ───H───X───@───X───H───X───@───X───H───────────M('result')───
                   │                   │                │
(1, 0): ───H───────@───H───X───H───X───H───X───H───M─────────────────
                   │
(2, 0): ───X───H───X─────────────────────────────────────────────────
Sampled results:
Counter({'01': 100})
Most common bitstring: 01
Found a match: True
```

Fig. 8 Implementation of Grover's algorithm using Google's Cirq

This algorithm uses the following gates: Measurement (M), X gate (X), Hadamard gate (H, the CNOT gate (X-@) and the Fredkin gate (X-@-@).

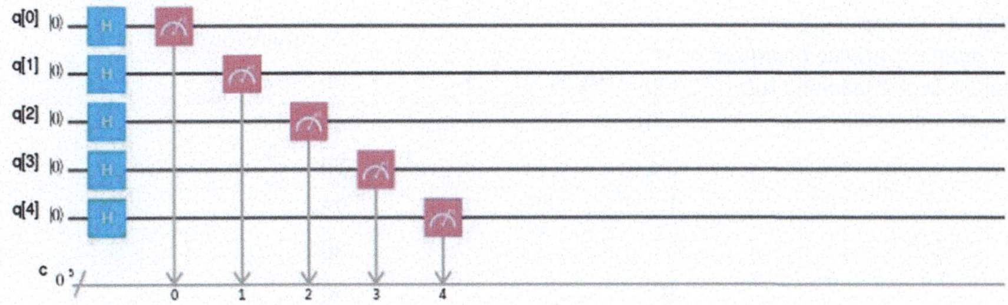

Fig. 9 A quantum circuit for generating a 5-bit random sequence using IBM's quantum experience

4.2 *Quantum Random Number Generator*

A random number generator (RNG) is a device that produces a random sequence of bits. RNG has applications in diverse fields such as cybersecurity, gaming and lottery (gambling). However, conventional RNG solutions can only offer pseudo-randomness. On the other hand, because of the true randomness of quantum measurement, a quantum random number generator (QRNG) offers a truly random number generation. Figure 9 shows a quantum circuit to generate a 5-bit random sequence.

4.3 *Differentially Private Quantum Machine Learning*

The concept of differentially private quantum machine learning was proposed in 2017 (Senekane et al. 2017). In this chapter, we demonstrate the implementation of a privacy-preserving quantum machine learning scheme on a NISQ platform. Both differential privacy and quantum machine learning were introduced in Sect. 2 of this chapter. As can be observed, quantum machine learning combines these two concepts, to produce a privacy-preserving scheme. Basically, the operation of a differentially private quantum machine learning scheme proposed in this chapter is as follows. It uses a classical dataset, adds the noise to the dataset, then encodes the result into the quantum states, and then implement on the quantum computer. The noise added depends on whether the differential privacy mechanism used satisfies -differential privacy definition or (ε, δ)-differential privacy definition (Dwork et al. 2014). Typically, Laplace mechanism (which adds Laplace noise to the data) is used as a ε-differential privacy mechanism, while Gaussian mechanism (which adds Gaussian noise to the data) is used as a (ε, δ)-differential privacy mechanism.

The differentially private quantum machine learning algorithm proposed in this chapter uses the ϵ-differential privacy, and applies Laplace noise to the input classical data (Senekane et al. 2017). In this case, the input dataset used is the Wisconsin Breast Cancer dataset (Learning 2019). Then using QISKit, a quantum support vector machine (QSVM) algorithm is designed, using values of ϵ ranging from 0.01 to 10. The results are summarized in Table 1.

Table 1 Performance of differentially private quantum support vector machine for various values of ϵ

ϵ	Accuracy (%)
Pure QSVM	90
0.01	60
0.1	70
1	75
5	80
10	90

5 Summary

In this chapter, Noisy Intermediate-Scale Quantum computing, which is one of the technologies driving the fourth Industrial Revolution, was discussed. Additionally, the chapter provided a succinct exposition of quantum models of computing, qubit technologies and NISQ computing disruptive players was done. Furthermore, the chapter discussed some applications of NISQ computing. Since NISQ computing is a rapidly growing field, it is virtually impossible to cover all the applications of NISQ computing. Finally, a differentially private quantum machine learning algorithm was demonstrated, using NISQ computing.

Quantum computing is already proving to be a disruptive technology of the fourth Industrial Revolution, and it already has the potential to be the most disruptive technology of the fourth Industrial Revolution. Its massive parallelism provides computational capabilities unmatched by the current computing technologies, resulting in the computational speedup, compared to its conventional computing counterpart. Already, quantum computing is employed in a plethora of applications, some of which were discussed in this chapter. These applications are in diverse fields such as cybersecurity, artificial intelligence, quantum chemistry and material science, and optimization.

Although quantum computing is already proving to be a disruptive technology, it is not without some challenges. Currently, quantum computing is implemented using the NISQ framework. In this framework, it is very challenging to demonstrate the computational speedup over conventional computing. Additionally, the NISQ circuits are constructed using noisy qubits, which are not error-corrected. The noisy nature of the NISQ circuits results in very shallow circuits, since all the computing operations must be computed before the qubits decohere (decoherence results in quantum states losing their quantumness; collapsing to classical states).

The NISQ framework is the only framework currently being implemented for quantum computing. A full-scale quantum computer is proving to be both conceptually and technologically challenging. The key challenge to the development of a fully functional quantum computer is decoherence. It is highly challenging to both create and isolate quantum states that can be used for quantum computing. Any interaction of the quantum state with the environment will alter such a state; resulting in errors in computation. In order to mitigate these errors, various error-correcting schemes have been developed. However, in order to function properly, these error-correcting codes require availability of a lot (in the order of thousands) of qubits. This requirement is both conceptually and technologically challenging.

Even though there are some serious challenges facing the development of a full-scale quantum computer, the future of this technology is far from being bleak. There are some promising developments both in the design of better error-correction codes and technologies that can generate and isolate several qubits. These conceptual and technological advances make significant contributions towards the realization of a full-scale quantum computer in the short-to-medium term.

References

Aaronson S (2013) Quantum computing since Democritus. Cambridge University Press
Adesso G, Ragy S, Lee AR (2014) Open Syst Inf Dyn 21:1440001
Aharonov D, Van Dam W, Kempe J, Landau Z, Lloyd S, Regev O (2008) SIAM Rev 50:755–787
Alicea J (2012) Rep Prog Phys 75:076501
Aspect A (2007) Nature 446:866
Barabasi S, Barrera J, Bhalani P, Dalvi P, Dimiecik R, Leider A, Mondrosch J, Peterson K, Sawant N, Tappert CC (2019) Future of information and communication conference. Springer, pp 547–563
Beenakker C (2013) Annu Rev Condens Matter Phys 4:113–136
Bennett CH, DiVincenzo DP (2000). Nature 404: 247
Bergholm V, Izaac J, Schuld M, Gogolin C, Killoran N (2018). arXiv preprint arXiv:1811.04968
Biamonte JD, Love PJ (2008) Phys Rev A 78:012352
Boixo S, Isakov SV, Smelyanskiy VN, Babbush R, Ding N, Jiang Z, Bremner MJ, Martinis JM, Neven H (2018) Nat Phys 14:595
Braunstein SL, Pati AK (2012) Quantum information with continuous variables. Springer Science & Business Media
Briegel HJ, Calarco T, Jaksch D, Cirac JI, Zoller P (2000) J Mod Opt 47:415–451
Castelvecchi D (2017) Nat News 541:9
Chaikin PM, Lubensky TC, Witten TA (1995) Principles of condensed matter physics, vol 1. Cambridge University Press
Childress L, Hanson R (2013) MRS Bull 38:134–138
Coles PJ, Eidenbenz S, Pakin S, Adedoyin A, Ambrosiano J, Anisimov P, Casper W, Chennupati G, Coffrin C, Djidjev H et al (2018). arXiv preprint arXiv:1804.03719
Collins GP (2006) Sci Am 294:56–63
Computing Z (2019) Zapata Computing. https://www.zapatacomputing.com/. Accessed 20 May 2019
Cross A (2018). APS Meeting Abstracts
Desurvire E (2009) Classical and quantum information theory: an introduction for the telecom scientist. Cambridge University Press
DiVincenzo DP (2000) Fortschritte der Physik: Progress Phys 48:771–783
Dowling JP, Milburn GJ (2003) Philos Trans Roy Soc Lond Ser A Math Phys Eng Sci 361:1655–1674
D-Wave (2019) D-Wave quantum computing company. https://www.dwavesys.com/. Accessed 20 May 2019
Dwork C (2006). Automata Lang Program 1–12
Dwork C (2011) Foundations of Computer Science (FOCS). In: 2011 IEEE 52nd Annual Symposium on (IEEE), pp 1–2
Dwork C, McSherry F, Nissim K, Smith A (2006) Theory of cryptography conference. Springer, pp 265–284

Dwork C, Roth A et al (2014) Found Trends Theor Comput Sci 9:211–407
Eriksson MA, Friesen M, Coppersmith SN, Joynt R, Klein LJ, Slinker K, Tahan C, Mooney P, Chu J, Koester S (2004) Quant Inf Process 3:133–146
Farhi E, Goldstone J, Gutmann S, Sipser M (2000). arXiv preprint quant-ph/0001106
Freedman MH, Larsen M, Wang Z (2002) Commun Math Phys 227:3
Gilyén A, Lloyd S, Tang E (2018). arXiv preprint arXiv: 1811.04909
Gleason NW (2018) Higher education in the era of the fourth industrial revolution. Springer
Gruska J (1999) Quantum computing, vol 2005. McGraw-Hill, London
Häner T, Steiger DS, Svore K, Troyer M (2018). Quant Sci Technol 3: 020501
Hayashi M (2017) Quantum information theory. Springer
IonQ (2019) IonQ https://ionq.co/. Accessed 20 May 2019
Killoran N, Izaac J, Quesada N, Bergholm V, Amy M, Weedbrook C (2019) Quantum 3:129
Kliesch M, Barthel T, Gogolin C, Kastoryano M, Eisert J (2011) Phys Rev Lett 107:120501
Knill E, Laflamme R, Milburn GJ (2001). Nature 409: 46
Lahtinen V, Pachos J (2017) Sci Post Phys 3:021
LaRose R (2019) Quantum 3:130
Learning UM (2019) Breast cancer Wisconsin (diagnostic) data set. https://archive.ics.uci.edu/ml/datasets/Breast+Cancer+Wisconsin+(Diagnostic). Accessed 15 March 2019
Lo HK, Spiller T, Popescu S (1998) Introduction to quantum computation and information. World Scientific
Loss D, DiVincenzo DP (1998) Phys Rev A 57:120
Maynard AD (2015) Nat Nanotechnol 10:1005
McClean JR, Kivlichan ID, Sung KJ, Steiger DS, Cao Y, Dai C, Fried E S, Gidney C, Gimby B, Gokhale P et al (2017). arXiv preprint arXiv: 1710.07629
Mermin ND (2007) Quantum computer science: an introduction. Cambridge University Press
Nayak C, Simon SH, Stern A, Freedman M, Sarma SD (2008) Rev Mod Phys 80:1083
Nielsen MA, Chuang I (2010) Quantum computation and quantum information. Cambridge University Press
Pachos JK (2012) Introduction to topological quantum computation. Cambridge University Press
Prawer S, Greentree AD (2008) Science 320:1601–1602
Preskill J (1998) California institute of technology, p 16
Preskill J (2018) Quantum 2:79
ProteinQure (2019) ProteinQure. https://www.proteinqure.com/ Accessed 20 May 2019
QBit (2019) 1QBit https://1qbit.com/. Accessed 20 May 2019
Raussendorf R (2009) Int J Quant Inf 7:1053–1203
Raussendorf R, Briegel HJ (2001) Phys Rev Lett 86:5188
Raussendorf R, Harrington J, Goyal K (2007) New J Phys 9:199
Razavy M (2013) Quantum theory of tunneling. World Scientific
Rigetti (2019) Rigetti https://www.rigetti.com/. Accessed 20 May 2019
Russell SJ, Norvig P (2016) Artificial intelligence: a modern approach. Pearson Education Limited, Malaysia
Saffman M (2016) J Phys B: Atom Mol Opt Phys 49:202001
Schuld M, Petruccione F (2018) Supervised learning with quantum computers, vol 17. Springer
Schwab K (2017) The fourth industrial revolution. Currency
Schwab K, Davis N (2018) Shaping the future of the fourth industrial revolution. Currency
Senekane M (2019) Mach Learn Knowl Extract 1:483–491
Senekane M, Mohapi L, Petruccione F (2013). Africon (IEEE), pp 1–4
Senekane M, Mafu M, Taele BM (2017) IEEE Africon. IEEE, pp 1432–1435
Serafini A (2017) Quantum continuous variables: a primer of theoretical methods. CRC Press
Sidebottom DL (2012) Fundamentals of condensed matter and crystalline physics: an introduction form students of physics and materials science. Cambridge University Press
Silva V (2018) Practical quantum computing for developers: programming quantum rigs in the cloud using Python, quantum assembly language and IBM QExperience. Springer

Sinayskiy I, Petruccione F (2012) Quantum Inf Process 11:1301–1309
Stanescu TD (2016) Introduction to topological quantum matter and quantum computation. CRC Press
Sutton RS, Barto AG (2018) Reinforcement learning: an introduction. MIT press
Tang E (2018). arXiv preprint arXiv: 1807.04271
Trauzettel B, Bulaev DV, Loss D, Burkard G (2007) Nat Phys 3:192
Um JS (2019) Drones as cyber-physical systems. Springer
Verstraete F, Wolf MM, Cirac JI (2009) Nat Phys 5:633
Walther P, Resch KJ, Rudolph T, Schenck E, Weinfurter H, Vedral V, Aspelmeyer M, Zeilinger A (2005) Nature 434:169
Weber J, Koehl W, Varley J, Janotti A, Buckley B, Van de Walle C, Awschalom DD (2010) Proc Natl Acad Sci 107:8513–8518
Weedbrook C, Pirandola S, García-Patron R, Cerf NJ, Ralph TC, Shapiro JH, Lloyd S (2012). Rev Modern Phys 84: 621
Wilde MM (2017) Quantum information theory. Cambridge University Press
Wille R, van Meter R, Naveh Y (2019) Design, automation and test in Europe conference and exhibition (DATE). IEEE, pp 1234–1240
Wittek P (2014) Quantum machine learning: what quantum computing means to data mining. Academic Press
Yanofsky NS, Mannucci MA (2008) Quantum computing for computer scientists. Cambridge University Press

The Disruptive 4IR in the Life Sciences: Metabolomics

Fidele Tugizimana, Jasper Engel, Reza Salek, Ian Dubery, Lizelle Piater, and Karl Burgess

Abstract A new era of systems biology is disruptively emerging, holistically describing biochemical events at both organismal and cellular level. In this new era, emerging '*-omics*' technologies have brought about a paradigm shift in biological sciences and research. Metabolomics, the youngest of the *omics* trilogy and defined as the qualitative and quantitative investigation of the entire metabolome of a biological system, has positioned itself as an indispensable methodology to investigate global biochemistry phenomena at a cellular level. Metabolomics is a multidisciplinary research field, involving a convergence of biology, chemistry, chemometrics, statistics and computer science. Metabolomics accordingly can provide unprecedented in-depth explanations and insights of the mechanisms responsible for various physiological conditions, given the innovative developments in analytical technologies (integrating artificial intelligence and machine learning), advancement in chemometric and statistical methods (big data analytics and management), and the integration of orthogonal biological approaches. Thus, the objective of this Chapter is to

Where authors are identified as personnel of the International Agency for Research on Cancer/World Health Organization, the authors alone are responsible for the views expressed in this article and they do not necessarily represent the decisions, policy or views of the International Agency for Research on Cancer/World Health Organization.

F. Tugizimana (✉) · I. Dubery · L. Piater
Department of Biochemistry, University of Johannesburg, Auckland Park, Johannesburg, South Africa
e-mail: ftugizimana@uj.ac.za; Fidele.Tugizimana@omnia.co.za

F. Tugizimana
International R&D, Omnia Group (Pty) Ltd, Bryanston, Johannesburg, South Africa

J. Engel
Biometris, Wageningen University & Research, Wageningen, Netherlands

R. Salek
International Agency for Research on Cancer, WHO, Lyon, France

K. Burgess
IQB3, University of Edinburgh, Edinburgh, Scotland, UK

W. Doorsamy et al. (eds.), *The Disruptive Fourth Industrial Revolution*, Lecture Notes in Electrical Engineering 674,
https://doi.org/10.1007/978-3-030-48230-5_10

provide an overview of 4IR in life sciences, illustratively pointing to some aspects in the metabolomics field. The latter, in its ontology, applies different 4IR technologies including big data analytics, machine learning, cloud computing, and artificial intelligence, amongst others. The momentum and maturation of metabolomics is undeniably evident, positively disruptive, and the field has visibly revolutionised the life sciences. The application of metabolomics spans a wide spectrum of the afore-said sciences, including biomedical technology, natural products, and plant biochemistry and—biotechnology research to name a few.

Keywords Metabolomics · Artificial intelligence (AI) · Big data · Cloud computing · Cloud metabolomics · Fourth industrial revolution—4IR · Machine learning

1 Introduction—The Dawn of a New Era

As early as 1957, Professor John McCarthy of Stanford University originally coined the phrase "artificial intelligence" (AI), which he described as the study of intellectual mechanisms for problem solving, be it in human being or machine. Going beyond the semantics, AI became pragmatically translated into an "epistemological engineering" as the emphasis was focussed on the generation and application of knowledge to solve real-world problems, with advanced development of computer software and programs manifesting intelligent behaviour in various tasks (Coles 1977; Karpatschof 1982). Fast-forward to the present times, AI with other emerging technologies like internet of things (IoT), machine learning, big data, cloud computing, virtual and augmented reality, is disruptively revolutionizing a wide spectrum of both fundamental research and applications, impacting substantially on all spheres of human society and everyday life (Makridakis 2017; Rose and Chilvers 2018; Wu and Zhao 2019). As reflected in other chapters, the momentous, non-linear innovation and advancements of these multiple technologies in the last two decades has led humanity into the fourth industrial revolution (4IR) era, and nobody really knows or can predict with certainty how this disruptive confluence of technologies will evolve or what its impact will be.

However, even though the future cannot be predicted with certainty, the 4IR reality has already begun, and with rapidly evolving cutting-edge technologies, it is inevitably and disruptively bringing the dawning of a new civilization, in which our biological limitations are transcended. As such, a fusion of real- and virtual reality is emerging, precision methodologies in all spheres of human society are rapidly being adopted to maximise the outcome, and automation and intelligent machines are gradually becoming next-generation systems in all aspects of life (Camacho et al. 2018; Kaur et al. 2017; Makridakis 2017; Tebani and Bekri 2019). In the life sciences, the 4IR technologies are inevitably revolutionizing fundamental and translational research and applications: AI and augmented intelligence, robotics, the IoT, nanotechnology, big data analytics are increasingly becoming drivers of transformative scientific achievements and breakthroughs, creating and shaping new horizons for life sciences. This can be exemplified by some of the recent headlines: AI finds autism-causing mutations in junk DNA (Zhou et al. 2019); AI paradigms in biotechnology (Goh et al. 2019); machine learning for biological networks (Camacho et al. 2018; Tugizimana et al. 2019a); metabolomics in a personalised world (Trivedi et al. 2017); innovative omics-based approaches in natural product studies as new strategies for drug discovery (Wolfender et al. 2019); machine learning in predicting drug-tissue relationships (Turki and Taguchi 2019), and AI and machine learning for precision agriculture (Qiu et al. 2019; Rose and Chilvers 2018), just to name a few in an exponentially growing list.

As illustrated in other chapters, the primary difference between previous revolutions and 4IR is the speed at which 'inevitable' transformation is taking place. As scientists, we are ever-changing and innovating to improve not just life, but the standards of science and technology, to contribute towards society. However, as well as generating solutions, scientific and technological advances also pose their own challenges. As illustrated in the above examples (and in other chapters), this new era is rapidly leading to paradigm shifts, creating a completely new *modus operandi*. Thus, to appreciate these new horizons in the life sciences, this chapter will explore and discuss the field of metabolomics. The 4IR technologies such as AI, machine learning, big data analytics, cloud computing to name a few, are integral components of the ontological fabric of a metabolomics workflow, and continuously shaping the growth and development of this field. Furthermore, on the other side of the same coin, metabolomics, an emergent and multidisciplinary scientific discipline, has disruptively positioned itself as one of the central pillars in systems biology approaches, and has increased in popularity and applicability across a vast array of fundamental and translational research domains (Trivedi et al. 2017). Thus, in this chapter, following the above introductory glimpse of a new era, a brief contextualised description of metabolomics is highlighted in the following section. Subsequent sections present and discuss the integration of AI and machine learning in metabolomics analytical platforms, big data metabolomics, and applications of metabolomics in the 4IR era.

2 A Primer on Metabolomics

The post-genomic era has rekindled interest in metabolism, with an eruption of systems biology methodologies, epistemologically encompassing the complementarity of hypothesis- and data-driven (deductive and inductive) reasoning and cognitive inquiry. This paradigm shift redefined how biological research is being carried out, with emerging novel possibilities to understand the nature of life at the molecular level (Kell and Oliver 2004; Ray 2010; Strange 2005). Biological science in the post-genomic era has consequently become a more data-intensive (big-data) science, driving the "cycle of knowledge" via a data-driven (inductive) reasoning to generate new hypotheses and insights. The cardinal *omics* layers, namely genomics, transcriptomics, proteomics and metabolomics, which collectively constitute systems biology, are *the* essential strategies driving the search to fully describe and understand the complex and dynamic metabolism of a biological system *in toto* (Döring et al. 2015; Goodacre 2005).

Metabolism, a cornerstone of life and characterized by an interconnected network of enzymatically-catalysed reactions, provides important cellular information regarding how environmental cues are translated into intracellular signals to coordinate biochemical processes such as nutrient utilisation, signalling or differentiation. Being a dynamic and integrated processes, metabolic regulation is underpinned by various interconnected translational and transcriptional mechanisms that consequently alter enzyme thermodynamics and kinetics (Fuhrer and Zamboni 2015; Ray 2010). This revival and emerging interest in metabolism rise from recognizing the etiolation of the field by the cloud of molecular biology (inclined towards the philosophical preference of deductive reasoning), and the awareness of the impossibility to answer many biological questions without comprehending the dynamics of metabolism. Recognising the reciprocal regulation of metabolism and other cellular processes—investigated through systems biology strategies—has disruptively revolutionised and is advancing our understanding of complex physiology, aiming at a comprehensive representation of biological systems in their respective ever-changing environments (Sévin et al. 2015).

Thus, the direct quantification of all metabolic fluxes (at all layers of biological information flow) has disruptively been the ideal of systems biology approaches. Proteomic and transcriptomic methodologies are generally applied and, although these offer genome-wide coverage; they do not necessarily describe the systemic outcome but rather provide information only relating to specific layers of regulation (Fuhrer and Zamboni 2015; Mathew and Padmanaban 2013; Tugizimana et al. 2013). The metabolome, on the other hand, is expectedly found to be more sensitive to perturbations in both enzyme activity and metabolic fluxes than transcriptome and proteome. Furthermore, changes in the metabolome are amplified relative to those in latter, and are arguably numerically more tractable (Kell et al. 2005; Richards et al. 2010). Hence, the qualitative and quantitative measurement of the dynamic multi-parametric metabolic reprogramming (related to pathophysiological stimuli or genetic—and epigenetic modification) of living systems reflect functional cues of the physiological state of the biological system being studied.

Semantically, metabolomics is a challenging discipline to define, due to the heterogeneity and dimensionality of the chemical space that it explores and the many analytical methods that are therefore required to detect and quantify the global 'metabolome'. There are many definitions of metabolomics in the literature, and the said term is often loosely cited to describe various experimental designs. In this chapter, metabolomics is considered as a multidisciplinary science that aims to define the entire complement of small molecular weight molecules, namely metabolites ($\leq$1500 Da in size), within a biological matrix of interest (Trivedi et al. 2017; Tugizimana et al. 2013). These compounds, as the end products of gene expression, allow insight into the biochemical phenotype and physiological state of a cell or tissue under given conditions (Beisken et al. 2015; Kell and Oliver 2016).

This youngest sibling of *omics* trilogy, metabolomics, differs disruptively from the classical or traditional targeted analytical methods in various fundamental aspects such as being a data-driven approach, and employing 4IR technologies to generate predictive models that intelligently grasp the silent characteristics of generated data (Grissa et al. 2016; Saccenti et al. 2014). Furthermore, as infographically depicted in a bird's-eye view (Fig. 1), the full complexity and multi-dimensionality of biochemical networks within a biological system (as decoded by systems biology approaches) are underlying dynamic frameworks that define physiological and phenotypic coherence. Hence, given that the biochemical actions of metabolites are far-reaching, including regulation of epigenetic mechanisms, post-translational modifications, involvement in protein transport and signal transduction, and active role in defence mechanisms, metabolomics can thus be seen as a powerful tool to interrogate cellular biochemistry at the systems level (Johnson et al. 2016; Likić et al. 2010). As such, the application of metabolomics as a global and data-driven (inductive) approach spans a great spectrum of life sciences including nutrition, therapeutics, diagnostics, surgical technology, natural products and plant biochemistry to name a few (Everett 2019; Phelps et al. 2018; Tebani and Bekri 2019; Tugizimana et al. 2019b).

Metabolomics is a multi-disciplinary skillset research field given that it converges from biology, chemistry, chemometrics, statistics and computer science (Beisken et al. 2015; Tugizimana et al. 2013). Metabolomics-generated biological insights and discoveries are increasingly rendered possible as the field matures and positions itself in the current innovative developments in analytical technologies, shaped by AI and machine learning, chemometric and statistical methodologies, and the integration of orthogonal biological approaches (Rai et al. 2019; Sen et al. 2019; Welch et al. 2019). Furthermore, the inherent sensitivity of the metabolome to alterations in the genome as well as environmental perturbations allows the measurement of subtle variations in biological pathways, thus generating important new mechanistic details (Johnson et al. 2016; Nagele 2014). Arguably, metabolomics is probably the most challenging and demanding of the *omics* fields, and this is due to the ontological complexity of the metabolome: highly dynamic (continuously changing at different rates), chemically diverse (dramatically different physicochemical properties, highly diverse and dynamic stereochemistries), wide range of metabolites and the intrinsic complexity of biological systems (biological cycles, organismal and cellular compartmentalisation) (Beisken et al. 2015; Heinig et al. 2013; Tugizimana et al. 2013). In order to attain its goal (i.e. holistic analysis of the metabolome, considering the complexity and chemo-diversity of the later), a wide range of chemistries, chemometrics methods, novel computational resources and innovative analytical tools that provide high degrees of sensitivity, selectivity and reproducibility are essential in metabolomics (Lee and Hu 2019; Miggiels et al. 2018; Misra 2018).

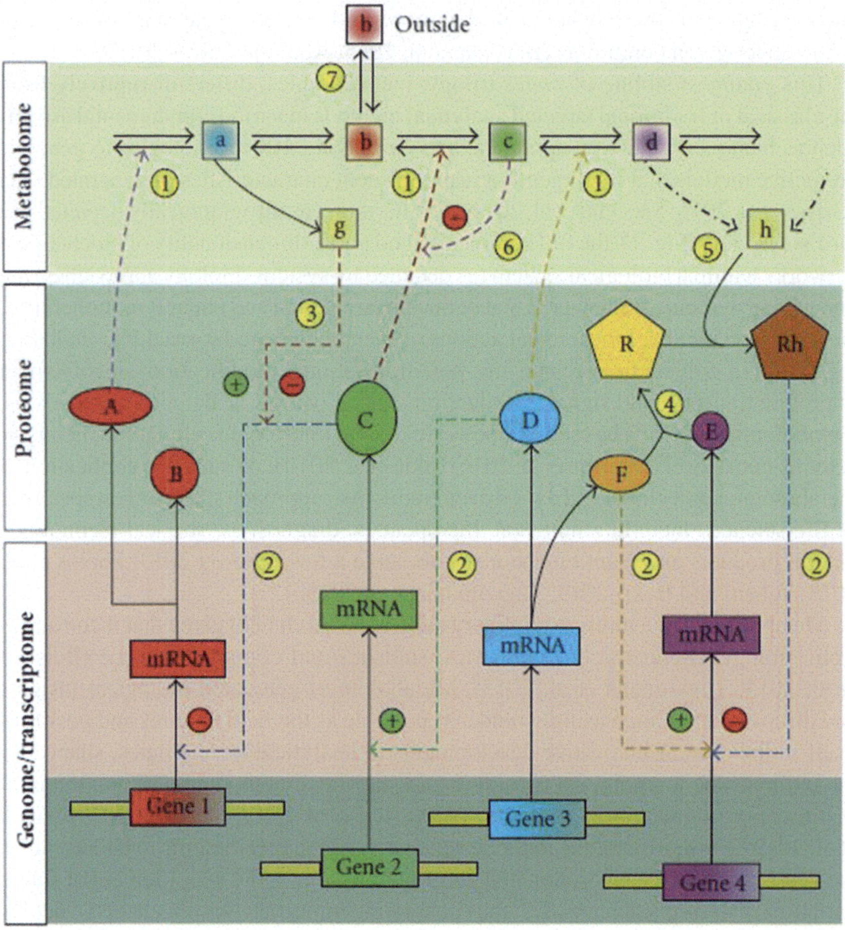

Fig. 1 A conceptualisation of biochemical networks showing genome, transcriptome, proteome and metabolome-level networks, pointing to their complexity and mutual interdependence. In biological systems, a large number of structurally and functionally diverse components (genes, proteins and metabolites) are involved in dynamic, linear and/or non-linear interactions, which in turn involve a range of time scales and interaction strengths. Direct conversions of species shown in solid lines, while some possible interactions (not necessarily one-step) are designated in dashed lines. Some of the types of interactions include: (1) enzyme catalysis, (2) posttranscriptional control of gene expression by proteins/protein complexes, including mechanisms that act on mRNAs (deadenylation, storage granulation) and mechanisms that act either directly or indirectly on DNA (histone modification, methylation), (3) effect of a metabolite on gene transcription mediated by a protein, (4) protein-protein interaction, (5) effect of a downstream ('reporter') metabolite on transcription through binding to a protein, (6) feedback inhibition/activation of an enzyme by a downstream metabolite, and (7) exchange of a metabolite with outside of the system (cell, organism). This schematic diagram (and descriptions) is from Likić et al. (2010)

3 Integration of AI and Machine Learning in Metabolomics Analytical Platforms

Metabolomics currently uses two main analytical platforms, mainly nuclear magnetic resonance spectroscopy (NMR) and mass spectrometry (MS). The latter is often coupled to a separation technique such as gas chromatography (GC), liquid chromatography (LC) or capillary electrophoresis (CE). It should, however, be kept in mind that all metabolites within a biological matrix cannot be detected using a single analytical platform, particularly pertaining to the diversity and complexity of the metabolome. Thus, the use of multi-dimensional techniques (e.g. LC × LC or GC × GC hyphenated to MS platforms) and integrated or combination of analytical approaches (e.g. GC-MS and LC-MS or different chromatographic column polarities) is beneficial to increase the coverage of the extracted metabolome. More details and description of these platforms are provided in the literature (Dunn and Ellis 2005; Miggiels et al. 2018; Zhang et al. 2012).

Liquid chromatography coupled with mass spectrometry has become the most suited analytical platform in metabolomics given both its high sensitivity and resolution. The last decade has seen innovative and disruptive technological developments in LC-MS-based systems ranging from advanced ultrahigh-performance liquid chromatographic (UHPLC) systems with improved peak separation to a newer generation of MS instruments with very high scan rates, sensitivity, and functionality and detection specificity (Putri et al. 2013). Recent capabilities of high-resolution accurate-mass MS (HRMS) to perform high-resolution full scan, reliable and sensitive quantitative analyses, have significantly impacted on LC-MS-based metabolomic analyses, thereby allowing a more detailed output pertaining to the composition of extracts under investigation. The current HRMS platforms are quadrupole-orbitrap-MS (Q-orbi-MS), ion trap-quadrupole-time-of-flight-MS (IT-Q-TOF-MS) and quadrupole-time-of-flight-MS (Q-TOF-MS) (Rochat 2016; Tugizimana et al. 2018).

There are several recently developed methodologies that attempt to perform metabolomics analysis without the time-consuming chromatographic step: matrix assisted laser desorption ionisation (MALDI)-TOF methods, flow injection MS. MALDI-TOF platform is a well-established technology, particularly in the field of imaging, where new methods provide the potential for a step change in *omics* capability. Based on spotting a dried sample on a target plate, along with an ionisation promoting compound (a matrix) MALDI uses a laser to ionise the sample and matrix, producing a cloud of gas-phase ions that can then be analysed (Balluff et al. 2014). Modern MALDI mass spectrometers have very fast lasers (2000 Hz), high resolution and the capability to perform tandem mass spectrometry. Additionally, recent advances in matrices for metabolomics have made MALDI a potential avenue for high throughput metabolome analysis (Heyman and Dubery 2015; Sans et al. 2018). Flow injection mass spectrometry, on the other hand, is simply conventional liquid chromatography mass spectrometry without the chromatography step, or column. Samples are injected into a liquid flow and directly introduced into the mass spectrometer. Two of the pioneering groups in this area are the Junot laboratory in France (Madalinski et al. 2008), and the Sauer/Zamboni research group in Zurich (Fuhrer and Zamboni 2015). While not quite as high throughput as MALDI, there are a plethora of pre-analytical steps that can be applied to improve coverage of the metabolome, especially as, with a sample analysis rate of less than a minute per injection, a typical day of analysis can comfortably include thousands of samples.

Finally, a very novel technology, 'acoustic mist ionization platform for direct and contactless ultrahigh-throughput MS analysis of liquid samples' (Sinclair et al. 2019), uses acoustic sampling techniques to infuse several samples per second. Using this platform, the authors are able to analyse samples at a rate of 100,000 per day. This technology makes large scale screening processes for biotechnology, using appropriate statistical experimental design feasible. Additionally, with this level of throughput, alternative extractions, large numbers of QCs and technical replicates can all be included to bolster the number of metabolites identified and the quality and reliability of the data (Sinclair et al. 2019). Furthermore, ion mobility spectrometry (IMS) technique has been explored and provides analytical advantages and opportunities. IMS is a rapid, gas-phase separation of ions based on their mobility in an electric field in the presence of a neutral buffer gas (Levy et al. 2019; May and McLean 2015). IMS technique has recently been integrated with MS-based analytical platforms, providing several benefits such as enhanced metabolome coverage, improved signal-to-noise and isomeric separation and increased confidence in metabolite identification and quantitation. The IMS coupled with direct analysis and imaging MS also provides enhanced coverage of the metabolome and a better description of spatial characterization of metabolites (Chouinard et al. 2019; Levy et al. 2019).

The disruptively increasing analytical power of these platforms rely on the continuous innovative advancements in the design and integration of analytical intelligence (AI), advanced automation and artificial intelligence (AI), pushing the boundaries of the traditional measurements. This can be illustrated by the recent release of: a cyclic ion-mobility mass spectrometry system from Waters corporation, with its multifunctional capabilities (multipass IM separation and IMS^n) (Giles et al. 2019); Shimadzu's new Nexera UHPLC series that incorporate artificial intelligence as analytical intelligence, which allows the system to detect and resolve analytical issues automatically, with integrated IoT for higher throughput analyses (https://www.shimadzu.com). Furthermore, Shimadzu corporation in partnership with Fujitsu are developing LC-MS technology that utilizes artificial intelligence for complete automation of peak picking in real-time, by applying deep learning algorithms (https://www.shimadzu.com). These advancements are allowing high-throughput analyses and enhanced coverage of the metabolome, which subsequently reveal unprecedented biological insights, shaping new horizons for the life sciences. It should be mentioned that, in a metabolomics workflow, this innovation in analytical technologies contributes (increasingly) to the generation of vast amount of data, which implies advanced data analytics, as discussed in the following section.

4 Big Data in Metabolomics: Current State, Challenges and Perspectives

Metabolomics studies, particularly untargeted approaches, generate complex and information-rich, high-dimensional data sets that are challenging to handle and fully mine information from (Goeddel and Patti 2012; Tugizimana et al. 2016). Furthermore, the amount of data produced is constantly rising at an ever-increasing pace. Also, because of the large number of advanced analytical platforms that are used, the data sets have very heterogeneous characteristics. Because of these reasons, untargeted metabolomics data, ontologically, can be placed under the umbrella of '**Big Data**' in terms of volume, velocity and variety (Gandomi and Haider 2015; May 2017). The availability of big data in metabolomics, but also in other fields, has introduced a new era of data-driven research where powerful computational tools, involving AI and machine learning algorithms, are used to find novel knowledge in the data (Biesecker 2013; Efron and Hastie 2016). However, it is worth pointing out that, despite advances in the development of (a tsunami of) data-mining tools and resources as described in the recent literature (Ebbels et al. 2019; Misra and Mohapatra 2019; Spicer et al. 2017), exploiting and extracting information (exhaustively) from metabolomic datasets (increasingly Big Data) remains a challenging endeavor.

Thus, dedicated mathematical modeling and/or chemometric algorithms are mandatory to cope with extracting relevant information from inherently complex metabolomic data sets. Various chemometric and bioinformatics tools and resources are continuously being developed (for mining metabolomics data), integrating mathematics, statistics and computer science (Boccard and Rudaz 2014; Cuperlovic-Culf 2018). More specifically, powerful algorithms are used to mine the data for patterns that are e.g. correlated to an experimental treatment. These results are then used to propose more specific hypotheses about the effect of the treatment, which may be tested in a follow-up study. Note that this approach is disruptively in contrast to traditional experiments where a small, specific, set of metabolites is analyzed based on an a priori specified hypothesis. A big advantage of data-driven research is that previously unknown or unexpected effects of the experimental treatment on the metabolome are being detected, leading to formulation of unprecedented hypotheses, with novel insights (Kell and Oliver 2004). This has disruptively positioned approaches such as untargeted metabolomics, in combination with big data analytics, at the forefront of science, creating a paradigm shift and new horizons for the life sciences.

4.1 Extracting Information from Metabolomics Big Data—Machine Learning Algorithms

The most time-consuming and challenging undertaking in untargeted metabolomics is extracting biologically relevant information from large data sets. The typical approach used in such analysis is well-describe in literature, and includes in the following steps: (i) processing (pre-processing—extracting features from raw instrumental data to a suitable 'clean' form; pre-treatment—normalization, scaling, centering and transformation to put all samples and variables on a comparable scale); (ii) chemometric, statistical analysis and modeling; (iii) annotation of the selected important features and (iv) interpretation and metabolic pathway and network analysis, leading to generation of research hypotheses and knowledge compilation, with integration of other *omics* for model reconstruction and system design (Fig. 2) (Brown et al. 2005; Goodacre et al. 2004; Tugizimana et al. 2016). It should be noted that the selected data processing strategy (pre-processing and pre-treatment methods) can greatly impact the outcome of subsequent analysis of the data and is therefore a crucial step in the metabolomics workflow (Gromski et al. 2015b; Tugizimana et al. 2016).

The data processing step involves removing artefacts from the data and thereby increase its quality and its biological information content (Madsen et al. 2010; Southam et al. 2017). Here, the instrumental raw data are transformed into a data matrix (defined and aligned peaks), which is subsequently, for example, adjusted for unwanted variation between samples (normalization); unwanted effects of highly abundant metabolites (scaling, transformation); distributional assumptions made during subsequent analysis (transformation). These processes involve a set of algorithms, some of which are machine learning based-methods (Figs. 2 and 3) (Beisken et al. 2015; Cuperlovic-Culf 2018). Subsequent mining of the pre-treated data can be

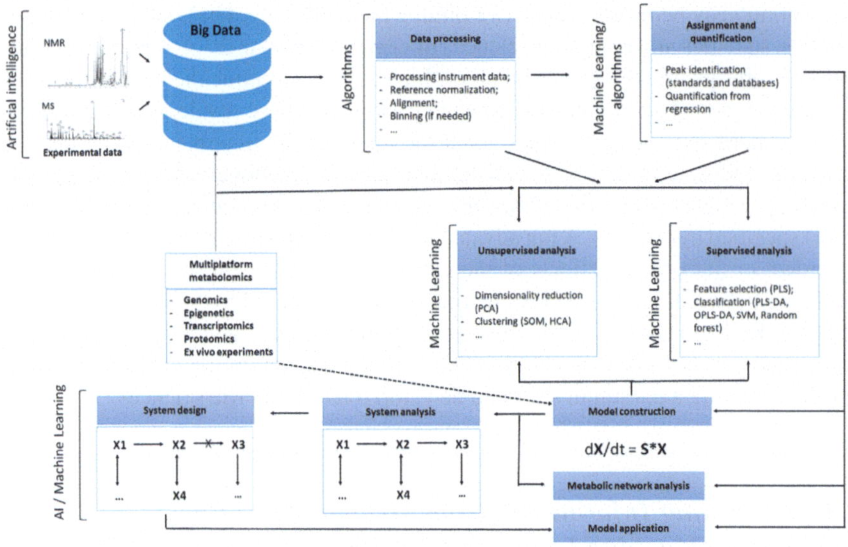

Fig. 2 Overview of a typical metabolomics data analysis workflow. As schematically depicted, metabolomics studies generate 'big data'. Handling and analysing these datasets follow a multistep workflow: from raw data pre-processing to data interpretation and/or integration. Each step of this workflow involves application of algorithms and machine learning (and mathematical modelling) methodologies

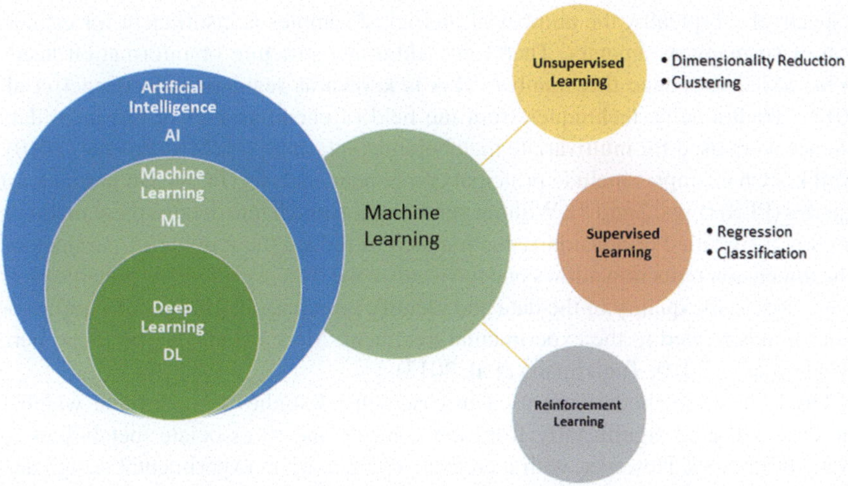

Fig. 3 Big data analytics and metabolomics data analysis. The diagram contextualises and describes big data analytics tools and methodologies. Artificial intelligence (AI) is a broader umbrella under which machine learning (ML) and deep learning (DL) are defined. DL is a subset of ML, and ML is a subset of AI. AI can be described here as the modelling of human mental functions by computer programs; i.e. any coding, modelling, algorithm that enables machine to mimic and develop (learn) human cognition or behaviour characteristics. ML, a subset of AI, refers to algorithms whose performance improves as exposed to more datasets, i.e. 'learning' from data, training a model from datasets. DL, a subset of ML inspired by our brain: multi-layered neural networks learn from vast amounts of data. In metabolomics data analysis ML and DL are used to mine and extract information from these complex big datasets. ML comprises different branches: unsupervised learning (algorithms for dimensionality reduction, clustering), supervised learning (regression and classification algorithms and modelling), and reinforcement learning

carried out either in a univariate or multivariate fashion. Univariate methods analyse the data in a metabolite-by-metabolite manner. This way changes to the experimental treatment in single metabolites are captured. Absence of such changes, however, does not mean that there are no multi-molecule combinations that clearly show an effect of the experimental treatment at a systems level (Saccenti et al. 2014). Therefore, multivariate techniques aim to take all metabolites into account simultaneously to detect single metabolite as well as multi-metabolite effects.

The large number of experimentally-generated metabolites, as measured in a typical untargeted approach for a (possibly relatively small) number of biological samples, poses many challenges for traditional statistical methods, which were developed for the opposite case (a few measurements on a large number of samples) (Efron and Hastie 2016; Engel et al. 2017). For example, classical multivariate data analysis relies on estimation of all pairwise correlations between metabolites. For a classical experiment with, for example, three metabolites this comes down to estimating three pairwise (partial) correlations. However, for modern experiments with 500 or 5000 metabolites already tens of thousands to millions of correlations need to be estimated, respectively. Typically, the number of biological samples is insufficient for estimation of so many parameters. Therefore, additional structure or information needs to be added to reduce this number. This is known as regularization (Engel et al. 2017). Traditionally, techniques from the field of chemometrics or chemical data science were used for multivariate metabolomic data analysis (Madsen et al. 2010). Well known examples include principal component analysis (PCA) and partial least squares (PLS) (Figs. 2 and 3). Without venturing too much into details, these methods employ a so-called dimension-reduction step to reduce the number of parameters. The dimension reduction allows one to visualize the data in easily interpretable plots to e.g. assess the quality of the data and identify groupings or clusters of samples or other trends related to the experimental treatment. More details can be found here (Madsen et al. 2010; Tugizimana et al. 2013).

Often, these methods were used in case-control studies to determine whether the cases differed significantly from the controls and to associate metabolites to these differences. However, with increasing complexity in experimental design and the number of metabolites that are measured the limitations of these methods have been realized (Engel et al. 2017; Gromski et al. 2015a). Also, at the present time, a metabolomics data set is often not studied in isolation, but information from other data sources is also considered. These other data sources can, for example, be additional metabolomics measurements from other analytical platforms or measurements on other levels of biological complexity such as genes or proteins (Fig. 1). This adds further complexity to data analysis, not only in terms of volume and variety of the data, but also because an integrative analysis where information from the different data sources is combined is required. These developments have disruptively resulted in a flurry of research activity and a multitude of improved dimension reduction methods have been proposed (Acar et al. 2015; Smilde et al. 2017).

Additionally, increasing use is made of modern methods for analysis of big data from fields such as statistical learning, machine learning (ML) and pattern recognition (Fig. 3). Note that, especially in the field of machine learning, often these methods have been developed for prediction, e.g. to accurately predict how likely someone is to suffer from a certain disease on the basis of information in his/her metabolite data. For data-driven research often prediction accuracy is not the most important goal. Rather, the researcher would like to be able to assess how the prediction actually came about. In other words, the researcher would like to identify which metabolites are important for identification of the disease. Therefore, especially the rising popularity of algorithms from fields such as statistical learning and ML that allow for such interpretation is expected to continue. Examples include the lasso, random forests (RF), support vector machines (SVM), and self-organizing maps (SOM) (Gromski et al. 2015a). Applications of these techniques in untargeted metabolomics can be found in numerous references such as (Gromski et al. 2015a; Lin et al. 2011). A review of the commonly used statistical and machine learning methods in metabolomics and their applications is beyond the scope of this chapter and can be found elsewhere (Cuperlovic-Culf 2018; Madsen et al. 2010; Saccenti et al. 2014).

Arguably, currently the most well-known and popular sub-field of ML is deep learning (Cuperlovic-Culf 2018) (Fig. 3). Here, deep neural networks are used for diverse big data applications ranging from autonomous driving to skin cancer diagnosis (Esteva et al. 2017). DL methods hold the promise of revolutionizing the future of *omics* studies (Grapov et al. 2018). They offer the elasticity to efficiently and successfully analyse and integrate a large volume of *omics* data of any type, including untargeted metabolomics data, given a large enough sample size and enough biological context (Grapov et al. 2018; Travers et al. 2018). Although interpretation of DL-output is not straightforward, a limited number of studies have demonstrated the potential that DL holds for analysis of untargeted metabolomics datasets and its effectiveness in integration with other data sources (genomic, proteomic and phenotypic data) (Alakwaa et al. 2018; Date and Kikuchi 2018; Grapov et al. 2018). Large sample size requirements, however, often hamper successful application of DL. The amount of data required can be reduced by integrating data from related experiments through transfer learning. Key to this is adoption of community-driven efforts for standardization and sharing of experimental data and results through controlled vocabulary and ontologies (see below) (Grapov et al. 2018; Travers et al., 2018).

4.2 Biological Interpretation of Metabolomics Data—Pathway and Network Analysis

Data mining typically identifies metabolites that are affected by the experimental treatment, but does not describe the underlying molecular mechanisms and biochemical processes that were affected. Metabolites are components of strongly connected pathways and therefore show a high degree of interconnectivity and correlation (Bartel et al. 2013). Thus, the observed differences should be functionally interpreted in light of the complete metabolome (Frainay and Jourdan 2016). A way to methodically model and describe the relationships between metabolites in the metabolome is to depict them as a graph or network, similar to how for instance social networks are visualized (Bartel et al. 2013; Cottret et al. 2010, 2018). Next, the graph can be investigated to identify sub-networks that were affected by the experimental treatment (Frainay and Jourdan 2016; Ren et al. 2015). Currently, global reconstructions of metabolic pathways are available in several publicly available databases such as KEGG or BioCyc and Recon3D (Brunk et al. 2018; Caspi et al. 2014). The biological networks contained in these can be used to identify those pathways involving the metabolites of interest (Cottret et al. 2010, 2018). However, the path that links the identified metabolites through the network may span multiple pathways, which is difficult to detect this way. Also, pathways can be defined differently in different databases. Therefore, it is useful to consider the complete metabolic network as a whole integrating all pathways into a single structure. To do so, use is made of powerful big data (ML) algorithms that were initially developed for constructing graphs of, for example, social networks and visualizing and analysing them (Figs. 2 and 3) (Cuperlovic-Culf 2018; Frainay and Jourdan 2016).

Unfortunately, pathway databases are far from complete, which makes any analyses based on them biased. Therefore, an alternative, data-driven, strategy is to estimate the metabolic network direction from the data (Bartel et al. 2013; Engel et al. 2017). Again, subsequently, (ML) algorithms can be applied to identify subnetworks or active modules that show significant differences as a result of the experimental treatment (Fig. 2). Although this approach typically fails in recovering the true underlying network structure, high overlap between the estimated networks and the metabolite reactions in databases have been observed, as well as novel candidates for pathway interactions (Bartel et al. 2013). In a simulation, the metabolic network can be defined and explained at various levels of detail. Network models can comprise kinetic description of defined pathways, enabling thus accurate prediction of the dynamic features of the metabolism under consideration. On the other hand, metabolic network models (e.g. genome scale models) can be computed involving stoichiometric representation with constraints and assumption of the steady state (Cuperlovic-Culf 2018).

Another form of network analysis algorithms is molecular networking. The latter is a mining and visualization strategy for nontargeted mass spectrometry (MS) data, and is increasingly becoming one of the key metabolite annotation and discovery strategies in metabolomics (Quinn et al. 2017). Molecular networking mathematically involves vector-based computational algorithms to associate the degree of similarity between every tandem-mass spectrum in a dataset. A molecular networking and data-sharing web-based platform, Global Natural Products Social Molecular Networking (GNPS) (http://gnps.ucsd.edu), is currently the world's largest repository and data exploitation resource for MS^n spectrometry data, and is increasingly being in use to decode the metabolomic 'dark matter' in different biological matrices (da Silva et al. 2018; Quinn et al. 2017). Different molecular networking tools are increasingly being developed and applied in different studies; and it may suffice here to point out the recent topic modeling, named MS2LDA (van der Hooft et al. 2016). This framework, inspired by machine learning algorithms for text processing, decomposes tandem mass spectra into patterns of co-occurring mass fragments and losses, designated as Mass2Motifs, which are generally molecular sub-structures. These motifs patterns are automatically extracted from complex MS/MS spectra using unsupervised text mining algorithms (van der Hooft et al. 2016). This approach allows the exploration of the 'fragmentome', increasing the coverage of annotated metabolites.

4.3 Cloud Metabolomics—e-Infrastructures for Data Analysis, Storage and Sharing

As illustrated in application section (under this chapter), the above highlighted data-driven mathematical (and ML-generated) models, describing the topology of biological network have disruptively revolutionized life sciences, providing unprecedented knowledge and insights, improving greatly our understanding of biological systems. However, as it can be deduced from the above paragraphs, a metabolomics data analysis pipeline (from pre-processing to network analysis) typically requires the use of various web services and specialized algorithms which have often been implemented in command-line driven software programs (Cottret et al. 2010, 2018). Therefore, such analysis is not straightforward for end users without sufficiently advanced computer skills and expert knowledge. An overview of the most widely used and freely available software tools can be found here (Spicer et al. 2017). The computational tools are often complex and require a diversity of tools and scripting languages. This, in combination with a lack of reporting on software versions and settings used, is one of the main contributors to the limited reproducibility of many metabolomics studies (Weber et al. 2017). Therefore, much community effort has been spent on developing e-infrastructure to overcome these difficulties (Fig. 4).

Powerful and user-friendly software is indispensable to make sense of the vast amount and variety of data generated in a typical metabolomics experiment and to ensure inter-laboratory reproducibility of the results (Weber et al. 2017). Recently, a growing number of users is starting to make use of workflow software which contain multiple interconnected algorithms covering several stages of data analysis, e.g. pre-processing, pre-treatment, mining and pathway analysis (Spicer et al. 2017). Because of this, users can perform their entire data analysis in only a limited number of programs. This removes problems with a lack of interoperability between tools and increases analysis reproducibility. Many workflows are offered as web-based services. MetaboAnalyst is such an online-based workflow suite. It consists of

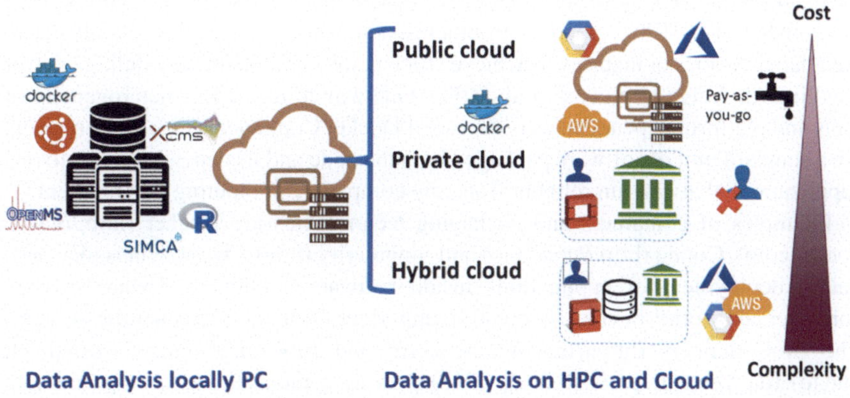

Fig. 4 Cloud metabolomics. Illustration of local high-performance computing (HPC) systems and public and hybrid cloud computing environments for the handling of metabolomics data analysis. Each cloud environment (public, private and hybrid) has its own merits and disadvantage, and the choice depends on the use-case. The complexity for setting up and usage of 'cloud metabolomics' increases from local HPC system to a hybrid integrated commercial system. The cost of public cloud environment is normally on a pay-as-you-go base, but requires the initial cost of purchase and set up on an HPC system. However, when it comes to sensitive and human clinical data, data privacy becomes an issue with public cloud environments

several modules including modules for data pre-treatment; data mining; and pathway analysis for functional interpretation of the data mining output. It is commonly used by metabolomics researchers with about 40000 jobs/month submitted to its servers in 2014 (Xia et al. 2015). XCMS Online is another popular web-based metabolomics data analysis suite with thousands of users (May 2017; Tautenhahn et al. 2012). It allows for pre-processing of LC/MS untargeted metabolomics data, statistical analysis/data mining, pathway analysis and multi-omic data integration. XCMS Online is a cloud-based solution which means that one can easily share datasets and data analysis results with collaborators or with the general public. MetExplore is a web server that offers the possibility to visualize metabolic networks and e.g. infer network properties of a list of metabolites, which, according to earlier data mining, appear to have been affected by the experimental treatment (Cottret et al. 2010, 2018). Also, MetExplore offers a collaborative environment for curation of metabolic networks.

Furthermore, besides increasing data size, usage of complex data analysis toolchains has necessitated to set up large-scale e-infrastructures and cloud computing environments for better handling of metabolomics analysis and integrating it with data i.e. locally and globally repositories. The High-Performance Computing (HPC) systems and cloud environments are currently the most common large-scale e-infrastructures used to date for data analysis (Fig. 4). However, installing and usage of different types of data analysis software, as required by the biologist, is often difficult and time-consuming due to; *incompatibility*, *scalability*, *security policies*, *resources and expertise* required to maintaining stable services over time. Another often limitation is the tool dependencies that are used in a set of data analysis workflows and their effective communication. Most currently commonly used tools are open source packages (Spicer et al. 2017), which often are not optimally written or set up to run on HPCs, cloud environments, or to work in a workflow mainly due to their dependencies or inability to achieve efficient job parallelism, scheduling or error handling (da Veiga Leprevost et al. 2017). One way to tackle this heterogeneity of tools and platform dependencies is the use of Docker Containers (Fig. 4). Containers are a standard unit of software packaged both the code and its dependencies into one "application" that can run reliably from any compatible computing environment.

Examples of containers and packaging technologies are Docker (http://www.docker.com), Conda (http://conda.io) and Singularity (https://www.sylabs.io/) facilitating deployment of applications inside software containers. Using software container and workflows have become an important component in scientific projects. The BioContainers (http://biocontainers.pro) and BioConda (http://bioconda.github.io) (da Veiga Leprevost et al. 2017) providing thousands of a large number of tools in Container that can run on a variety of desktops, local workstations, HPC and cloud environments. By combining different tools in a workflow environment, creating complex workflow, it makes it easer and possible to effectively connect the compatible tools to perform complex data analysis. Common workflow environments are Taverna (Li et al. 2010), Galaxy (Afgan et al. 2018), Snakemake (Köster and Rahmann 2012) and more recently Nextflow (Di Tommaso et al. 2017).

The PhenoMeNal (Phenome and Metabolome aNalysis) consortium (http://phenomenal-h2020.eu/) (Peters et al. 2019) was one of the first initiatives that took advantage of current technologies and set up a large-scale e-infrastructure, which supportively allows workflow based metabolomics analysis pipelines. The PhenoMeNal architecture is based on light-weight microservices, a software engineering methodology that complex applications are broken into a collection of smaller components that communicate over a network (Peters et al. 2019). In PhenoMeNal, Docker (https://www.docker.com/) containers were used to encapsulate the metabolomics software tools with an automated solution for installing tool based on Virtual Research Environment (VRE) utilizing Kubernetes (https://kubernetes.io/) as container orchestrator over multiple compute nodes (Peters et al. 2019). All the tools are available in the GitHub public repository with continuous integration testing implemented (Moreno et al. 2018). To utilise and to connect the tools into a workflow, Galaxy was used in the PhenoMeNal VRE. The PhenoMeNal project has a web portal for launching VREs on a selection of cloud providers, such as Amazon Web Services and Google Cloud Platform, or alternatively, can be set up locally and run on OpenStack-based HPC system (Moreno et al. 2018). Resource in such a system can be configured on-demand for example processing and memory needs, storage and software tools based on the data analysis planed, paying only for the time the resources were needed.

The PhenoMeNal tool can be easily used on a variety of cloud infrastructures (Peters et al. 2019). This offers a compelling approach to metabolomics data analysis where users can easily scale up computational power and storage requirements based on the size of a dataset (Gao et al. 2019). As above-mentioned, the size of metabolomics data sets as well as the pace at which they are being generated are dramatically increasing. An example of big data in metabolomics is the National Phenome Centre where an average 1.5 Petabytes of metabolomics data is generated on a yearly basis (Gao et al. 2019). This increase makes it cumbersome to carry out analysis on a standard desktop computer. Recently, Gao et al. (2019) demonstrated that data that took four days to process on a standard desktop computer was processed in ten minutes in the cloud. Also, the cloud-based analysis was more cost effective compared to the use of desktop computers (Gao et al. 2019). Another advantage of cloud-based approaches is that it allows for easier sharing of data and data analysis results facilitating collaborating between different laboratories in a standardized and reproducible manner.

Data deposition in an open-access repository (Figs. 4 and 5) is an important final step in a metabolomics experiment to ensure that a study's findings can be reproduced or re-used in new studies (Haug et al. 2017). Because of this several open-access metabolomics data repositories have emerged, such as MetaboLights and Metabolomics Workbench (Salek et al. 2013; Sud et al. 2016). In order to enable exchange and re-use of data and meta-data stored in such databases, they are stored in community-agreed standards. These standards were first defined by the Metabolomics Standards Initiative (MSI) and later extended by the Coordination of Standards in Metabolomics (COSMOS) initiative (Haug et al. 2017; Salek et al. 2015a, b). The reporting standards set by the MSI are nowadays backed by a rich set of ontologies to ensure that the consistent vocabulary is used. To structure data according to MSI-standards and ontologies, the ISA-TAB format is often

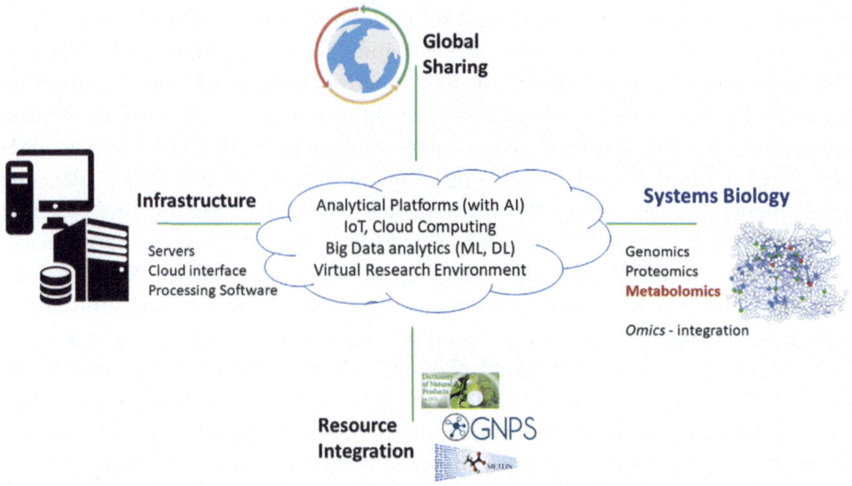

Fig. 5 Systems Biology layers and 4IR technologies. *Omics* approaches (systems biology layers) are being driven and shaped by 4IR technologies, but also, as feedback, these approaches push the boundaries of 4IR technologies. The centrality of 4IR technologies in systems biology approach is illustrated by the usage of AI, IoT, big data analytics and cloud environments in *omics* fundamental research and applications. Ability to integrate multiple resources, data sharing and interconnectedness with local infrastructure are rendered possibly by 4IR tools and disruptively advancing systems biology research and applications

used, for example, by MetaboLights. The COSMOS has promoted the use of open data formats such as mzML to encode primary research data rather than storing the data in vendor-specific formats to ensure that users can effortlessly access the data (Haug et al. 2017). The number of metabolomics studies deposited to databases is rapidly increasing. MetabolomeXchange (http://www.metabolomexchange.org/) combines data from four data repositories including MetaboLights and Metabolomics Workbench. Currently, (28 May 2019) 1627 metabolomics datasets are available through MetabolomeXchange, including in many cases raw experimental data. This increasing volume of metabolomics data that is shared around the globe in public repositories will disruptively enable analysis previously not possible such as integrative or meta-analysis of several metabolomics studies (Haug et al. 2017). This is a key to learning even more about biological systems, new horizons in the life sciences.

5 Applications of Metabolomics in the 4IR Era

Metabolomics, a disruptive technology and an '*omics*' branch utilising disruptive 4IR tools, offers exciting opportunities in systems biology and has revolutionized a broad range of life sciences, both fundamental and translational work, including biomedical technology (e.g. diagnostics and therapeutics) and plant sciences (Alexander et al. 2017; Kell and Oliver 2016; Pereira et al. 2009; Rai et al. 2019). The term 'metabolomics' was coined in 1998 and some recent publications have provided a reflection on the trajectories of this concept that comes of age (Alseekh and Fernie 2018; Kell and Oliver 2016); and the International Metabolomics Society with its affiliates (http://metabolomicssociety.org/) play key role in promoting the growth and application of metabolomics in the life sciences. Undoubtedly, metabolomics has disruptively positioned itself as one of the key pillars in systems biology. The latter embraces Aristotle's wisdom that "the whole is greater than the sum of its parts," and is disruptively being driven by 4IR technologies (Fig. 5). As reflected in Sect. 2 of this chapter, this special position of metabolomics among the modern *omics* disciplines of the systems biology approach rises from the fact that the metabolome, being the endpoint of biological events, carries imprints of environmental and genetic factors. Thus, one of the most biologically best descriptions of metabolism is the metabolic fluxes it generates, which represent the integrated output of the molecular machinery and biochemical characteristics of a biological system (Likic et al. 2010; Beisken et al. 2015).

Although a detailed account of systems biology in 4IR era is beyond the scope of this chapter, it suffices to highlight that *omics* approaches (genomics, proteomics, metabolomics, etc.), layers of systems biology, are disruptively revolutionising biology and life sciences in general, advancing our understanding of biological processes (Likic et al. 2010). The generation of *omics* datasets, a disruptive explosion of biological "big data", handling and extracting information from such datasets are rendered possible by emerging 4IR technologies (Fig. 5): artificial- and analytical intelligence (AI and AI)-equipped analytical platforms (as illustrated in Sect. 3 of this chapter), internet of things (IoT) and cloud environments (as illustrated in Sect. 4). These 4IR-driven systems biology layers (*omics* approaches, individually and integrated) are increasingly enabling us to explore and characterise the 'moliverse', a molecular space, of a biological system in unprecedented ways, revealing modular organisations and interactome networks of the system's chemical space (Doring et al. 2015; Johnson et al. 2016). Despite the challenges and gaps that still need attention (e.g. *omics* integration, decoding 'dark matter' in metabolism and detailed mapping of biochemical fluxes), such increasingly comprehensive biological insights are disruptively impacting life sciences, as illustrated in the following paragraphs, focusing on metabolomics.

The field of metabolomics has matured and, being a multidisciplinary science, metabolomics has disruptively pushed traditional boundaries of scientific endeavour, contributing significantly in shaping data-driven discoveries and knowledge generation trajectories. The list of metabolomics applications has grown exponentially and is still expanding as new insights open to more questions and 'dark matter', and the rapidly increasing development of novel technologies and innovation ecosystem. It may suffice here to highlight few domains in which this multidisciplinary science, metabolomics, has been applied, and further information on applications are given in the cited literature.

Since its early development, metabolomics has been applied in biomarker screening (or discovery) for diseases or follow drug efficacy (Everett 2019; Trivedi et al. 2017). This has led to unprecedented detailed insights into disease conditions and development, with significant contribution in identifying novel (potential) diagnostic and treatment-related markers for pathological processes affecting the heart (Deidda et al. 2015; Lance et al. 2018), lung diseases (du Preez and Loots 2014; Stringer et al. 2016), human immunodeficiency virus (HIV) infection (Hollenbaugh et al. 2016; Sitole et al. 2013), cancer (Armitage and Barbas 2014), diabetes (Zhao et al. 2010) and autoimmune diseases (Kang et al. 2015), to name a few. This positioned metabolomics as a revolutionary tool in biomedical research and bridging the gap between pharmaceutical development and disease prognosis, diagnosis and/or treatment. Furthermore, metabolomics has contributed in development of (medical) tools that are disruptively changing the medical space, e.g. the surgical intelligent knife. The latter distinguishes normal, borderline and malignant tissues, allowing the surgeon to tailor the operation to the patient in real-time (Phelps et al. 2018). With its applications in drug research and development (Kell and Goodacre 2014; Tolstikov 2016) and in nutrition (Tebani and Bekri 2019), in integration to biomedical research, metabolomics is and will play a key role in establishment of precision and personalized medicine (Beger et al. 2016; Trivedi et al. 2017).

In stem cell research, nanotechnology and the development of micro- and nanoscale surfaces that guide stem cell persistence and differentiation via mechanotransduction are one of the key technologies driving the development of new therapeutics (McMurray et al. 2011; Sazonova et al. 2011). Electron beam lithography can be used to make defined structures (with scalable levels of order and disorder) on the nanoscale, and these patterns can be translated to lower cost biocompatible materials such as PDMA or PEEK via injection moulding or (in the case of titanium surfaces) by mask etching. More disordered surfaces can be produced by HF etching or biomimetic materials can be produced from templates from biological materials such as nacre (McMurray et al. 2011). Metabolomics has been a key technology tracking the effectiveness of stem cell differentiation across these surfaces, and has even guided discovery of new biocompatible compounds that drive stem cell differentiation into osteoblasts (Seras-Franzoso et al. 2014; Tsimbouri et al. 2012).

Metabolomics has also been used extensively in biotechnology, especially the optimisation of strains and the development of bioprocesses. Both intracellular metabolomics (fingerprinting) and extracellular metabolomics (footprinting) are extremely valuable in bioprocesses and the development of high performing strains (Causon and Hann 2016; Kultschar et al. 2019). The intracellular metabolome allows pathways to be directly monitored for the pool sizes of particular intermediates, highlighting carbon sinks into side reactions, or enzymes providing a bottleneck to efficient production, or, using stable isotopes incorporated into the feedstocks, can allow flux analysis to be performed, providing quantitative information on the fate of carbon or nitrogen sources (Junker 2014; Sauer 2004). Extracellular metabolomics can provide a global picture of the output of a bioprocess, useful for modelling (Kell et al. 2005). Additionally, combined with the intracellular metabolome, it can provide a clear avenue for optimisation of the feedstock, especially for timings in fed batch fermentations (Causon and Hann 2016; Yang et al. 2014).

In plant sciences research and agriculture, metabolomics has increasingly been used in a wide spectrum of studies including relating genotype and biochemical phenotype (Fiehn et al. 2000; Rai et al. 2019), metabolic pathway studies (Bernard et al. 2014; Kang et al. 2019), silent phenotypes of mutations (Sweetlove et al. 2014), plant-pathogen interactions (Balmer et al. 2013; Tugizimana et al. 2019b) and plant priming (Balmer et al. 2015; Tugizimana et al. 2018). Such fundamental and translational research are shaping and contributing significantly to agriculture industry and the emerging 'agriculture 4.0' (Rose and Chilvers 2018), considering the alarming climate change and the global population explosion and concomitant demand on food supply (Prado et al. 2018). Although only a few such recent examples are presented, the possibilities and applications in the said fields are now coming to the fore and these include both targeted and non-targeted approaches. Here, apart from run-of-the-mill investigations relating to how plants respond to stressed (biotic and abiotic) versus normal or primed states (Mhlongo et al. 2018; Tugizimana et al. 2018), metabolomics applications pertain to plant/crop improvement strategies (e.g. cultivar/variety-specific quality, taste, ripening, nutrient and antioxidant content, yield) (Kumar et al. 2017) and food safety/risk assessment, where plant engineering or genetically modified (GM) crops (resistance, tolerance, improved nutrients, increased biosynthetic products, shelf-life considerations etc.) may also be included (Christ et al. 2018; Hong et al. 2016). Unquestionably, metabolomic comparison of the latter to natural variation (substantial equivalence) could generate significant insight/evidence in the ongoing controversy relating to GM food products (Christ et al. 2018; Simó et al. 2014), or snapshot plants/crops with improved traits during breeding strategies and cultivar selection (Kumar et al. 2017; Zeiss et al. 2018).

In addition to plants, livestock also significantly partake in the science of agriculture. As such, metabolomics studies pertaining to characteristics such as product (meat and milk) quality, animal health and even methane production have revealed metabolites and biomarkers which could be related to a particular trait. Agricultural considerations should also not lose sight of the negative impact of antimicrobial resistance in livestock, and here future metabolomics is expected to enable a more in-depth understanding (Kong et al. 2018; Prado et al. 2018). Lastly, in recent years a trend in affluent communities is developing to follow a healthier diet by including organic foods, however, much debate is made regarding the accuracy of labelling and, of course, whether such products are indeed not merely conventionally produced. Using machine-learning algorithms to profile metabolomics data, Kessler et al. (2015) demonstrated that organic and conventional wheat could be catalogued for a specific year and cultivar. Furthermore, the results prompted the authors to coin the term "metabolite-agnostic", since the strategy did not depend on accurate metabolite identification nor known biomarkers, but rather on data sets which were reliably annotated.

6 Conclusion and Perspectives

Metabolomics, the youngest in the *omics* trilogy, has become a central pillar in systems biology, revolutionizing the life sciences in different domains, as highlighted in this chapter. This multidisciplinary science, utilizing 4IR technologies, has matured rapidly in the last two decades, growing in its popularity and applicability. The increasing use and exploration of cutting edge analytical technologies to the analysis of metabolites, coupled to big data analytics (involving machine learning algorithms) are continuously and increasingly helping in illuminating the 'dark matter' of dynamic metabolism in biological system. The applications of metabolomics span a wide spectrum of life sciences from medical fields to agriculture.

The advancement and (disruptive) contribution of metabolomics in the 4IR era is indeed worth appreciating (as highlighted in this chapter), but at the same time, it is still a tip of an iceberg, considering the bottlenecks in metabolomics workflow, the unexplored areas and increasing questions and challenging arising from new insights. Furthermore, the amount of metabolomics data is expected to grow exponentially; therefore, more than ever capabilities are needed to scale up storage and analysis for *omics* data; and considering also the integration of other *omics* datasets for a comprehensive systems biology framework. In order to meet with these growing demands, metabolomics databases hosting both experimental and reference spectra need to be compatible with cloud computing technologies. Furthermore, with a 4IR-driven momentum, ongoing efforts from the global *omics* research community are paving the realization of integrated-omics frameworks, which will reveal novel biological insights.

References

Acar E, Bro R, Smilde AK (2015) Data fusion in metabolomics using coupled matrix and tensor factorizations. Proc IEEE 103:1602–1620. https://doi.org/10.1109/JPROC.2015.2438719

Afgan E, Baker D, Batut B, van den Beek M, Bouvier D, Čech M et al (2018) The Galaxy platform for accessible, reproducible and collaborative biomedical analyses: 2018 update. Nucleic Acids Res 46:W537–W544. https://doi.org/10.1093/nar/gky379

Alakwaa FM, Chaudhary K, Garmire LX (2018) Deep learning accurately predicts estrogen receptor status in breast cancer metabolomics data. J Proteome Res 17:337–347. https://doi.org/10.1021/acs.jproteome.7b00595

Alexander J, Gildea L, Balog J, Speller A, Mckenzie J, Muirhead L et al (2017) A novel methodology for in vivo endoscopic phenotyping of colorectal cancer based on real-time analysis of the mucosal lipidome: a prospective observational study of the iKnife. Surg Endosc 31:1361–1370. https://doi.org/10.1007/s00464-016-5121-5

Alseekh S, Fernie AR (2018) Metabolomics 20 years on: what have we learned and what hurdles remain? Plant J 94:933–942. https://doi.org/10.1111/tpj.13950

Armitage EG, Barbas C (2014) Metabolomics in cancer biomarker discovery: Current trends and future perspectives. J Pharm Biomed Anal 87:1–11. https://doi.org/10.1016/j.jpba.2013.08.041

Balluff B, Carreira RJ, Mcdonnell LA (2014) Mass Spectrometry imaging in proteomics and metabolomics. Comprehensive analytical chemistry. Elsevier, pp 159–185. https://doi.org/10.1016/b978-0-444-62651-6.00007-6

Balmer D, Flors V, Glauser G, Mauch-Mani B (2013) Metabolomics of cereals under biotic stress: current knowledge and techniques. Front. Plant Sci. 4:1–12. https://doi.org/10.3389/fpls.2013.00082

Balmer A, Pastor V, Gamir J, Flors V, Mauch-Mani B (2015) The 'prime-ome': towards a holistic approach to priming. Trends Plant Sci 20:443–452. https://doi.org/10.1016/j.tplants.2015.04.002

Bartel J, Krumsiek J, Theis FJ (2013) Statistical methods for the analysis of high-throughput metabolomics data. Comput Struct Biotechnol J 4:1–9. https://doi.org/10.5936/csbj.201301009

Beger RD, Dunn W, Schmidt MA, Gross SS, Kirwan JA, Cascante M et al (2016) Metabolomics enables precision medicine: "A White Paper, Community Perspective". Metabolomics 12:149. https://doi.org/10.1007/s11306-016-1094-6

Beisken S, Eiden M, Salek RM (2015) Getting the right answers: understanding metabolomics challenges. Expert Rev Mol Diagn 15:97–109. https://doi.org/10.1586/14737159.2015.974562

Bernard T, Bridge A, Morgat A, Moretti S, Xenarios I, Pagni M (2014) Reconciliation of metabolites and biochemical reactions for metabolic networks. Brief Bioinform 15:123–135. https://doi.org/10.1093/bib/bbs058

Biesecker LG (2013) Hypothesis-generating research and predictive medicine. Genome Res 23:1051–1053. https://doi.org/10.1101/gr.157826.113

Boccard J, Rudaz S (2014) Harnessing the complexity of metabolomic data with chemometrics. J Chemom 28:1–9. https://doi.org/10.1002/cem.2567

Brown M, Dunn WB, Ellis DI, Goodacre R, Handl J, Knowles JD et al (2005) A metabolome pipeline: from concept to data to knowledge. Metabolomics 1:39–51. https://doi.org/10.1007/s11306-005-1106-4

Brunk E, Sahoo S, Zielinski DC, Altunkaya A, Mih N, Gatto F et al (2018) Recon3D: a resource enabling a three-dimensional view of gene variation in human metabolism. Nat Biotechnol 36:272–281. https://doi.org/10.1038/nbt.4072.Recon3D

Camacho DM, Collins KM, Powers RK, Costello JC, Collins JJ (2018) Next-generation machine learning for biological networks. Cell 173:1581–1592. https://doi.org/10.1016/j.cell.2018.05.015

Caspi R, Altman T, Billington R, Dreher K, Foerster H, Fulcher CA et al (2014) The MetaCyc database of metabolic pathways and enzymes and the BioCyc collection of pathway/genome databases. Nucleic Acids Res 42:459–471. https://doi.org/10.1093/nar/gkt1103

Causon TJ, Hann S (2016) Review of sample preparation strategies for MS-based metabolomic studies in industrial biotechnology. Anal Chim Acta 938:18–32. https://doi.org/10.1016/j.aca.2016.07.033

Chouinard CD, Nagy G, Smith RD, Baker ES (2019) Ion mobility-mass spectrometry in metabolomic, lipidomic, and proteomic analyses. Comprehensive analytical chemistry. Elsevier B.V., pp 123–159. https://doi.org/10.1016/bs.coac.2018.11.001

Christ B, Pluskal T, Aubry S, Weng JK (2018) Contribution of untargeted metabolomics for future assessment of biotech crops. Trends Plant Sci 23:1047–1056. https://doi.org/10.1016/j.tplants.2018.09.011

Coles LS (1977) The application of artificial intelligence to medicine. Futures 9:315–323. https://doi.org/10.1016/0016-3287(77)90097-0

Cottret L, Wildridge D, Vinson F, Barrett MP, Charles H, Sagot M-F et al (2010) MetExplore: a web server to link metabolomic experiments and genome-scale metabolic networks. Nucleic Acids Res 38:W132–W137. https://doi.org/10.1093/nar/gkq312

Cottret L, Frainay C, Chazalviel M, Cabanettes F, Gloaguen Y, Camenen E et al (2018) MetExplore: collaborative edition and exploration of metabolic networks. Nucleic Acids Res 46:W495–W502. https://doi.org/10.1093/nar/gky301

Cuperlovic-Culf M (2018) Machine learning methods for analysis of metabolic data and metabolic pathway modeling. Metabolites 8:4. https://doi.org/10.3390/metabo8010004

da Silva RR, Wang M, Nothias L-F, van der Hooft JJJ, Caraballo-Rodríguez AM, Fox E et al (2018) Propagating annotations of molecular networks using in silico fragmentation. PLoS Comput Biol 14:e1006089. https://doi.org/10.1371/journal.pcbi.1006089
Date Y, Kikuchi J (2018) Application of a deep neural network to metabolomics studies and its performance in determining important variables. Anal Chem 90:1805–1810. https://doi.org/10.1021/acs.analchem.7b03795
da Veiga Leprevost F, Grüning BA, Alves Aflitos S, Röst HL, Uszkoreit J, Barsnes H et al (2017) BioContainers: an open-source and community-driven framework for software standardization. Bioinformatics 33:2580–2582. https://doi.org/10.1093/bioinformatics/btx192
Deidda M, Piras C, Bassareo PP, Cadeddu Dessalvi C, Mercuro G (2015) Metabolomics, a promising approach to translational research in cardiology. IJC Metab Endocr 9:31–38. https://doi.org/10.1016/j.ijcme.2015.10.001
Di Tommaso P, Chatzou M, Floden EW, Barja PP, Palumbo E, Notredame C (2017) Nextflow enables reproducible computational workflows. Nat Biotechnol 35:316–319. https://doi.org/10.1038/nbt.3820
Döring M, Brüninghaus A, Kollek R, Petersen I (2015) Systems-oriented approaches in biology: system biologist's narratives of present, past, and future. Contextualizing systems biology. Springer International Publishing, Cham, pp 119–146. https://doi.org/10.1007/978-3-319-17106-7_3
Dunn WB, Ellis DI (2005) Metabolomics: current analytical platforms and methodologies. Trends Anal Chem 24:285–294. https://doi.org/10.1016/j.trac.2004.11.021
du Preez I, Loots DT (2014) Can metabolomics improve tuberculosis diagnostics? Metabolomics 10:877–886. https://doi.org/10.1007/s11306-014-0633-2
Ebbels TMD, Pearce JTM, Sadawi N, Gao J, Glen RC (2019) Big data and databases for metabolic phenotyping. The handbook of metabolic phenotyping. Elsevier, pp 329–367. https://doi.org/10.1016/b978-0-12-812293-8.00011-6
Efron B, Hastie T (2016) Computer age statistical inference: algorithms, evidence, and data science. https://doi.org/10.1017/cbo9781316576533
Engel J, Buydens L, Blanchet L (2017) An overview of large-dimensional covariance and precision matrix estimators with applications in chemometrics. J Chemom 31:1–19. https://doi.org/10.1002/cem.2880
Esteva A, Kuprel B, Novoa RA, Ko J, Swetter SM, Blau HM et al (2017) Dermatologist-level classification of skin cancer with deep neural networks. Nature 542:115–118. https://doi.org/10.1038/nature21056
Everett JR (2019) Applications of metabolic phenotyping in pharmaceutical research and development. Elsevier Inc. https://doi.org/10.1016/b978-0-12-812293-8.00014-1
Fiehn O, Kopka J, Dörmann P, Altmann T, Trethewey RN, Willmitzer L (2000) Metabolite profiling for plant functional genomics. Nat Biotechnol 18:1157–1161. https://doi.org/10.1038/81137
Frainay C, Jourdan F (2016) Computational methods to identify metabolic sub-networks based on metabolomic profiles. Brief Bioinform 1–14. https://doi.org/10.1093/bib/bbv115
Fuhrer T, Zamboni N (2015) High-throughput discovery metabolomics. Curr Opin Biotechnol 31:73–78. https://doi.org/10.1016/j.copbio.2014.08.006
Gandomi A, Haider M (2015) Beyond the hype: big data concepts, methods, and analytics. Int J Inf Manage 35:137–144. https://doi.org/10.1016/j.ijinfomgt.2014.10.007
Gao J, Sadawi N, Karaman I, Pearce JTM, Moreno P, Larsson A et al (2019) Metabolomics in the cloud: scaling computational tools to big data. arXiv Prepr 1904:1–25. http://arxiv.org/abs/1904.02288
Giles K, Ujma J, Wildgoose J, Pringle S, Richardson K, Langridge D et al (2019) A cyclic ion mobility-mass spectrometry system. Anal Chem. https://doi.org/10.1021/acs.analchem.9b01838
Goeddel LC, Patti GJ (2012) Maximizing the value of metabolomic data. Bioanalysis 4:2199–2201. https://doi.org/10.4155/bio.12.210
Goh W, Bin W, Sze CC (2019) AI paradigms for teaching biotechnology. Trends Biotechnol 37:1–5. https://doi.org/10.1016/j.tibtech.2018.09.009

Goodacre R (2005) Making sense of the metabolome using evolutionary computation: seeing the wood with the trees. J Exp Bot 56:245–254. https://doi.org/10.1093/jxb/eri043

Goodacre R, Vaidyanathan S, Dunn WB, Harrigan GG, Kell DB (2004) Metabolomics by numbers: acquiring and understanding global metabolite data. Trends Biotechnol 22:245–252. https://doi.org/10.1016/j.tibtech.2004.03.007

Grapov D, Fahrmann J, Wanichthanarak K, Khoomrung S (2018) Rise of deep learning for genomic, proteomic, and metabolomic data integration in precision medicine. Omi A J Integr Biol 22:630–636. https://doi.org/10.1089/omi.2018.0097

Grissa D, Pétéra M, Brandolini M, Napoli A, Comte B, Pujos-Guillot E (2016) Feature selection methods for early predictive biomarker discovery using untargeted metabolomic data. Front Mol Biosci 3:1–15. https://doi.org/10.3389/fmolb.2016.00030

Gromski PS, Xu Y, Hollywood KA, Turner ML, Goodacre R (2015b) The influence of scaling metabolomics data on model classification accuracy. Metabolomics 11:684–695. https://doi.org/10.1007/s11306-014-0738-7

Gromski PS, Muhamadali H, Ellis DI, Xu Y, Correa E, Turner ML et al (2015a) A tutorial review: metabolomics and partial least squares-discriminant analysis—a marriage of convenience or a shotgun wedding. Anal Chim Acta. https://doi.org/10.1016/j.aca.2015.02.012

Haug K, Salek RM, Steinbeck C (2017) Global open data management in metabolomics. Curr Opin Chem Biol 36:58–63. https://doi.org/10.1016/j.cbpa.2016.12.024

Heinig U, Gutensohn M, Dudareva N, Aharoni A (2013) The challenges of cellular compartmentalization in plant metabolic engineering. Curr Opin Biotechnol 24:239–246. https://doi.org/10.1016/j.copbio.2012.11.006

Heyman HM, Dubery IA (2015) The potential of mass spectrometry imaging in plant metabolomics: a review. Phytochem Rev. https://doi.org/10.1007/s11101-015-9416-2

Hollenbaugh JA, Montero C, Schinazi RF, Munger J, Kim B (2016) Metabolic profiling during HIV-1 and HIV-2 infection of primary human monocyte-derived macrophages. Virology 491:106–114. https://doi.org/10.1016/j.virol.2016.01.023

Hong J, Yang L, Zhang D, Shi J (2016) Plant metabolomics: an indispensable system biology tool for plant science. Int J Mol Sci 17:767. https://doi.org/10.3390/ijms17060767

Johnson CH, Ivanisevic J, Siuzdak G (2016) Metabolomics: beyond biomarkers and towards mechanisms. Nat Rev Mol Cell Biol 17:451–459. https://doi.org/10.1038/nrm.2016.25

Junker BH (2014) Flux analysis in plant metabolic networks: Increasing throughput and coverage. Curr Opin Biotechnol 26:183–188. https://doi.org/10.1016/j.copbio.2014.01.016

Kang J, Zhu L, Lu J, Zhang X (2015) Application of metabolomics in autoimmune diseases: insight into biomarkers and pathology. J Neuroimmunol 279:25–32. https://doi.org/10.1016/j.jneuroim.2015.01.001

Kang KB, Ernst M, Hooft JJJ, Silva RR, Park J, Medema MH et al (2019) Comprehensive mass spectrometry-guided phenotyping of plant specialized metabolites reveals metabolic diversity in the cosmopolitan plant family Rhamnaceae. Plant J. https://doi.org/10.1111/tpj.14292

Karpatschof B (1982) Artificial intelligence or artificial signification? J Pragmat 6:293–304. https://doi.org/10.1016/0378-2166(82)90005-4

Kaur J, Rahat B, Thakur S, Kaur J (2017) Trends in precision medicine. Elsevier Inc. https://doi.org/10.1016/b978-0-12-809411-2.00015-5

Kell DB, Goodacre R (2014) Metabolomics and systems pharmacology: why and how to model the human metabolic network for drug discovery. Drug Discov Today 19:171–182. https://doi.org/10.1016/j.drudis.2013.07.014

Kell DB, Oliver SG (2004) Here is the evidence, now what is the hypothesis? The complementary roles of inductive and hypothesis-driven science in the post-genomic era. BioEssays 26:99–105. https://doi.org/10.1002/bies.10385

Kell DB, Oliver SG (2016) The metabolome 18 years on: a concept comes of age. Metabolomics 12:1–8. https://doi.org/10.1007/s11306-016-1108-4

Kell DB, Brown M, Davey HM, Dunn WB, Spasic I, Oliver SG (2005) Metabolic footprinting and systems biology: the medium is the message. Nat Rev Microbiol 3:557–565. https://doi.org/10.1038/nrmicro1177

Kessler N, Bonte A, Albaum SP, Mäder P, Messmer M, Goesmann A, Niehaus K, Langenkämper G, Nattkemper TW (2015) Learning to classify organic and conventional wheat: a machine learning driven approach using the MeltDB 2.0 metabolomics analysis platform. Front Bioeng Biotech 3

Kong Z, Hart M, Liu H (2018) Paving the way from the lab to the field: using synthetic microbial consortia to produce high-quality crops. Front Plant Sci 9:1–5. https://doi.org/10.3389/fpls.2018.01467

Köster J, Rahmann S (2012) Snakemake-a scalable bioinformatics workflow engine. Bioinformatics 28:2520–2522. https://doi.org/10.1093/bioinformatics/bts480

Kultschar B, Dudley E, Wilson S, Llewellyn CA (2019) Intracellular and extracellular metabolites from the cyanobacterium chlorogloeopsis fritschii, PCC 6912, during 48 hours of UV-B exposure. Metabolites 9:74. https://doi.org/10.3390/metabo9040074

Kumar R, Bohra A, Pandey AK, Pandey MK, Kumar A (2017) Metabolomics for plant improvement: status and prospects. Front Plant Sci 8:1–27. https://doi.org/10.3389/fpls.2017.01302

Lance N, Damian O, Michael MC, Olga JM, Meredith I, Naggie S et al (2018) Metabolites predict cardiovascular disease events in persons living with HIV: a pilot case—control study. Metabolomics 14:1–6. https://doi.org/10.1007/s11306-018-1318-z

Lee M, Hu T (2019) Computational methods for the discovery of metabolic markers of complex traits. Metabolites 9:66. https://doi.org/10.3390/metabo9040066

Levy AJ, Oranzi NR, Ahmadireskety A, Kemperman RHJ, Wei MS, Yost RA (2019) Recent progress in metabolomics using ion mobility-mass spectrometry. TrAC Trends Anal Chem 116:274–281. https://doi.org/10.1016/j.trac.2019.05.001

Likić VA, McConville MJ, Lithgow T, Bacic A (2010) Systems biology: the next frontier for bioinformatics. Adv Bioinform 2010:1–10. https://doi.org/10.1155/2010/268925

Lin X, Wang Q, Yin P, Tang L, Tan Y, Li H et al (2011) A method for handling metabonomics data from liquid chromatography/mass spectrometry: combinational use of support vector machine recursive feature elimination, genetic algorithm and random forest for feature selection. Metabolomics 7:549–558. https://doi.org/10.1007/s11306-011-0274-7

Li P, Dada JO, Jameson D, Spasic I, Swainston N, Carroll K et al (2010) Systematic integration of experimental data and models in systems biology. BMC Bioinform 11:582. https://doi.org/10.1186/1471-2105-11-582

Madalinski G, Godat E, Alves S, Lesage D, Genin E, Levi P et al (2008) Direct introduction of biological samples into a LTQ-Orbitrap hybrid mass spectrometer as a tool for fast metabolome analysis. Anal Chem 80:3291–3303. https://doi.org/10.1021/ac7024915

Madsen R, Lundstedt T, Trygg J (2010) Chemometrics in metabolomics–a review in human disease diagnosis. Anal Chim Acta 659:23–33. https://doi.org/10.1016/j.aca.2009.11.042

Makridakis S (2017) The forthcoming Artificial Intelligence (AI) revolution: its impact on society and firms. Futures 90:46–60. https://doi.org/10.1016/j.futures.2017.03.006

Mathew AK, Padmanaban VC (2013) Metabolomics: the apogee of the omics trilogy. Int J Pharm Pharm Sci 5:45–48

May M (2017) Big data, big picture: metabolomics meets systems biology. Science 356(80):646–648. https://doi.org/10.1126/science.356.6338.646

May JC, McLean JA (2015) Ion mobility-mass spectrometry: time-dispersive instrumentation. Anal Chem 87:1422–1436. https://doi.org/10.1021/ac504720m

McMurray RJ, Gadegaard N, Tsimbouri PM, Burgess KV, McNamara LE, Tare R et al (2011) Nanoscale surfaces for the long-term maintenance of mesenchymal stem cell phenotype and multipotency. Nat Mater 10:637–644. https://doi.org/10.1038/nmat3058

Mhlongo MI, Piater LA, Madala NE, Labuschagne N, Dubery IA (2018) The chemistry of plant-microbe interactions in the rhizosphere and the potential for metabolomics to reveal signaling related to defense priming and induced systemic resistance. Front Plant Sci 9:1–17. https://doi.org/10.3389/fpls.2018.00112

Miggiels P, Wouters B, van Westen GJP, Dubbelman AC, Hankemeier T (2018) Novel technologies for metabolomics: more for less. TrAC Trends Anal Chem 2018:1–9. https://doi.org/10.1016/j.trac.2018.11.021

Misra BB (2018) New tools and resources in metabolomics: 2016–2017. Electrophoresis 39:909–923. https://doi.org/10.1002/elps.201700441

Misra BB, Mohapatra S (2019) Tools and resources for metabolomics research community: a 2017–2018 update. Electrophoresis 40:227–246. https://doi.org/10.1002/elps.201800428

Moreno P, Pireddu L, Roger P, Goonasekera N, Afgan E, Beek M, van den et al (2018) Galaxy-Kubernetes integration: scaling bioinformatics workflows in the cloud. bioRxiv 1–3. https://doi.org/10.1101/488643

Nagele T (2014) Linking metabolomics data to underlying metabolic regulation. Front Mol Biosci 1:1–6. https://doi.org/10.3389/fmolb.2014.00022

Pereira H, Martin J-F, Joly C, Sébédio J-L, Pujos-Guillot E (2009) Development and validation of a UPLC/MS method for a nutritional metabolomic study of human plasma. Metabolomics 6:207–218. https://doi.org/10.1007/s11306-009-0188-9

Peters K, Bradbury J, Bergmann S, Capuccini M, Cascante M, de Atauri P et al (2019) PhenoMeNal: processing and analysis of metabolomics data in the cloud. Gigascience 8:1–12. https://doi.org/10.1093/gigascience/giy149

Phelps DL, Balog J, Gildea LF, Bodai Z, Savage A, El-Bahrawy MA et al (2018) The surgical intelligent knife distinguishes normal, borderline and malignant gynaecological tissues using rapid evaporative ionisation mass spectrometry (REIMS). Br J Cancer. https://doi.org/10.1038/s41416-018-0048-3

Prado RM do, Porto C, Nunes E, Aguiar CL de, Pilau EJ (2018) Metabolomics and agriculture: what can be done ? mSystems 3:2–4. https://doi.org/10.1128/msystems.00156-17

Putri SP, Yamamoto S, Tsugawa H, Fukusaki E (2013) Current metabolomics: technological advances. J Biosci Bioeng 116:9–16. https://doi.org/10.1016/j.jbiosc.2013.01.004

Qiu Q, Sun N, Bai H, Wang N, Fan Z, Wang Y et al (2019) Field-based high-throughput phenotyping for maize plant using 3D LiDAR point cloud generated with a "Phenomobile". Front Plant Sci 10:1–15. https://doi.org/10.3389/fpls.2019.00554

Quinn RA, Nothias LF, Vining O, Meehan M, Esquenazi E, Dorrestein PC (2017) Molecular networking as a drug discovery, drug metabolism, and precision medicine strategy. Trends Pharmacol Sci 38:143–154. https://doi.org/10.1016/j.tips.2016.10.011

Rai A, Yamazaki M, Saito K (2019) A new era in plant functional genomics. Curr Opin Syst Biol 15:58–67. https://doi.org/10.1016/j.coisb.2019.03.005

Ray LB (2010) Metabolism is not boring. Science 330(80):1337. https://doi.org/10.1126/science.330.6009.1337

Ren S, Hinzman AA, Kang EL, Szczesniak RD, Lu LJ (2015) Computational and statistical analysis of metabolomics data. Metabolomics 11:1492–1513. https://doi.org/10.1007/s11306-015-0823-6

Richards SE, Dumas M-E, Fonville JM, Ebbels TMD, Holmes E, Nicholson JK (2010) Intra- and inter-omic fusion of metabolic profiling data in a systems biology framework. Chemom Intell Lab Syst 104:121–131. https://doi.org/10.1016/j.chemolab.2010.07.006

Rochat B (2016) From targeted quantification to untargeted metabolomics: why LC-high-resolution-MS will become a key instrument in clinical labs. TrAC Trends Anal Chem 84:151–164. https://doi.org/10.1016/j.trac.2016.02.009

Rose DC, Chilvers J (2018) Agriculture 4.0: broadening responsible innovation in an era of smart farming. Front Sustain Food Syst 2:1–7. https://doi.org/10.3389/fsufs.2018.00087

Saccenti E, Hoefsloot HCJ, Smilde AK, Westerhuis JA, Hendriks MMWB (2014) Reflections on univariate and multivariate analysis of metabolomics data. Metabolomics 10:361–374. https://doi.org/10.1007/s11306-013-0598-6

Salek RM, Haug K, Steinbeck C (2013) Dissemination of metabolomics results: role of MetaboLights and COSMOS. Gigascience 2:8. https://doi.org/10.1186/2047-217X-2-8

Salek RM, Arita M, Dayalan S, Ebbels T, Jones AR, Neumann S et al (2015a) Embedding standards in metabolomics: the metabolomics society data standards task group. Metabolomics 11:782–783. https://doi.org/10.1007/s11306-015-0821-8

Salek RM, Neumann S, Schober D, Hummel J, Billiau K, Kopka J et al (2015b) COordination of Standards in MetabOlomicS (COSMOS): facilitating integrated metabolomics data access. Metabolomics 11:1587–1597. https://doi.org/10.1007/s11306-015-0810-y

Sans M, Feider CL, Eberlin LS (2018) Advances in mass spectrometry imaging coupled to ion mobility spectrometry for enhanced imaging of biological tissues. Curr Opin Chem Biol 42:138–146. https://doi.org/10.1016/j.cbpa.2017.12.005

Sauer U (2004) High-throughput phenomics: experimental methods for mapping fluxomes. Curr Opin Biotechnol 15:58–63. https://doi.org/10.1016/j.copbio.2003.11.001

Sazonova OV, Lee KL, Isenberg BC, Rich CB, Nugent MA, Wong JY (2011) Cell-cell interactions mediate the response of vascular smooth muscle cells to substrate stiffness. Biophys J 101:622–630. https://doi.org/10.1016/j.bpj.2011.06.051

Sen P, Orešič M, Sen P, Orešič M (2019) Metabolic modeling of human gut microbiota on a genome scale: an overview. Metabolites 9:22. https://doi.org/10.3390/metabo9020022

Seras-Franzoso J, Tsimbouri PM, Burgess KV, Unzueta U, Garcia-Fruitos E, Vazquez E et al (2014) Topographically targeted osteogenesis of mesenchymal stem cells stimulated by inclusion bodies attached to polycaprolactone surfaces. Nanomedicine 9:207–220. https://doi.org/10.2217/nnm.13.43

Sévin DC, Kuehne A, Zamboni N, Sauer U (2015) Biological insights through nontargeted metabolomics. Curr Opin Biotechnol 34:1–8. https://doi.org/10.1016/j.copbio.2014.10.001

Simó C, Ibáñez C, Valdés A, Cifuentes A, García-Cañas V (2014) Metabolomics of genetically modified crops. Int J Mol Sci 15:18941–18966. https://doi.org/10.3390/ijms151018941

Sinclair I, Bachman M, Addison D, Rohman M, Murray DC, Davies G et al (2019) Acoustic mist ionization platform for direct and contactless ultrahigh-throughput mass spectrometry analysis of liquid samples. Anal Chem 91:3790–3794. https://doi.org/10.1021/acs.analchem.9b00142

Sitole LJ, Williams AA, Meyer D (2013) Metabonomic analysis of HIV-infected biofluids. Mol Biosyst 9:18–28. https://doi.org/10.1039/c2mb25318f

Smilde AK, Måge I, Næs T, Hankemeier T, Lips MA, Kiers HAL et al (2017) Common and distinct components in data fusion. J Chemom 31:1–20. https://doi.org/10.1002/cem.2900

Southam AD, Weber RJM, Engel J, Jones MR, Viant MR (2017) A complete workflow for high-resolution spectral-stitching nanoelectrospray direct-infusion mass-spectrometry-based metabolomics and lipidomics. Nat Protoc 12:310–328. https://doi.org/10.1038/nprot.2016.156

Spicer R, Salek RM, Moreno P, Cañueto D, Steinbeck C (2017) Navigating freely-available software tools for metabolomics analysis. Metabolomics 13:1–16. https://doi.org/10.1007/s11306-017-1242-7

Strange K (2005) The end of "naive reductionism": rise of systems biology or renaissance of physiology? AJP Cell Physiol 288:C968–C974. https://doi.org/10.1152/ajpcell.00598.2004

Stringer KA, McKay RT, Karnovsky A, Quémerais B, Lacy P (2016) Metabolomics and its application to acute lung diseases. Front Immunol 7. https://doi.org/10.3389/fimmu.2016.00044

Sud M, Fahy E, Cotter D, Azam K, Vadivelu I, Burant C et al (2016) Metabolomics workbench: an international repository for metabolomics data and metadata, metabolite standards, protocols, tutorials and training, and analysis tools. Nucleic Acids Res 44:D463–D470. https://doi.org/10.1093/nar/gkv1042

Sweetlove LJ, Obata T, Fernie AR (2014) Systems analysis of metabolic phenotypes: what have we learnt? Trends Plant Sci 19:222–230. https://doi.org/10.1016/j.tplants.2013.09.005

Tautenhahn R, Patti GJ, Rinehart D, Siuzdak G (2012) XCMS online: a web-based platform to process untargeted metabolomic data. Anal Chem 84:5035–5039. https://doi.org/10.1021/ac300698c

Tebani A, Bekri S (2019) Paving the way to precision nutrition through metabolomics. Front Nutr 6:1–10. https://doi.org/10.3389/fnut.2019.00041

Tolstikov V (2016) Metabolomics: bridging the gap between pharmaceutical development and population health. Metabolites 6:20. https://doi.org/10.3390/metabo6030020

Travers C, Himmelstein DS, Beaulieu-Jones BK, Do BT, Way GP et al (2018) Opportunities and obstacles for deep learning in biology and medicine. https://doi.org/10.1098/rsif.2017.0387

Trivedi DK, Hollywood KA, Goodacre R (2017) Metabolomics for the masses: the future of metabolomics in a personalized world. New Horizons Transl Med. 3:294–305. https://doi.org/10.1016/j.nhtm.2017.06.001

Tsimbouri PM, McMurray RJ, Burgess KV, Alakpa EV, Reynolds PM, Murawski K et al (2012) Using nanotopography and metabolomics to identify biochemical effectors of multipotency. ACS Nano 6:10239–10249. https://doi.org/10.1021/nn304046m

Tugizimana F, Djami-Tchatchou AT, Fahrmann JF, Steenkamp PA, Piater LA, Dubery IA (2019a) Time-resolved decoding of metabolic signatures of in vitro growth of the hemibiotrophic pathogen Colletotrichum sublineolum. Sci Rep 9:3290. https://doi.org/10.1038/s41598-019-38692-7

Tugizimana F, Piater LA, Dubery IA (2013) Plant metabolomics: a new frontier in phytochemical analysis. S Afr J Sci 109:18–20. https://doi.org/10.1590/sajs.2013/20120005

Tugizimana F, Steenkamp P, Piater L, Dubery I (2016) A conversation on data mining strategies in LC-MS untargeted metabolomics: pre-processing and pre-treatment steps. Metabolites 6:1–18. https://doi.org/10.3390/metabo6040040

Tugizimana F, Mhlongo M, Piater L, Dubery I (2018) Metabolomics in plant priming research: the way forward? Int J Mol Sci 19:1759. https://doi.org/10.3390/ijms19061759

Tugizimana F, Djami-Tchatchou AT, Steenkamp PA, Piater LA, Dubery IA (2019b) Metabolomic analysis of defense-related reprogramming in sorghum bicolor in response to colletotrichum sublineolum infection reveals a functional metabolic web of phenylpropanoid and flavonoid pathways. Front Plant Sci 9:1–20. https://doi.org/10.3389/fpls.2018.01840

Turki T, Taguchi YH (2019) Machine learning algorithms for predicting drugs–tissues relationships. Expert Syst Appl 127:167–186. https://doi.org/10.1016/j.eswa.2019.02.013

van der Hooft JJJ, Wandy J, Barrett MP, Burgess KEV, Rogers S (2016) Topic modeling for untargeted substructure exploration in metabolomics. Proc Natl Acad Sci 113:13738–13743. https://doi.org/10.1073/pnas.1608041113

Weber RJM, Lawson TN, Salek RM, Ebbels TMD, Glen RC, Goodacre R et al (2017) Computational tools and workflows in metabolomics: an international survey highlights the opportunity for harmonisation through Galaxy. Metabolomics 13:1–5. https://doi.org/10.1007/s11306-016-1147-x

Welch JD, Kozareva V, Ferreira A, Vanderburg C, Martin C, Macosko EZ (2019) Single-cell multi-omic integration compares and contrasts features of brain cell identity. Cell 1–15. https://doi.org/10.1016/j.cell.2019.05.006

Wolfender J-L, Litaudon M, Touboul D, Queiroz EF (2019) Innovative omics-based approaches for prioritisation and targeted isolation of natural products—new strategies for drug discovery. Nat Prod Rep. https://doi.org/10.1039/c9np00004f

Wu J, Zhao Y (2019) Machine learning technology in the application of genome analysis: a systematic review. Gene 705:149–156. https://doi.org/10.1016/j.gene.2019.04.062

Xia J, Sinelnikov IV, Han B, Wishart DS (2015) MetaboAnalyst 3.0—making metabolomics more meaningful. Nucleic Acids Res 43:W251–W257. https://doi.org/10.1093/nar/gkv380

Yang D, Fan X, Shi X, Lian S, Qiao J, Guo R (2014) Metabolomics reveals stage-specific metabolic pathways of microbial communities in two-stage anaerobic fermentation of corn-stalk. Biotechnol Lett. https://doi.org/10.1007/s10529-014-1508-3

Zeiss DR, Mhlongo MI, Tugizimana F (2018) Comparative metabolic phenotyping of tomato (Solanum lycopersicum) for the identification of metabolic signatures in cultivars differing in resistance to ralstonia solanacearum. Int J Mol Sci 19:1–17. https://doi.org/10.3390/ijms19092558

Zhang A, Sun H, Wang P, Han Y, Wang X (2012) Modern analytical techniques in metabolomics analysis. Analyst 137:293–300. https://doi.org/10.1039/c1an15605e

Zhao X, Fritsche J, Wang J, Chen J, Rittig K, Schmitt-Kopplin P et al (2010) Metabonomic fingerprints of fasting plasma and spot urine reveal human pre-diabetic metabolic traits. Metabolomics 6:362–374. https://doi.org/10.1007/s11306-010-0203-1

Zhou J, Park CY, Theesfeld CL, Wong AK, Yuan Y, Scheckel C et al (2019) Whole-genome deep-learning analysis identifies contribution of noncoding mutations to autism risk. Nat Genet 51. https://doi.org/10.1038/s41588-019-0420-0

Virtual and Augmented Reality in Surgery

Osama Halabi, Shidin Balakrishnan, Sarada Prasad Dakua, Nassir Navab, and Mohammed Warfa

Abstract To provide an overview of the novel applications of Augmented Reality (AR) and Virtual Reality (VR) technologies within the realm of Surgery. This chapter will clarify the concepts associated with AR and VR in the context of surgery, and address implementation approaches of AR and VR in surgical pre-operative planning and surgical training. Some examples of state-of-the-art solutions and system architecture developed by the editors will be described and discussed.

Keywords Virtual reality · Augmented reality · Mixed reality · Surgery · Image-guided surgery · Segmentation · Registration

1 Introduction

The scope of the chapter will be limited to novel applications of AR and VR in the healthcare sector, specifically for surgical pre-operative planning and surgical education. Our focus is to highlight the potential applications of AR and VR on all

O. Halabi
Department of Electrical Engineering, Qatar University, PO Box 2713, Doha, Qatar
e-mail: ohalabi@qu.edu.qa

S. Balakrishnan (✉) · S. P. Dakua
Department of Surgery, Hamad Medical Corporation, PO Box 3050, Doha, Qatar
e-mail: sbalakrishnan1@hamad.qa

S. P. Dakua
e-mail: sdakua@hamad.qa

N. Navab
Department of Informatics, Technische Universität München, Munich, Germany
e-mail: navab@cs.tum.edu

M. Warfa
Department of Radiology, Wake Forest Baptist Medical Center, Winston-Salem, NC, USA
e-mail: mwarfa@wakehealth.edu

W. Doorsamy et al. (eds.), *The Disruptive Fourth Industrial Revolution*, Lecture Notes in Electrical Engineering 674,
https://doi.org/10.1007/978-3-030-48230-5_11

human organs, however, due to page constraints, we have included only a few organs in this chapter such as liver, heart, etc.

Today, enhanced/augmented surgical visualization could be in the form of 2D/3D projections of data on a conventional 2D/3D screen/handheld display device or holographic projections directly onto the patient's body/surgical site. The kind of data visualized depends on the surgical procedure; it could be just a simple numerical overlay of patient data, or a complex display of the reconstructed organs and tissues. The reconstruction and display of data (preoperative patient imaging data, anatomic models, surgical instruments etc.) on the surgeon's visual field using the concepts of simulated reality ("Augmented Reality" (AR) or "Virtual Reality" (VR)) creates an enhanced, intuitive and immersive surgical environment. The result is increased simplicity and accuracy during the actual procedure, and hence lowered distractions, negative outcomes and time under anesthesia. Such enhanced visualization aids both pre-operative planning (virtual reconstruction and exploration of patient-specific anatomy, planning optimal surgical route/approach, teaching surgical residents in preparation for the procedure) and intra-operative augmentation (updated patient statistics during surgery, and improved visualization, orientation and navigation within the surgical site). The advent, rapid advancement and, miniaturization of virtual and augmented reality systems and the technological advances in the processing power of microcomputers has resulted in a feasible, usable and realistic simulation and augmented visualization of the surgical environment. Using AR and VR, current cutting-edge research focuses on the acquisition, manipulation, and visualization of patient-data/images in real-time.

2 Virtual and Augmented Reality Technologies for Visualization

Virtual Reality (VR) and Augmented Reality (AR) technologies are the current trends and affecting nearly every industry and set to revolutionize the way we view and interact with the surrounding. These technologies have potential applications in medicine and considered disruptive for many medical specialties by creating a new paradigm shift in how medical care is provided.

2.1 Defining Reality and Virtuality

The next-generation interaction and communication environments are mixed between sufficient reality and virtual spaces. People are starting to talk about a new term, Mixed Reality (MR). There is some confusion about MR, VR, and AR. To clear this confusion, the "virtuality continuum" will be used to describe the difference between these different terms. The concept of the Reality-Virtuality (RV) continuum

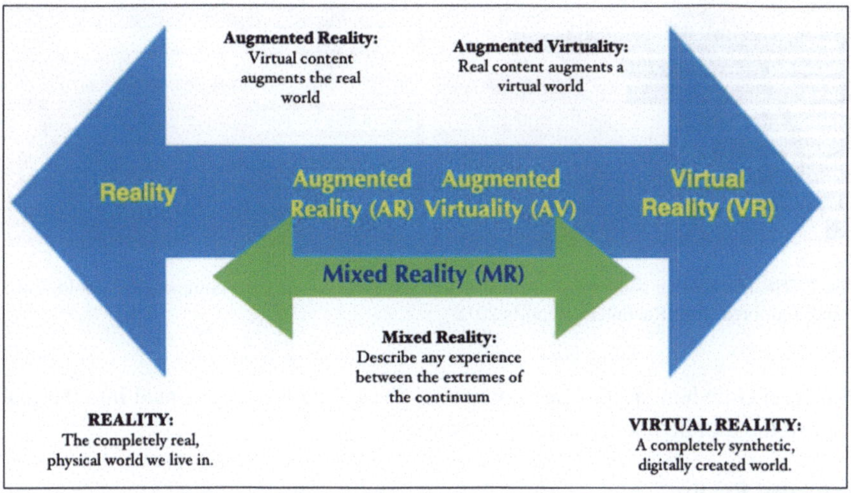

Fig. 1 The reality-virtuality continuum concept created by Milgram and Kishimo (1994)

was first introduced by Paul Milgram and Kishino (1994; Milgram et al. 1995). Figure 1 illustrates the concept where the real worlds are shown at one end of the continuum and the virtual completely digital worlds are at the opposite extremum. MR environments represent anywhere between the real and virtual world. It blends real and virtual worlds to create complex environments where physical and digital elements can interact in real time. On the other hand, the AV is a virtual world with elements of the real world introduced into it, similar to the AR where the real world is augmented with virtual elements.

VR is an entirely computer-generated view of a world in 3D where everything in the view is purely virtual. Hence, everything the user sees is synthesized and not real (Milgram and Kishimo 1994). Immersion or sense of being there is a key element in VR to enable high fidelity interaction with the user to make him believe that the virtual world is real. In VR the virtual world can provide a virtual experience of either amplified or subdued versions of reality. Besides, VR has to provide multimodal interaction feedback to the wearer's visual, auditory, olfactory, haptic, tactile, etc. to get closer to real experiences. Consumer low-cost VR Head Mounted Displays (HMD) are commercially available from HTC, Oculus, and Sony, among others with high-resolution displays and tracking capabilities of the user motion and actions.

Conversely, AR takes simulation and synthesized worlds and combine them with real-world objects and interactions. It enables the user to see their native real-world surroundings while placing 2D/3D images within it. The onboard camera can provide continuous view of the real world and can be displayed on a mobile device or an unobtrusive HMD. A popular consumer application of AR technology has been

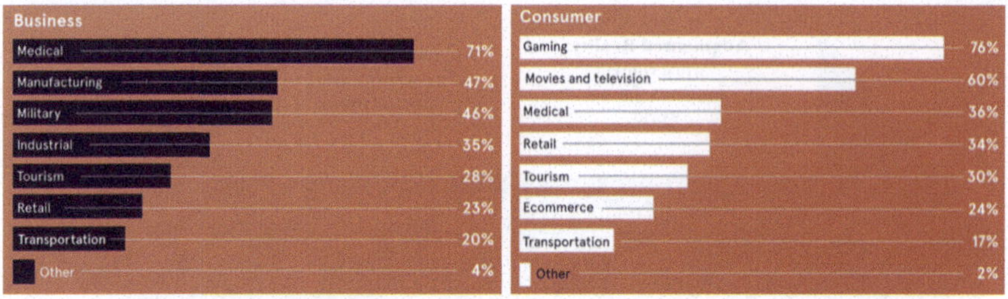

Fig. 2 The market that is the most promising for XR when XR product company stakeholder was asked) (adopted from Raconteur Reports 2018

Pokémon Go where avatars and contextual game data was augmented into the real view of the mobile phone camera (Heine 2016).

Extended Reality
Extended Reality (XR) is a new term that used as the umbrella for VR, AR, and MR, as well as all future realities that might emerge from these technologies. Therefore, XR covers the full spectrum between reality and virtuality. XR describes all the development of computer-generated environments that either fully immersed the user in virtual worlds or create a mixed interactive experience by merging the physical and the virtual world. XR is expected to impact different industry and Medical business is on the top industry use-cases that will be impacted by XR technologies. According to the report published in THE TIMES by Racounteur Reports (2018), the Medial applications are the top-use case for XR business and runup to third for consumer applications, see Fig. 2.

2.2 *Display Technologies*

AR system uses a camera to track the subject's view to be merged with the virtual objects. The devices have on a display unit to visualize the computer-generated enhancements overlaid on the actual physical world. Generally, Head Mounted Displays (HMDs) are used for this purpose, though conventional 2D/3D screens/monitors (e.g. tablet-screens, iPads, computer monitors, etc.) can also be used as a cheaper and available alternative to display AR/VR applications using built-in or connected sensors such as gyroscopic and motion sensors and its camera. Paul Milgram was the first to categorize the types of AR and MR displays to see-through and monitor-based AR displays. In his paper (Milgram et al. 1995). Now a days the displays are broadly classified into two types, Optical See-Through (OST) or Video-See Through (VST) (Monsky et al. 2019). VST displays project a simulated video stream of the real-world environment with computer-generated overlays onto the visual field of the user; i.e. the user sees what the video stream shows him as opposed to the real world. Hence, if using VR during surgery, the surgical site is not in view and thus, the VR modality cannot be used during live surgery to optically augment and enhance the information available to a surgeon. However, VST-HMDs can be extensively used for surgical training, to allow junior surgeons objectively and repetitively practice surgical skills. HTC Vive and Odyssey from Samsung are examples of VR headsets equipped with cameras which is can be used for creating AR experience.

The OST enables the observer to view reality directly through optical elements such as holographic waveguides and other systems that enable graphical overlay on the real world. An example of this type is the Magic Leap One, HoloLens from Microsoft, and Google Glass. If used for surgery, world-anchoring the virtual projections (organs, tissues etc.) locks them in place on a real-world scenery (patient's body), effectively making the surgical site a virtual monitor, so the surgeon can freely look around and even interact with the patient/surgical staff while the virtual object stays put at the surgical site. Easy visualization and assimilation of the medical information in real-time are crucial for the success of a simulated-reality based surgical planning and/or training module. The greatest advantage of an OST device over VST in a surgical setting is that the surgeon can visualize the enhanced information directly onto the surgical field (patient's body). They can synchronize their body movements and surgical actions in response to the enhanced/augmented image they see through the OST-HMD directly on the surgical site. Additionally, it is possible in OST to world-anchor the enhanced visualizations/projections, to minimize spatial disorientation and contextual-loss of information at the surgical site. All of these advantages make OST-HMDS within the realm of AR/VR, as the most attractive option for the development of modules in surgical simulations, anatomic training, pre-operative planning, and intra-operative navigation. We reviewed the commercially available VR and AR headsets on the market and proposed classification to three categories for VR, namely VR headsets for consoles and PC, Standalone VR headsets, and Mobile headsets and AR headsets. Table 1 shows the main technical specification for each device. For current models on the market, two forms of tracking are used: "outside-in" and "inside-out." The VR displays use external fixed optical tracking for accurate positioning and tracking within a volume in outside-in tracking. Meanwhile, the HMD relies on an onboard camera in inside-out tracking to detect fiducial markers within the same area.

3 Visualization Technology in Surgical Workflow

To incorporate AR in surgery there are three main technical aspects related to tracking, registration, and calibration that need to be considered (El-Hariri et al. 2018; Chen et al. 2015). Figure 3 illustrates the complete framework for AR medical system with the necessary technologies.

Table 1 Technical specification of wearable VR and AR displays on the market today

Devices	Company	Resolution	Refresh rate	FOV	Tracking
VR headsets for consoles and PC					
Oculus Rift	Oculus	2160 × 1200	90 Hz	110	Positional, outside-in, 6DOF controller
HTC Vive Pro	HTC	2,880 × 1,600	90 Hz	110	Positional and location, outside-in, 6DOF
Sony PlayStation VR	Sony	1020 × 1080	120 Hz, 90 Hz	100	Positional 360, Inside-out 6DOF
Samsung Odyssey+	Samsung	1400 × 1600	90 Hz	110	Positional and location, Inside-out,6DOF
Oculus Rift S	Oculus	1280 × 1440 per eye	80 Hz	>110	Inside-out, 6DOF
Standalone VR headsets					
Oculus Quest	Oculus	1,440 × 1,660 per eye	72 Hz	100	Inside-out (6DOF)
Oculus Go	Oculus	2560 × 1440	72 Hz	110	Inside-out (3DOF)
Mirage Solo	Lenovo	2560 × 1440	75 Hz	110	Inside-out (6DOF)
Mobile VR headsets					
Gear VR	Samsung	2560 × 1440	60 Hz	101	–

(continued)

Table 1 (continued)

VR headsets for consoles and PC					
Devices	Company	Resolution	Refresh rate	FOV	Tracking
Daydream View	Google	Depends on device	Depends on device	90	–
Google Cardboard	Google	Depends on device	Depends on device	45	–
Zeiss One Plus	Zeiss	Depends on device	Depends on device	100	–

AR headsets	
Devices	Specifications
Microsoft HoloLens 2	See-through holographic lens, 2 K pixel per eye, Qulacoom Snapdragon 850, hand tracking, eye tracking, voice control, Wi-Fi, Bluetooth
Magic Leap	Nvidia Parker processor, GPU NVidia Pascal with 256 cores, 8 GB, 128 GB built-in storage, Wi-Fi, Bluetooth, Controller with 6DoF tracking and haptics
Meta	2550 × 1440 resolution, 60 Hz refresh rate, 90° FOV, four speaker near-ear audio system, 720p front-facing camera, sensor array for hand interactions and positional tracking
Epson Moverio BT-300FPV	1280 × 720 pixels, 23° angle of view, 80° virtual screen size, 2 GB memory, 6 h battery life
Google Glass Enterprise Edition	8-megapixel camera 640 × 360 optical display, multi-touch gesture touchpad. 802.11ac Wi-Fi and Bluetooth 5, 3 GB of RAM and 32 GB of onboard storage

(continued)

Table 1 (continued)

VR headsets for consoles and PC					
Devices	Company	Resolution	Refresh rate	FOV	Tracking
Vuzix Blade AR	8 Megapixel camera, 720p 30fps or 1080p 24fps, Quad core ARM CPU, Wi-Fi and Bluetooth wireless, touch pad with gesture, head motion trackers				
Optinvent Ora-2	dual core processor w/GPU, 42 pixel/deg, 5 M Pixels camera, microphone, sound, inertial sensors, Wi-Fi, Bluetooth, GPS, track pad, ambient light sensor, and a high capacity rechargeable battery				
ODG R-7	Qualcomm Snapdragon 805 2.7 GHz quad-core Processor, 3 GB RAM, 64 GB storage, dual 720p see through, Bluetooth				

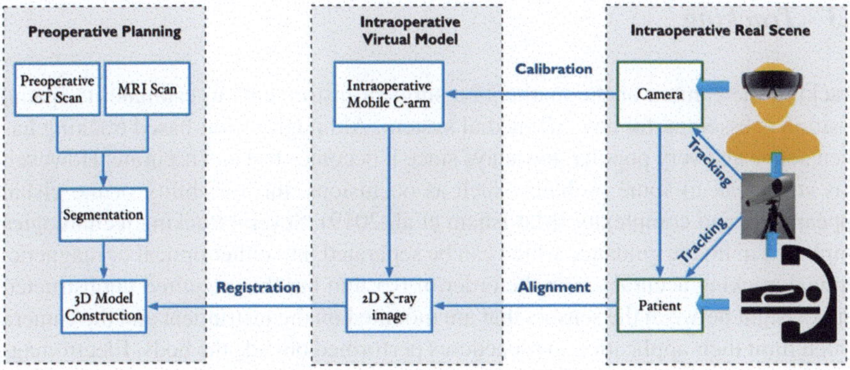

Fig. 3 The framework for the complete AR medical system

3.1 Registration

The patient data is usually taken from computed tomography (CT), ultrasound imaging, or magnetic resonance imaging (MRI) to construct the 3D volume data for preoperative planning. The patient real-world coordinate must be registered with the 3D data to have both data and real patient in the same perspective (Li et al. 2017a, b, c). State-of-art resignation techniques is described in the Sect. 4.2 image registration section.

3.2 Calibration

The AR display fuses the virtual objects with the real world, thereby requiring transformation between the real world and the camera coordinate. Perspective projection is used to map the 3D points to the 2D reference plane where the transformation between the image and the real-world coordinates is represented by a projection matrix. The estimation of the parameters for a pinhole model of the projection matrix is the process of calibration of the camera. In practice, the preoperative images in HMD are aligned with the patient and then calibration is required. The adjustment between the positioning of the patient in relation to the HMD where the patient's or HMD's motion does not influence the interaction between the device coordinate and the patient coordinate (Chen et al. 2015).

3.3 Tracking

Tracking the camera or the markers on surgical instruments to estimate its spatial position is essential for any AR medial system. Although vision-based tracking has been active and very popular nowadays since it becomes fast and accurate. However, it is still weak to some problems such as occlusions, the variability of the visual appearance, and complexity (Mat Isham et al. 2019). Several tracking technologies employed in image guidance which can be separated into either optical or magnetic. Optical tracking accuracy is in the order of 0.5 mm but they required unobstructed line-of-sight between the sensors that are mounted on the instrument and the camera which limit their application to procedures performed outside the body. Electromagnetic tracking does not suffer from such limitation, however, they cannot be used with the presence of ferromagnetic materials in the vicinity of the field.

3.4 Advantages of Using AR Over VR for Visualization in Surgery Using HMD

Easy visualization and assimilation of the medical information in real-time are crucial for the success of a simulated-reality based surgical planning and/or training module. The greatest advantage of an AR-based HMD over VR in a surgical setting is that the surgeon is able to visualize the enhanced information directly on-to the surgical field (patient's body). He can synchronize his body movements and surgical actions in response to the enhanced/augmented image he sees through the HMD (OST-HMD) directly on the surgical site. Additionally, it is possible in AR to world-anchor the enhanced visualizations/projections, to minimize spatial disorientation and contextual-loss of information at the surgical site.

In a surgical setting, HMDs are greatly preferred as an option for enhanced display over conventional monitors, which are the current workhorses. HMDs enable the surgeon directly to interact with the projections very intuitively, without losing their spatial orientation and while maintaining hand-eye coordination. There is no need to gaze away from the surgical site. This is in contrast to a conventionally used 2-D monitor, where the surgeon would have to continuously switch his gaze and bodily orientation between the augmented image-overlay on the monitor and then re-orient his body, hand-eye coordination and surgical actions to perform the procedure on the surgical site.

4 Related State of Art Work

4.1 Image Segmentation

On radiological images, the liver is usually considered to be difficult to separate from the neighboring organs (e.g. heart, kidney, muscle, and stomach) due to similar gray level intensities. The major challenges in its segmentation are variability in shape, size, and other similar adjacent\neighboring organs with similar gray level intensity. Moreover, the blurry edges and low contrast usually characterize CT images making the segmentation more challenging. Furthermore, there are other issues such as partial volume effects resulting from patient movement, spatial averaging, and reconstruction artifacts. Several methods have been reported in the literature related to liver segmentation on CT datasets. Conceptually, the methods are either pixel driven/intensity-based (e.g. thresholding and morphological filtering, etc.) or model-based (e.g. active contours with level sets or snake, and so forth). There are some simple standard classical techniques that include thresholding and region-based techniques (Kennedy et al. 1989; Grady 2006) that are quite user friendly. Fuzzy segmentation techniques are preferred for multi-channel image segmentation, though they still can be applied for single-channel images (Ruan et al. 2002; Patwardhan et al. 2005). There are also statistical approaches reported for robust segmentation such as reported in (Zhang et al. 2010; Nuzillard and Lazar 2007), however, these techniques typically involve some form of constraints, sometimes leading to under-segmentation results.

Although manual segmentation is mostly preferred by the clinicians, this is quite tedious and time-consuming, therefore a robust computer-aided liver (tumor\vessel) segmentation has been a core topic over the years. There have been a lot of attempts\published segmentation methods, however, a proven and clinically accepted method is yet to be established because of the challenges, mainly: (1) low contrast, (2) similar gray level intensity between the neighboring muscles\tissues, (3) unclear lesion boundary, (4) variation in lesion shape, etc. Tumors from CT can only be identified if lesions are segmented. The literature of segmentation is huge (Mharib et al. 2012): there have been proposed many automatic and semiautomatic segmentation methods. A few of them are discussed here. Knowledge of local organs' shape is essential in order to compare with each other; the 3D affine invariant shape parameterization approach by Linguraru et al. (2012) first generates a regular sampling of the liver and then various 3D surface features are compared using this parameter space point-to-point. Another method by Seo (2005) uses a multi-stage approach to sequentially segment liver, hepatic vessels, and then tumor using an optimal threshold value.

There are many such automatic segmentation methods; however, a pseudo-segmentation may have resulted since there is no provision for correction during its execution. On the other hand, a semi-automatic method is more suitable and reliable because the output contour can be guided as per the desire of the user's clinical requirement (Heimann et al. 2009; Dakua 2013). Chartrand et al. (2017)

propose a semiautomatic method for liver segmentation; they first approximately model a liver and then keep deforming it by a Laplacian mesh optimization approach until the liver is precisely segmented. The model proposed by Peng et al. utilizes level sets that incorporate the likelihood energy with that of edge. The likelihood of energy is then minimized to get the liver segmented. Zhang et al. (2017) propose a semiautomatic method to segment the liver in clinical cases based on Couinaud's theory. Watershed method with active contour has also been used in the past to study hepatic metastatic lesions on CT slices (Yim and Foran 2003). Zhao et al. report a region-growing method that uses shape constraints so that the region growing does not leak into surrounding tissues (Zhao et al. 2006). Mahr et al. (1999) in their study of different segmentation approaches find region-growing, graph-based and snakes as the potential futures of all investigations on liver segmentation. The main challenge in liver vessel segmentation is that all bright structures are not liver vessels and the parts inside the liver are to be segmented. There are many vessel segmentation methods too. Due to the introduced artifacts, for instance during the acquisition, image features are usually hidden or distorted in the normal image scale/resolution. There are certain techniques that have been proven to have high resistance to noise and have high processing speed; some of them include: Wavelet, Ridgelet, Curvelet, and Contourlet transforms, where they are embraced in medical image segmentation (AlZubi et al. 2011). In order to visualize the complete anatomy such as liver vessels, there have been also many attempts. An active contour method by Shang et al. (2011) exploits the Gaussian mixture model to segment the thick vessel first and then based on the centerline of vascular vector field, weak and thin vessels are segmented. There are many methods (Sato et al. 1998) to the date on vessel segmentation; one may have a look at the paper (Kirbas and Quek 2005) to get a comprehensive review. For the sake of the present discussion flow, the input images, their contrast enhancement and liver segmentations are provided in Fig. 4.

4.2 *Image Registration/Fusion*

In recent years, the development of augmented reality (AR) is remarkably noticeable. The display resolution is dramatically improved and the cost is dropped to consumer-level from USD 18,000 to USD1000 or less. Virtual reality provides an immersive environment that can enhance surgical simulators but it is not appropriate for intraoperative image guidance since it isolates the surgeon forms the surrounding. Meanwhile, AR provides the optimal in the case of both virtual and real worlds by integrating the virtual and real objects. These integrations occur in real time; the virtual and real objects interactively run by aligning with each other (Der et al. 2013). Therefore, AR visualization can provide significant advantages for image-guided surgical systems. AR visualization provides the ability to superimpose the virtual CT data onto the patient's anatomy. This has been presented in (Navab et al. 2007), where AR is performed after the registration of the preoperative CT with the intra-operative US. Liao et al. (2010) describe a navigation system with augmented

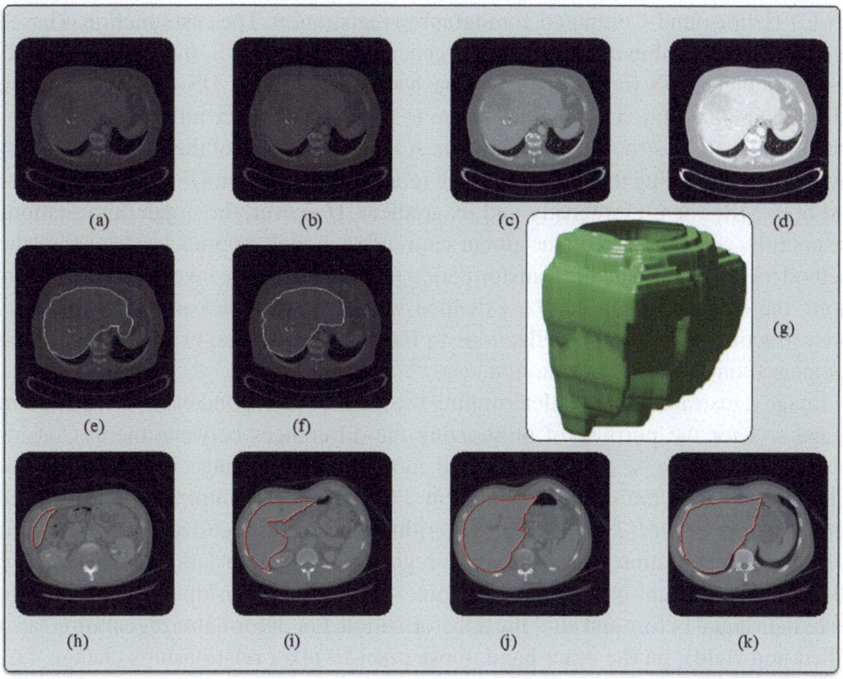

Fig. 4 **a** Input liver CT image. **b–d** Contrast enhancement of the input liver images. **e–k** 2D and 3D Liver segmentations

reality for MRI guidance where a half-slivered mirror helps in superimposing the images superpose onto the surgical area.

A recent work by Zhang et al. (2017) has described a see-through surgical guidance system using AR, however, they also use the half-slivered mirror as a display. Such an AR display suffers from brightness and vividness issues, in addition to the need for designing special heavy structures for holding the mirror. Recent advancement in AR display brings the attention of its potential to be considered for the visualization of guidance due to its high image resolution and brightness. For example, the new Hololens 2 from Microsoft provides 43° by 29° field of view (FOV) (Microsoft HoloLens 2019). The other state-or-art AR head mounted display (HMD) is Magic Leap One (2005) with 40° by 30° provides almost the same resolution with hand interaction and positional tracking. The hand tracking in Magic Leap Oe seems to be more successful than Hololens 2 with the ability to detect the position of the fingers of the user in space. Standard similarity measures such as the Sum of Squared Difference (SSD), Normalized Cross Correlation (NCC) or Mutual Information (MI) are limited in performance for robust registration due to the different nature of intensity values and structures that are in-discernible in one of the two image sets (e.g. because of shadowing artifacts). The Linear Correlation of Linear Combination (LC2) similarity measure was introduced as a method specially designed for

US-CT (Ultrasound-Computed Tomography) registration. The cost function is based on local linear combinations of patches generated from the US, from CT, and a CT-based simulated US image. This approach was adapted for US-MRI registration, introducing an additional weighting term to allow for a larger influence of patches with high levels of structural information. A main advantage of the LC2 metric is its invariance concerning the locally varying relationship between US image intensities and both MRI (or CT) intensity and its gradient. However, the organ deformations are not fully compensated by the global semi-affine model. Approaches to expand the method towards de-formable transformation models need to be investigated. Furthermore, the US simulation can be extended with more complex models for effects like refraction and multiple reflections to increase robustness, especially real-time compensation of respiratory motion.

Image registration aims at determining the spatial correspondence of two or more image sets for the purpose of minimizing the differences between them. Considering two image sets, a static $S(x)$ and moving $M(x^*)$, image registration algorithms try to find the optimal transform $T(x^*)$, which minimizes the difference between $S(x)$ and $M(T(x^*))$. Such algorithms can be categorized into rigid and deformable registration. For rigid image registration, the translation and rotation of all image pixels is uniform, such that all pixel-to-pixel relationships within the image set remain equal before and after the transformation. For deformable registration (also called non-rigid), on the other hand, those pixel-to-pixel relationships change, i.e., when two images are on the same reference frame of coordinates, a pixel on one image set does not necessarily represent the same anatomical structure of the other although the pixel coordinates of both image sets are same. Therefore, deformable image registration can account for local distortions, occurring since organs and tissue are non-rigid structures and subject to deformation.

In this application, we need the exact registration of pre-operative CT volumes with intra-operative US images. Additional to the real-time constraint, many factors can complicate accurate registration: Especially for tissue located in thorax and abdomen, like the liver, the respiratory motion leads to organ tissue deformation due to the contraction of the diaphragm. Dissimilarities can occur due to inter- and intra-fractional anatomical variations from the pre-operative image set; that is variations between different treatment sessions and during single treatment sessions, respectively. Soft tissue deformation can be caused by plain physiological changes such as bladder filling, patient weight loss, tumor shrinkage/growth or differences in patient positioning. The registration must be independent of the presence of medical instruments in the US images or varying application of contrast agent. Furthermore, the multimodal images taken at different time frames makes the registration process quite challenging. There is a greater number of degrees of freedom in the deformable registration to tackle the aforementioned local distortions between anatomical structures. However, this also typically requires more computational power and/or time which inherently influences the real-time capabilities of such registration algorithms.

A deformable registration algorithm consists of three main components, the objective function, the optimization method, and the deformation model. The objective function requires a definition of similarity between two image sets. Similarity measures can be intensity-based, feature-based or a combination of the two. Depending on the task at hand, the type of images and the expected kind and amplitude of misalignment, a different similarity metric may be appropriate for accurate registration. Intensity-based objective functions define the degree of similarity by using image intensity values. For single-modality images, this approach can be robust as the assumption is met, that the intensity values of the same anatomical area are similar among images. Feature-based objective functions require the definition of image features which can be independent of image intensity. However, the construction of features can be difficult and time-consuming while introducing inter- and intra-observer dependencies. Recent advances in unsupervised Deep Learning have reduced those disadvantages and allow for feature creation without manual input but require large datasets and ground truth for training.

While from a medical perspective, different modalities complement each other for diagnostic and interventional purposes, the registration of CT and US images is a complex task. This is due to the information represented in the different modalities originating from different physical processes and properties. Various direction-dependent artefacts and the speckle noise affect the US intensities that represent the changes in acoustic impedance. Intensities in CT images, on the other hand, provide a representation of X-ray attenuation. This is approximately proportional to the tissue density which again is proportional to acoustic impedance. However, the correlation is influenced by inherent differences of the imaging method: The acquisition of US images is user-dependent and provides a limited field of view and is prone to image quality reduction due to overlying structures, subcutaneous fat or gas-containing organs. Furthermore, ultrasound is usually acquired in arbitrary planes, which are not parallel to the ones typically visualized by physicians from CT imaging (axial, coronal, or sagittal).

Multiple prior studies have already been reported regarding image fusion in non-rigid, deformable structures like the liver and prostate. However, these studies always come up with the problems of accurate image registration and segmentation, which they have attempted to solve using various computing/mathematical techniques (Lee 2014; Mauri et al. 2015; Esser et al. 2016; Li et al. 2017a, b, c; Seror 2017; Venkatesan et al. 2011; Kadoury et al. 2016). Standard similarity measures such as the Sum of Squared Difference (SSD), Normalized Cross Correlation (NCC) or Mutual Information (MI) are limited in performance for robust registration due to the diverse nature of grey-scale values and anatomies that are indiscernible in one of the two image sets (e.g. because of shadowing artefacts). The Linear Correlation of Linear Combination (LC2) similarity measure was introduced as a method specially designed for US-CT registration. The cost function is based on local linear combinations of patches generated from the US, from CT and from a CT-based simulated US image (Wein et al. 2008). The reconstruction has to be estimated at every iteration during optimization, however, this also means that the algorithm can search for the best alignment using the best reconstruction. Large variations in image intensities can be adjusted by the local linear combination. This approach was adopted by (Wein et al. 2013) for US-MRI registration, introducing an additional weighting term to allow for a larger influence of patches with high levels of structural information. A main advantage of the LC2 metric is its invariance concerning the locally varying relationship between US image intensities and both MRI (or CT) intensity and its gradient. However, the organ deformations are not fully compensated by the global semi-affine model.

4.3 Tool Tracking

Since the literature on tool tracking is huge, we have included only a few methods in this chapter due to page constraints. The methods are quite diverse, therefore, it is quite difficult to generalize them if a particular category of methods produces the best results, rather, we have included the recent state of art methods for the readers knowledge and information. A method based on edge information is proposed by Kim and Park (2004) to assist video coding, compensation of motion, and motion estimation for Moving Picture Experts Group (MPEG) 7 and 4. This method utilizes the human visual perception providing edge information. Markov random field model and posterior probability estimation technique are used by Subudhi et al. (2011) for temporal segmentation and spatiotemporal spatial segmentation. The color descriptors, as well as the gradient descriptors incorporation with the generalized Hough transform can track an object. Duffner and Garcia (2017) present such an algorithm that uses probabilistic segmentation for foreground and background color distributions. Correlation framework (CF) is used to model a tracker that maximizes the margin between the surrounding background and target (Li et al. 2017a, b, c). In this method, the background information is exploited effectively. The CF is trained first by multilevel scale supervision making it sensitive to the target scale variation. Then a single framework is developed from these two individual modules simplifying the tracking model. A combination of blob detection and fuzzy morphological operation is proposed by Mahalingam et al. for object tracking (Mahalingam and Subramoniam 2018). A specific type of correlation particle filter is designed by combining a particle filter with a correlation filter tracking the object (Zhang et al. 2018).

Kernelized Correlation Filters (KCF) is used to analyze the video frames that are extracted from the video sequences (Guo et al. 2018). A three-dimensional trajectory of a cabbage butterfly is plotted when combining KCF and Background Subtraction (BS). The KCF-BS algorithm obtains the coordinates of the target centroid in two video sequences tracking the butterfly in video frames (Guo et al. 2018). The object tracking in satellite videos is also one of the hot topics in computer vision. Du et al. (2018) propose such an algorithm that fuses a three-frame-difference method and the KCF tracker. Sometimes multiple positions' detections and alternate templates help in tracking (Liu et al. 2018). Liu et al. (2018) propose such a method that is based on a correlation filter. The estimated target speed repositions the detection position by an optical flow method with a template update mechanism. Sometimes histogram becomes exceptionally pivotal in tool tracking. Histograms of RGB, orient, and motion integrate forming a statistical model to track the target (Liu and Feng 2018). Du et al. (2018) demonstrate a method for object tracking that is iterative graph-based. Here the mid-level visual cues represent target parts, whereas the super-pixel representation exploits the local appearance variations. Three key steps constitute the method: 1—the target is selected from the background, 2—majority voting classifies the associate parts between consecutive frames, 3—majority voting of matching results estimate the target state. A unified energy minimization framework consisting of the target part selection, part matching, and state estimation incorporates the structural information in local parts variations.

5 Surgical Planning and Navigation

The recent advances in VR/AR headset technologies pave the way to deliver what has been promised by extended reality technologies to enable physicians to work in the native 3D world and understand the organ anatomy noninvasively. The new class of HMD devices have high-resolution, low-cost, comfortable and light-weight, and can display high-quality clinical data at fast response time that can be used for extended longer periods with less motion sickness. The medical community is starting to embrace XR technology for prior purposes. The Ture 3D system developed by Echopixel is one of the first 3D cardiovascular disease pre-procedural planning to obtain US approval. Food and Drug Administration which is included in a DICOM workstation diagnostic grade. The system uses a 3D interface where the user wearing polarized glasses identical to that used in 3D movie theaters and uses a handheld wand to control images. The system used to visualize arteries in patients with pulmonary atresia with major aortopulmonary collateral arteries (Chan et al. 2013). The cardiologists were able to interoperate the data faster in True 3D display in 13 min compared to 22 min in the traditional readout method. In 2016 computer-generated holography coupled with 3D transeophageal echocardiographic was created in (Bruckheimer and Rotschild 2016). The system used to identify anatomical landmarks within the hologram and typical cardiac imaging that was the first of its kind to demonstrate feasibility in the laboratory for cardiac catheterization. Conventional surgery for chronic full occlusion of the coronary artery is highly dependent on 2D X-ray scans and the expertise of the physician, which can contribute to misidentification of the coronary artery or inaccurate location of the coronary artery stenosis (Ha and Hong 2016). A system was proposed to fuse the 3D CT angiography model with X-ray images to provide 3D anatomical information to the surgeon (Jeon et al. 2016).

Halabi and Halwani (Halabi and Halwani 2018) presented a preoperative surgical planning system for the design and implementation of haptic virtual fixtures (VFs). The approach may be very useful for surgical procedures involving pathway navigation. Preoperative surgical preparation has been applied for the heart catheterisation operation. Upon segmentation and triangulation, the CT scanner images were used to create a 3D model. The research explains how surgeon can establish their desired path by selecting the midpoints in the correct order directly on the heart 3D model

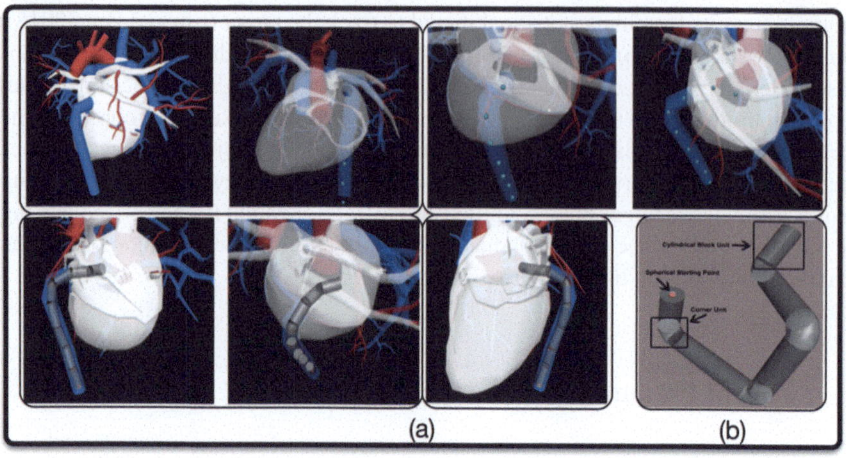

Fig. 5 **a** Pre-operative and operational stages. The operator can adjust the camera view and visualize the 3D model. **b** Assembled path of VF elements

during the preoperative planning stage with the help of the haptic interface device, as shown in Fig. 5a–d. The user is given the option to change the model's transparency and adjust the camera. Such methods help the user imagine the world in 3D better. Thus, in the 3D model, VFs are generated dynamically along the midpoints and aligned within the path as shown in Fig. 5e–g from different camera perspectives. The VFs are generated to fit within the surrounding diameters of the blood vessel after forming the midpoints, see Fig. 5.

VR/AR technology provides great assistance in surgical planning and navigation. VR can be very beneficial in pre-operative planning with highly immersive visualization and interaction. Meanwhile, AR is more appropriate for intra-operative guidance due to the fact that relevant 3D anatomy can be superimposed on the actual view of the patient. In this manner, AR is better suited than VR in intra-operative guidance. AR has been used in many surgical planning applications. AR was used in stereotactic surgery where radiographic images were superimposed on the patient to enable accurate and safer neuro-navigation (Parkhomenko et al. 2018). AR fused images to improve the intra-procedural guidance for it provides a 3D depiction of the patient's specific anatomy and pathology. The pathologists utilized AR for gross and histopathological assessment with even the ability to perform examination remotely (Hanna et al. 2018).

6 Augmented and Virtual Reality in Surgical Education

As expected in any field where technical skills are passed down to novices by experts, surgical education follows the "apprenticeship" model. This model, however, depends on the subjective capability of the expert to impactfully impart his surgical knowledge and skills to a training surgeon and is thus inefficient and sub-optimal. Moreover, the assessment of the actual transfer of proficiency is subjective rather than objective as it stands today, relying greatly on the opinion of the field-expert rather than objective metrics or assessment. It is here that the use of AR/VR coupled with known procedural metrics and benchmarks could supplement the existing surgical educational landscape in terms of imparting and assessing surgical proficiencies.

Fida et al. have shown in 2018 that the most common surgical setting for AR/VR enhanced image-guided open surgical training using OST-HMDs is in the specialties of Hepatobiliary, Pancreatic, Gastrointestinal and Urogenital surgeries (Fida et al. 2018). These studies indicate a definitive clinical need for enhanced surgical visualization and training, and an undeniable possible solution in the form of AR. Simulation of procedures in AR/VR can be used to augment a surgeon's training while also providing an objective measurement of their surgical proficiency. All of the above would significantly improve surgical times, outcomes and accuracy.

Benninger et al. showed the utilization of an ultrasound probe as a visual stethoscope to generate 3D images projected on a Google Glass HMD (Benninger 2015). They were thus able to supplement anatomical learning by mastering understanding the body's static and dynamic stereo-structural design. Borgmann et al. used Google Glass to assess the feasibility, safety, and usefulness for 10 different types of procedures in Urological Surgery by augmenting the visualization with CT pre-operative CT images (Borgmann et al. 2017). Volonté et al. successfully attempted to enhance visualization in surgery by using OsiriX software to manipulate pre-operative CT images to generate 3-D volume-rendered overlays that were superimposed on the patients during surgery (Volonté et al. 2011). Volente's study was performed on multiple laparoscopic procedures, including cholecystectomies, liver resections, abdominal explorations, and distal pancreatic resections. Pratt et al. have reported their pioneering efforts in applying VR to reconstructive Plastic Surgery by using the Microsoft HoloLens to assist in the "accurate identification, dissection and execution" of vascular pedunculated flaps (Pratt et al. 2018). Real-time augmentation techniques in surgery by overlaying 3D holograms of anatomic structures on HMDs using anatomic reference points (object-anchored) on the patients are also being studied. Gavaghan et al. have shown the use of an overlay projector device to enable portable projection of anatomical images directly onto the patient's body which is neither hindered by the animate or inanimate objects in the environment nor afflicted by lowered "perspective effects" (Gavaghan et al. 2011). Sauer et al. have explored the use of Microsoft HoloLens in Surgery to aid visualization complex Hepatobiliary viscera and assess the impact on the understanding of vascular anatomy of the region (Sauer et al. 2017). Similar studies on AR applications is being done in many other surgical fields, including Acute Care surgery (Jeon et al. 2014), Vascular surgery (Cheng et al. 2014), Neurosurgery (Hooten et al. 2014), Urology (Borgmann et al. 2017; Dickey et al. 2016; Hamacher et al. 2016), Maxillofacial Surgery (Badiali et al. 2014), ENT (Ear, Nose and Throat) Surgery (Piromchai et al. 2015) and Orthopedic Surgery (Ponce et al. 2014).

7 Augmented Reality and Virtual Reality in Image Guided Robotic Surgery

Minimally invasive surgical skills require a complex finesse as well as time-dependency owing to obscurity in visualizing the operating field, reduced operative space, and a resulting dearth of visual cues and tactile feedback available to the surgeon. The proficiency of the psychomotor skills of a surgeon can be vastly improved using VR/AR, and this can greatly impact the overall perioperative outcome for the patient, surgeon, and hospital. Figure 8, adopted from (Pandya 2004), shows examples of data displayed as VR view during Image Guided Surgery, and a 3D geometric depiction of the same data displayed on live video view (AR View).

In Urology, AR/VR has been used to perform and augment image guided surgery, especially some critical surgical sub-steps of robot-assisted partial nephrectomy, such as the dissection of the renal artery, as shown by Kobayashi et al. (2018). Such augmentation techniques have provided surgeons better visualization of the conventional operating field, better control over inefficient tool motions (such as "insert," "pull," and "rotate" motions), as well as better control over vascular structures. Another study involving pediatric orthopedic surgeons using AR/VT found similar improvements in surgical outcomes through augmenting the information available to the surgeon about the positioning and motion of his surgical tools, leading to better accuracy and reduced procedural time (Docquier et al. 2016; Sugano 2003) (Fig. 6).

In addition to surgical procedures, AR/VT techniques could also be used to augment surgical imaging for pre-planning invasive procedures. As mentioned earlier, imaging modalities generally used for procedural planning include MRI, CT, and Ultrasound. In Urology, biopsies of the prostate gland are being undertaken by fusion of preoperative MRI with intraoperative real-time ultrasound in AR, enabling surgeons to accurately sample tissue for better prognosis, patient-safety, and outcome

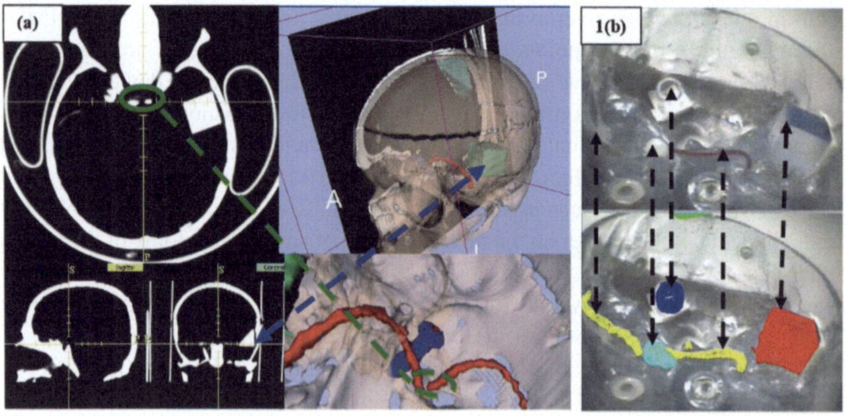

Fig. 6 **a** Virtual data (VR view) displayed during Image Guided Surgery. **b** 3D geometric data of the same case registered and displayed on live video view (AR View) (Adapted from Pandya 2004)

(Tadayyon et al. 2011). For any endovascular/urological procedures making use of a guidewire, tracking the motion of this guidewire can help prevent catastrophic sequelae of vascular rupture or other unintended anatomical violations (Ozkan et al. 2017). In addition to these, applications of AR/VR in surgical navigation systems are being potentially explored in other surgical fields, including ear-nose-and-throat (ENT) surgery (Pruliere-Escabasse and Coste 2010), craniomaxillofacial surgery (Liu et al. 2016), cardiothoracic surgery (Engelhardt et al. 2016), and orthopedic surgery (Niehaus et al. 2017).

8 Challenges in the Comprehension of Visual Information During Surgery

A majority of studies exploring the application of enhanced visualization in surgery indicate a definitive clinical need for enhanced surgical visualization and training, and an undeniable possible solution in the form of AR. However, the common denominator to most of these studies is that they are all specialty-specific. There does not seem to be any study that looks into creating a common institution-wide infrastructure that would service all specialties with generic yet adaptable algorithm.

Additionally, while advances in technology allow accurate depiction of data with low latency, the data segmentation, 3D asset generation, holographic module deployment and anatomic anchoring in most of the studies were done manually using an on-site technical assistance, which could be considered a limitation and bottleneck for the application of these study-outcomes in a clinical setting. The current technique for the rendering of 3D models (generally based on volumetric rendering) could also be improved and optimized. Another common limitation seems to be the lack of re-alignment of the 3D holographic assets to compensate for any body movement or tissue deformation (occurring due to respiratory action, surgical manipulation, etc.) once it has been registered/anchored onto the surgical site. In the realm of clinical setting, optimizing the above-mentioned factors could lead to accurate, real-time acquisition and display of augmented information with low latency.

Since there are multiple ways of displaying the augmented information (Fig. 7), other clinical challenges to usage of OST-HMDs could be inattentional blindness (where the surgeon becomes blinded to an object which was covered by the 3D overlay), information overload (where the overlay outputs more information than what the surgeon can comprehend meaningfully), and overlay distraction. There could also be ergonomic issues such as heavyweight of the HMD, wear comfort as well as possible simulator sickness.

The challenges of adopting different extended reality modalities to pre-operative planning, intraoperative guidance, and medical education depend on the inherent advantages and limitations. VR blocks the real view completely meanwhile AR is transparent and the real is still accessible with digital additions. VR is more suitable for training and education where students can interact in a fully immersive simulation

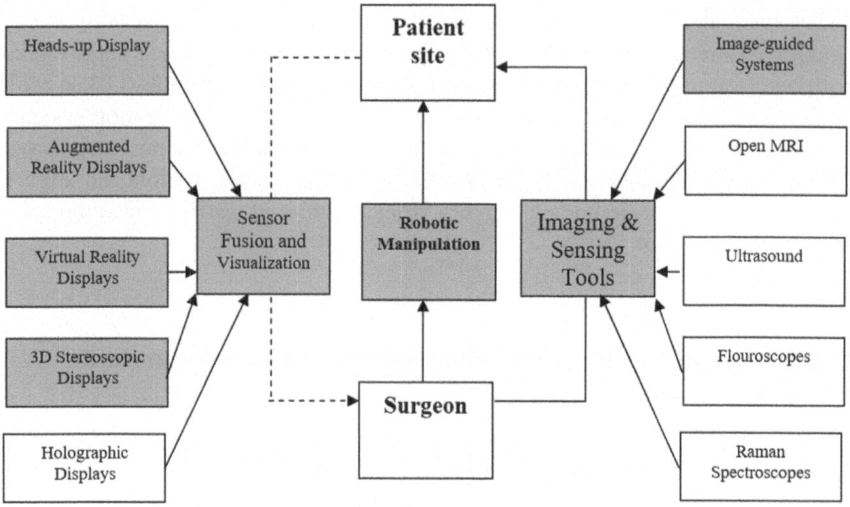

Fig. 7 The tools are available to the surgeon (right) and the display modalities where the information can be visualized (left) (Adapted from Pandya 2004)

with isolation from outside distractions. AR allows the practitioners to maintain a realistic view of the patient hence the ability to interact with the patient. However, the most critical aspect is the accuracy of registration and tracking as this significantly affects the intraoperative navigation and guidance accuracy. Therefore, improvements in all various components of AR is required. The reconstruction techniques and registration methods have to be improved in terms of precision and automation. The tracking technique needs to be more accurate and provide real-time feedback. The displacement and deformation of the organs during intraoperative procedure can affect the accuracy of registration, therefore different techniques are needed to reduce such effects. Display technologies are the largest cost and design constraint because that it is the most demanding aspect of any XR system. XR device is needed to deliver high visual quality, low cost and weight, processing speed, and interactivity. For intraoperative guidance the display must provide high visual quality to match the human visual system (HVS) capabilities. Normal HVS acuity is approximately 9000 by 8100 pixels/eye with 150° to 170° FOV (Silva et al. 2018). Achieving such a requirement is still not economically feasible with current display technologies. As a result, compromising pixel density, FOV, and brightness is a common practice between manufacturers. For example, Oculus Rift has 4.9 (arc-min) angular resolution and 132.2° FOV compared to 1.4, and 43° in Hololens 2 which means that the VR display is still better than the state-of-art AR display.

Another technical challenge that need improvement is the accommodation which is the perceived depth at close distance due to disparity in focal depth. Accommodation is necessary to enable the user to focus on digital objects and instruments at the same time. For example, accommodation conflict of surgical guidance and augmented digital enhancements within the action space and personal space can cause discomfort at close working distances (Silva et al. 2018).

9 Conclusions

This work provides an extensive survey of state of art augmented and virtual reality methods that are being used or considered to be used as popular visualization techniques in clinical practice, especially in Surgery. Furthermore, various associated processes used in AR and VR are also described in detail. The popularity of these visualization techniques and the potential challenges faced by these techniques while using in clinical practice have also been discussed. Finally, we believe that addressing these challenges would further broaden the scope of these visualization techniques to be practically used and truly used in clinical practice.

Acknowledgements This work was partly supported by NPRP Grant #NPRP 11S-1219-170106 from the Qatar National Research Fund (a member of Qatar Foundation).

References

AlZubi S, Islam N, Abbod M (2011) Multiresolution analysis using wavelet, ridgelet, and curvelet transforms for medical image segmentation. Int J Biomed Imaging. https://doi.org/10.1155/2011/136034

Badiali G, Ferrari V, Cutolo F et al (2014) Augmented reality as an aid in maxillofacial surgery: validation of a wearable system allowing maxillary repositioning. J Craniomaxillofac Surg 42: 1970–1976. https://doi.org/10.1016/j.jcms.2014.09.001

Benninger B (2015) Google glass, ultrasound and palpation: the anatomy teacher of the future? Clin Anat 28: 152–155. https://doi.org/10.1002/ca.22480

Borgmann H, Rodriguez Socarras M, Salem J et al (2017) Feasibility and safety of augmented reality-assisted urological surgery using smartglass. World J Urol 35: 967–972. https://doi.org/10.1007/s00345-016-1956-6

Bruckheimer E, Rotschild C (2016) Holography for imaging in structural heart disease. EuroIntervention 12:X81–X84. https://doi.org/10.4244/EIJV12SXA15

Chan F, Aguirre S, Bauser-Heaton H et al (2013) Head tracked stereoscopic pre-surgical evaluation of major Aortopulmonary collateral arteries in the newborns

Chartrand G, Cresson T, Chav R et al (2017) Liver segmentation on CT and MR using Laplacian mesh optimization. IEEE Trans Biomed Eng. https://doi.org/10.1109/TBME.2016.2631139

Chen X, Xu L, Wang Y et al (2015) Development of a surgical navigation system based on augmented reality using an optical see-through head-mounted display. J Biomed Inform 55:124–131. https://doi.org/10.1016/j.jbi.2015.04.003

Cheng I, Shen R, Moreau R et al (2014) An augmented reality framework for optimization of computer assisted navigation in endovascular surgery. Conf Proc. Annual Int Conf IEEE Eng Med Biol Soc Eng Med Biol Soc Conf 2014: 5647–5650. https://doi.org/10.1109/EMBC.2014.6944908

Dakua SP (2013) Performance divergence with data discrepancy: a review. Artif Intell Rev. https://doi.org/10.1007/s10462-011-9289-8

Yang M Der, Chao CF, Huang KS et al (2013) Image-based 3D scene reconstruction and exploration in augmented reality. Autom Constr. https://doi.org/10.1016/j.autcon.2012.09.017

Dickey RM, Srikishen N, Lipshultz LI et al (2016) Augmented reality assisted surgery: a urologic training tool. Asian J Androl 18: 732–734. https://doi.org/10.4103/1008-682X.166436

Docquier P-L, Paul L, TranDuy K (2016) Surgical navigation in paediatric orthopaedics. EFORT Open Rev. https://doi.org/10.1302/2058-5241.1.000009

Du B, Sun Y, Cai S, et al (2018) Object tracking in satellite videos by fusing the Kernel correlation filter and the three-frame-difference algorithm. IEEE Geosci Remote Sens Lett. https://doi.org/10.1109/LGRS.2017.2776899

Du D, Wen L, Qi H et al (2018) Iterative graph seeking for object tracking. IEEE Trans Image Process. https://doi.org/10.1109/TIP.2017.2785626

Duffner S, Garcia C (2017) Fast pixelwise adaptive visual tracking of non-rigid objects. IEEE Trans Image Process. https://doi.org/10.1109/TIP.2017.2676346

El-Hariri H, Pandey P, Hodgson AJ, Garbi R (2018) Augmented reality visualisation for orthopaedic surgical guidance with pre- and intra-operative multimodal image data fusion. Healthc Technol Lett 5:189–193. https://doi.org/10.1049/htl.2018.5061

Engelhardt S, De Simone R, Al-Maisary S et al (2016) Accuracy evaluation of a mitral valve surgery assistance system based on optical tracking. Int J Comput Assist Radiol Surg 11:1891–1904. https://doi.org/10.1007/s11548-016-1353-z

Esser M, Schneeweiss S, Nikolaou K, Horger M (2016) Complexity of ultrasound and CT fusion examinations: are they feasible in the daily routine? Am J Roentgenol. https://doi.org/10.2214/AJR.16.16246

Fida B, Cutolo F, di Franco G et al (2018) Augmented reality in open surgery. Updates Surg. https://doi.org/10.1007/s13304-018-0567-8 [doi]

Gavaghan KA, Peterhans M, Oliveira-Santos T, Weber S (2011) A portable image overlay projection device for computer-aided open liver surgery. IEEE Trans Biomed Eng 58: 1855–1864. https://doi.org/10.1109/TBME.2011.2126572

Grady L (2006) Random walks for image segmentation. IEEE Trans Pattern Anal Mach Intell. https://doi.org/10.1109/TPAMI.2006.233

Guo YY, He DJ, Liu C (2018) Target tracking and 3D trajectory acquisition of cabbage butterfly (*P. rapae*) based on the KCF-BS algorithm. Sci Rep. https://doi.org/10.1038/s41598-018-27520-z

Ha H-G, Hong J (2016) Augmented reality in medicine. Hanyang Med Rev 36:242. https://doi.org/10.7599/hmr.2016.36.4.242

Halabi O, Halwani Y (2018) Design and implementation of haptic virtual fixtures for preoperative surgical planning. Displays 54:9–19. https://doi.org/10.1016/j.displa.2018.07.004

Hamacher A, Kim SJ, Cho ST et al (2016) Application of virtual, augmented, and mixed reality to urology. Int Neurourol J 20: 172–181. https://doi.org/10.5213/inj.1632714.357

Hanna MG, Ahmed I, Nine J et al (2018) Augmented reality technology using microsoft HoloLens in anatomic pathology. Arch Pathol Lab Med 142:638–644. https://doi.org/10.5858/arpa.2017-0189-OA

Heimann T, Van Ginneken B, Styner MA et al (2009) Comparison and evaluation of methods for liver segmentation from CT datasets. IEEE Trans Med Imaging. https://doi.org/10.1109/TMI.2009.2013851

Heine C (2016) Pokémon go is inspiring small retailers. So has augmented reality gone mainstream? Adweek

Hooten KG, Lister JR, Lombard G et al (2014) Mixed reality ventriculostomy simulation: experience in neurosurgical residency. Neurosurgery 10(Suppl 4): 576–581; discussion 581. https://doi.org/10.1227/NEU.0000000000000503

Jeon Y, Choi S, Kim H (2014) Evaluation of a simplified augmented reality device for ultrasound-guided vascular access in a vascular phantom. J Clin Anesth 26: 485–489. https://doi.org/10.1016/j.jclinane.2014.02.010

Jeon S, Hwangbo S, Hong J (2016) A surgical navigation system to assist in chronic total occlusion intervention. In: 2016 13th international conference on ubiquitous robots and ambient intelligence. URAI

Kadoury S, Fecamp V, Gerard M, Soulez G (2016) Fiducial-less fusion imaging of 3D ultrasound with prior CT/MRI in liver interventions. J Vasc Interv Radiol. https://doi.org/10.1016/j.jvir.2015.12.080

Kennedy DN, Filipek PA, Caviness VS (1989) Anatomic segmentation and volumetric calculations in nuclear magnetic resonance imaging. IEEE Trans Med Imaging

Kim BG, Park DJ (2004) Unsupervised video object segmentation and tracking based on new edge features. Pattern Recognit Lett. https://doi.org/10.1016/j.patrec.2004.07.009

Kirbas C, Quek F (2005) A review of vessel extraction techniques and algorithms. ACM Comput Surv. https://doi.org/10.1145/1031120.1031121

Kobayashi L, Zhang XC, Collins SA et al (2018) Exploratory application of augmented reality/mixed reality devices for acute care procedure training. West J Emerg Med 19: 158–164. https://doi.org/10.5811/westjem.2017.10.35026

Lee MW (2014) Fusion imaging of real-time ultrasonography with CT or MRI for hepatic intervention. Ultrasonography. https://doi.org/10.14366/usg.14021

Li Y, Liang W, An H, et al (2017a) C-arm based image-guided percutaneous puncture of minimally invasive spine surgery

Li Z, Cui L, Wang J, Li S (2017b) Fusion of real-time us with CT images to guide selective lumbosacral nerve root interventional therapy. Ultrasound Med Biol. https://doi.org/10.1016/j.ultrasmedbio.2017.08.1624

Li J, Zhou X, Chan S, Chen S (2017c) Robust object tracking via large margin and scale-adaptive correlation filter. IEEE Access. https://doi.org/10.1109/ACCESS.2017.2778740

Liao H, Inomata T, Sakuma I, Dohi T (2010) 3-D augmented reality for MRI-guided surgery using integral videography autostereoscopic image overlay. IEEE Trans Biomed Eng 57:1476–1486. https://doi.org/10.1109/TBME.2010.2040278

Linguraru MG, Richbourg WJ, Liu J et al (2012) Tumor burden analysis on computed tomography by automated liver and tumor segmentation. IEEE Trans Med Imaging. https://doi.org/10.1109/TMI.2012.2211887

Liu S, Feng Y (2018) Real-time fast moving object tracking in severely degraded videos captured by unmanned aerial vehicle. Int J Adv Robot Syst. https://doi.org/10.1177/1729881418759108

Liu TJ, Ko AT, Tang YB et al (2016) Clinical application of different surgical navigation systems in complex craniomaxillofacial surgery: the use of multisurface 3-dimensional images and a 2-plane reference system. Ann Plast Surg 76:411–419. https://doi.org/10.1097/sap.0000000000000429

Liu G, Liu S, Muhammad K et al (2018) Object tracking in vary lighting conditions for fog based intelligent surveillance of public spaces. IEEE Access. https://doi.org/10.1109/ACCESS.2018.2834916

Magic Leap Magic Leap One (2005). https://www.magicleap.com. Accessed 20 Sept 2005

Mahalingam T, Subramoniam M (2018) A robust single and multiple moving object detection, tracking and classification. Appl Comput Inform. https://doi.org/10.1016/j.aci.2018.01.001

Mahr A, Levegrün S, Bahner ML et al (1999) Usability of semiautomatic segmentation algorithms for tumor volume determination. Invest Radiol. https://doi.org/10.1097/00004424-199902000-00007

Mat Isham MI, Mohamed F, Siang CV, et al (2019) A framework of ultrasounds image slice positioning and orientation in 3D augmented reality environment using hybrid tracking method. In: 2018 IEEE Conference on Big Data Analysis ICBDA 2018, pp. 105–110. https://doi.org/10.1109/ICBDAA.2018.8629589

Mauri G, Cova L, De Beni S et al (2015) Real-time US-CT/MRI image fusion for guidance of thermal ablation of liver tumors undetectable with US: results in 295 cases. Cardiovasc Intervent Radiol. https://doi.org/10.1007/s00270-014-0897-y

Mharib AM, Ramli AR, Mashohor S, Mahmood RB (2012) Survey on liver CT image segmentation methods. Artif Intell Rev

Microsoft HoloLens (2019) HoloLens 2. In: Microsoft HoloLens. https://www.microsoft.com/en-us/hololens. Accessed 20 Sept 2005

Milgram P, Kishimo F (1994) A taxonomy of mixed reality. IEICE Trans Inf Syst 77:1321–1329

Milgram P, Takemura H, Utsumi A, Kishino F (1995) Augmented reality: a class of displays on the reality-virtuality continuum. In: Das H (ed). pp 282–292

Monsky WL, James R, Seslar SS (2019) Virtual and augmented reality applications in medicine and surgery-the fantastic voyage is here. Anat Physiol Curr Res 9:1–5. https://doi.org/10.4172/2161-0940.1000313

Navab N, Traub J, Sielhorst T et al (2007) Action- and workflow-driven augmented reality for computer-aided medical procedures. IEEE Comput Graph Appl 27:10–14. https://doi.org/10.1109/MCG.2007.117

Niehaus R, Schilter D, Fornaciari P et al (2017) Experience of total knee arthroplasty using a novel navigation system within the surgical field. Knee 24:518–524. https://doi.org/10.1016/j.knee.2016.10.021

Nuzillard D, Lazar C (2007) Partitional clustering techniques for multi-spectral image segmentation. J Comput. https://doi.org/10.4304/jcp.2.10.1-8

Ozkan E, Tanner C, Kastelic M et al (2017) Robust motion tracking in liver from 2D ultrasound images using supporters. Int J Comput Assist Radiol Surg 12:941–950. https://doi.org/10.1007/s11548-017-1559-8

Pandya A (2004) Medical augmented reality system for image-guided and robotic surgery: development and surgeon factors analysis. Wayne State University

Parkhomenko E, Safiullah S, Walia S et al (2018) Mp26-20 virtual-reality projected renal models with urolithiasis as an educational and preoperative planning tool for nephrolithotomy: a pilot study. J Urol 199. https://doi.org/10.1016/j.juro.2018.02.878

Patwardhan SV, Dai S, Dhawan AP (2005) Multi-spectral image analysis and classification of melanoma using fuzzy membership based partitions. Comput Med Imaging Graph. https://doi.org/10.1016/j.compmedimag.2004.11.001

Piromchai P, Avery A, Laopaiboon M et al (2015) Virtual reality training for improving the skills needed for performing surgery of the ear, nose or throat. Cochrane Database Syst Rev (9). https://doi.org/10.1002/14651858.CD010198.pub2

Ponce BA, Menendez ME, Oladeji LO et al (2014) Emerging technology in surgical education: combining real-time augmented reality and wearable computing devices. Orthopedics 37: 751–757. https://doi.org/10.3928/01477447-20141023-05

Pratt P, Ives M, Lawton G et al (2018) Through the HoloLens looking glass: augmented reality for extremity reconstruction surgery using 3D vascular models with perforating vessels. Eur Radiol Exp 2. https://doi.org/10.1186/s41747-017-0033-2

Pruliere-Escabasse V, Coste A (2010) Image-guided sinus surgery. Eur Ann Otorhinolaryngol Head Neck Dis 127:33–39. https://doi.org/10.1016/j.anorl.2010.02.009

Raconteur Reports (2018) XR For business

Ruan S, Moretti B, Fadili J, Bloyet D (2002) Fuzzy Markovian segmentation in application of magnetic resonance images. Comput Vis Image Underst. https://doi.org/10.1006/cviu.2002.0957

Sato Y, Nakajima S, Shiraga N et al (1998) Three-dimensional multi-scale line filter for segmentation and visualization of curvilinear structures in medical images. Med Image Anal. https://doi.org/10.1016/S1361-8415(98)80009-1

Sauer IM, Queisner M, Tang P et al (2017) Mixed reality in visceral surgery: development of a suitable workflow and evaluation of intraoperative use-cases. Ann Surg 266: 706–712. https://doi.org/10.1097/SLA.0000000000002448

Seo KS (2005) Automatic hepatic tumor segmentation using composite hypotheses. In: Lecture Notes in Computer Science (including subseries Lecture Notes in Artificial Intelligence and Lecture Notes in Bioinformatics)

Seror O (2017) Novel guidance for percutaneous treatment of hepatic tumors: fusion images and cone-beam CT. Ultrasound Med Biol. https://doi.org/10.1016/j.ultrasmedbio.2017.08.1473
Shang Y, Deklerck R, Nyssen E et al (2011) Vascular active contour for vessel tree segmentation. IEEE Trans Biomed Eng. https://doi.org/10.1109/TBME.2010.2097596
Silva JJNA, Southworth M, Raptis C, Silva JJNA (2018) Emerging applications of virtual reality in cardiovascular medicine. JACC Basic Transl Sci 3:420–430. https://doi.org/10.1016/j.jacbts.2017.11.009
Subudhi BN, Nanda PK, Ghosh A (2011) A change information based fast algorithm for video object detection and tracking. IEEE Trans Circ Syst Video Technol. https://doi.org/10.1109/TCSVT.2011.2133870
Sugano N (2003) Computer-assisted orthopedic surgery. J Orthop Sci 8:442–448. https://doi.org/10.1007/s10776-002-0623-6
Tadayyon H, Lasso A, Kaushal A et al (2011) Target motion tracking in MRI-guided transrectal robotic prostate biopsy. IEEE Trans Biomed Eng 58:3135–3142. https://doi.org/10.1109/tbme.2011.2163633
Venkatesan AM, Kadoury S, Abi-Jaoudeh N et al (2011) Real-time FDG PET guidance during biopsies and radiofrequency ablation using multimodality fusion with electromagnetic navigation. Radiology. https://doi.org/10.1148/radiol.11101985
Volonté F, Pugin F, Bucher P et al (2011) Augmented reality and image overlay navigation with OsiriX in laparoscopic and robotic surgery: not only a matter of fashion. J Hepatobiliary Pancreat Sci 18:506–509. https://doi.org/10.1007/s00534-011-0385-6
Wein W, Brunke S, Khamene A et al (2008) Automatic CT-ultrasound registration for diagnostic imaging and image-guided intervention. Med Image Anal. https://doi.org/10.1016/j.media.2008.06.006
Wein W, Ladikos A, Fuerst B et al (2013) Global registration of ultrasound to MRI using the LC2 metric for enabling neurosurgical guidance. In: Lecture notes in computer science (including subseries Lecture Notes in Artificial Intelligence and Lecture Notes in Bioinformatics)
Yim PJ, Foran DJ (2003) Volumetry of hepatic metastases in computed tomography using the watershed and active contour algorithms
Zhang X, Tian J, Deng K et al (2010) Automatic liver segmentation using a statistical shape model with optimal surface detection. IEEE Trans Biomed Eng. https://doi.org/10.1109/TBME.2010.2056369
Zhang Q, Fan Y, Wan J, Liu Y (2017) An efficient and clinical-oriented 3D liver segmentation method. IEEE Access. https://doi.org/10.1109/ACCESS.2017.2754298
Zhang X, Chen G, Liao H (2017b) High-quality see-through surgical guidance system using enhanced 3-D autostereoscopic augmented reality. IEEE Trans Biomed Eng 64:1815–1825. https://doi.org/10.1109/TBME.2016.2624632
Zhang T, Liu S, Xu C et al (2018) Correlation particle filter for visual tracking. IEEE Trans Image Process. https://doi.org/10.1109/TIP.2017.2781304
Zhao B, Schwartz LH, Jiang L et al (2006) Shape-constraint region growing for delineation of hepatic metastases on contrast-enhanced computed tomograph scans. Invest Radiol. https://doi.org/10.1097/01.rli.0000236907.81400.18

Osama Halabi (first author) has been active in academia for the last 17 years. Presently, he is working as an assistant professor in the Department of Computer Science at Qatar University. Prior to joining Qatar University, he has participated in various organizations in various capacities such as Fujitsu Endowed Chair (Fujitsu Co.,) at Japan Advanced Institute of Science and Technology from 2001 to 2003, researcher in Virtual Systems Laboratory at Gifu University, Japan, from 2003 to 2006, assistant professor at Iwate University, Japan, from 2006 to 2010. He received M.Sc. degree in computer science from Shanghai University, and a Ph.D. degree in Information Science from Japan Advanced Institute of Science and Technology (JAIST) in 1998, and 2001, respectively. His principal research interests include virtual reality (VR), augmented reality

(AR), haptic interface, human-centered computing, robotics, interactive systems, laser graphics, and computer graphics. He has intensive experience in software development for haptic interfaces and virtual realit, especially for medical training and simulation. Dr. Halabi has participated in many QNRF and Qatar university research projects and published more than 50-refereed technical papers in various international conferences and journals, two books, and one patent.

Shidin Balakrishnan (corresponding author) is a Physician-Scientist working in the Department of Surgery at Hamad Medical Corporation (HMC), Qatar since 2015. He graduated from Weill Cornell Medicine as a Doctor of Medicine (MD) in 2012, following which he completed a research internship at the Qatar Robotic Surgery Center in the Qatar Science and Technology Park where he worked on surgical-simulation projects. At HMC, Dr. Shidin has participated in multiple awarded QNRF and institutional surgical translational research projects to build technology systems that help surgical practitioners perform cognitive tasks more intuitively. He provides a unique blend of medical background, statistical expertise, and research interest and experience in conducting surgico-technological research that is crucial to the successful delivery of multidisciplinary projects. Dr. Balakrishnan's research-focus has thus far resulted in 10+ conference/journal publications and patent filings, and R&D output.

Sarada Prasad Dakua is an Academic Research Scientist in the Department of Surgery, Hamad Medical Corporation (HMC). Prior to joining HMC in 2017, he has worked over six years in Qatar Robotic Surgery Centre, Qatar Science and Technology Park (Qatar Foundation) as a Research Executive. He has 13+ years of experience in computer vision and image processing (medical image segmentation, image registration, image analysis, image compression, object motion tracking, late potential detection in ECG, IoT, etc.). He has been involved in a couple of NPRP projects as PI and LPI. Furthermore, he has been awarded a few grants from Hamad Medical Corporation. He has 60+ scientific publications including peer-reviewed journals, conferences, books and book chapters. His other activities include teaching, supervising interns, reviewing Ph.D. theses, journal and conference manuscripts, etc. He is a regular reviewer of more than 30 peer-reviewed international journals. He has also participated in more than 40 national and international conferences in different capacities such as TPC member, chair, advisor, and program committee member. He holds a Ph.D. degree from the Indian Institute of Technology Guwahati in medical image processing and analysis.

Nassir Navab has been active in the research field for 30 years. He is currently a full professor and the Chair for Computer Aided Medical Procedures, Department of Informatics, Technical University of Munich, Germany, as well as a Part-time Research Professor at the Computer Science Department, Johns Hopkins University, Baltimore, USA. Before joining Technical University of Munich, he worked for Siemens Corporate Research (SCR), Princeton, the USA for 10 years from 1994 to 2003, was a Postdoctoral Fellow at MIT, Boston, USA (1993–1994) and received his Ph.D. degree from INRIA, Sophia Antipolis, France in 1993. Prof. Navab is a founder of the IEEE International Symposium on Mixed and Augmented Reality (ISMAR) community and served on its steering committee for 8 years until 2008. Since 2006 he has been a fellow and member of the board of directors of the MICCAI Society, the organizer of the world's leading conference on medical image computing and computer assisted intervention. He is also on the editorial board of many international journals, such as IEEE TMI, Medical Image Analysis, Medical Physics and International Journal of Computer Vision. He has authored hundreds of scientific publications, has received more than 9200 citations and holds more than 60 international patents. His research interests include, but are not limited to Medical Image Analysis, Reconstruction, Machine Learning for Medical Applications, Robot-guided Surgery, Ultrasound Imaging, Computer Vision and Augmented Reality.

Mohamed Warfa is a radiology fellow at Wake Forest University in North Carolina, USA. He completed his medical school training at Weill Cornell Medical College and his residency at Hamad Medical Corporation, Doha, Qatar. Dr. Warfa is an abdominal imager and is interested in the utilization of VR/Augmented Reality in interventional oncology.

The manufacturer's authorised representative in the EU is Springer Nature Customer Service Centre GmbH, Europaplatz 3, 69115 Heidelberg, Germany. If you have any concerns regarding our products, please contact ProductSafety@springernature.com

Printed and bound by CPI Group (UK) Ltd, Croydon, CR0 4YY
30/09/2026
02225648-0009